Earth in Our Care

Earth in Our Care

Ecology, Economy, and Sustainability

CHRIS MASER

RUTGERS UNIVERSITY PRESS

NEW BRUNSWICK, NEW JERSEY, AND LONDON

The publication of this book was supported, in part, by a grant
(#GF-18-2008-318) from Global Forest Science,
a nonprofit conservation institute.

LIBRARY OF CONGRESS CATALOGING-IN-PUBLICATION DATA

Maser, Chris.
Earth in our care : ecology, economy, and sustainability / Chris Maser.
    p. cm.
Includes bibliographical references and index.
ISBN 978–0–8135–4559–2 (hardcover : alk. paper)
1. Sustainable development.   2. Nature—Effect of human beings on.   I. Title.
HC79.E5M3595 2009
338.9'27—dc22   2008040065

A British Cataloging-in-Publication record for this book is available
from the British Library.

Visit our Web site: http://rutgerspress.rutgers.edu

Manufactured in the United States of America

If nothing else, nature is persistent. I have seen the tender fiddle neck of a fern push its way, cell by cell, up through asphalt to grow in a city street, where, on a cold morning, there shone heaven's own rainbow in a drop of dew clinging to a frond. As a boy, I watched the roadside ditch in which I played carry away pebble and grain, as it wore its way down into the soil. The Grand Canyon began in the same way, millions of years ago.

Over the millennia, water dissolved its way deep into the crust of the Earth and, through the long reach of time, carved an ever-changing mural of exquisite beauty that today enriches the life of every person who pauses to marvel at the creative forces of nature. Thus it is that I, with a great deal of humility, dedicate this book to planet Earth for gracing my life with such magnificent beauty for so many, many years.

This book is also lovingly dedicated to the memory of Blessing, a magnificent crow that my wife, Zane, befriended and fed. Despite missing a foot and part of her leg, Blessing integrated the three spheres of life with grace and ease–the atmosphere (air), the litho-hydrosphere (the rock that constitutes the restless continents and the water that surrounds them), and the biosphere (all life sandwiched in the middle). Gliding on the restless breezes, she would land to partake of nourishment from the Earth. English-walnut halves were her favorite. Having eaten, she would load her beak with all the goodies it could hold and, rising once again, would navigate through the air to her favorite branch, where she would either finish her meal or stash her booty to be enjoyed at a later time. Then, she was off for a refreshing drink and splash in the birdbath before once again riding a zephyr to some cherished place.

CONTENTS

Foreword by Okechukwu Ukaga   ix

Preface   xi

Acknowledgments   xvii

Introduction   1

1   Of Ignorance and Knowledge   5

2   Our Ever-Changing Landscape Patterns   25

3   How Species Enrich Our Lives and the World   56

4   The Never-Ending Stories of Cause, Effect, and Change   86

5   Act Locally and Affect the Whole World   117

6   Repairing Ecosystems   142

7   Where Do We Go from Here?   198

Appendix: Common and Scientific
  Names of Plants and Animals   207

Notes   217

Glossary   247

Index   265

FOREWORD

Earth's social, environmental, and economic fabric is being threatened from all sides by such challenges as global warming, violence, poverty, and general environmental degradation caused by unsustainable use of Earth's resources. Nations in the West, whose economies became industrialized early, bear the brunt of responsibility for damage done to the environment so far. Nevertheless, as densely populated countries, like China, India, and Brazil, quickly transition to technology-based, consumer economies, demands for Earth's resources might reach a breaking point.

As Chris Maser puts it, fulfilling our obligation as environmental trustees of Earth as a biological living trust requires fundamental changes in our social consciousness and cultural norms. To meet these challenges, we need to fundamentally reframe our way of thinking. Instead of arbitrarily delineating our seamless world into discrete parts, we need a holistic approach—one that acknowledges the interconnectedness of causes and effects, actions and consequences. Knowledge of systems is essential if we are to pass a habitable, healthy planet to future generations. Proper trusteeship is critical to maintaining the Earth's ability to produce, nourish, and maintain life. Without it, we risk becoming the originators of our own demise.

But what is systems thinking, and how can we harness it to put our planet on a sustainable course? Systems thinking goes to and deals with the root cause of a problem; it is the opposite of symptomatic thinking, which deals with the world in piecemeal fashion. In this book, Maser reviews some of the factors that relate to the workings, services, and resilience of our planet—from nature's biophysical principles to the role of ignorance and knowledge, to the tradeoffs of every decision and action, to ever-changing landscape patterns, to the never-ending cycles of cause and effect, and so on. In doing so, he makes a unique and simultaneous use of both scientific and philosophical reasoning in articulating how the Earth works according to the immutable biophysical laws that govern it. The book goes beyond superficial recommendations, however, to call for self-conquest as a fundamental foundation for social-environmental sustainability; through self-conquest individuals can bring their thoughts and behaviors in line with biophysical laws and, acting locally, can then change the world for the better.

Sustainability has become an increasingly urgent global imperative. With that goal in mind, *Earth in Our Care* opens a new chapter in our search for practical solutions to environmental problems, which must include changing the way we think—raising the level of our consciousness—about the Earth as a living system and our place in it.

Okechukwu Ukaga, Ph.D.

Executive Director and Extension Professor

Northeast Minnesota Sustainable Development Partnership

University of Minnesota

PREFACE

The mind creates the abyss, and the heart crosses it.–Sri Nisargadatta, an
Indian spiritual teacher and philosopher of Advaita (Nondualism)

In a world exploding in the fire of ethnic and religious hatreds, I see fear and its
grisly gang of distrust, divisiveness, separation, slander, reprisal, greed, fraud, dis-
tortion, and duplicity slithering through the dark halls of governments in each of
the four hemispheres. It matters not which hemisphere you choose; each has its
despots with fingers on the trigger as they suck the life energy from the people in
a bid for the power of control. In their anxiety about life's uncertainties and the
irrational fear of the future it spawns, their sense of security depends on this control to suppress the imagined portents of personal annihilation.

In such a world, it is difficult to remain consciously aware of the miraculous
beauty of form and function that surrounds us. I am particularly blessed in that I
have been privileged to travel in many lands, near and afar, from ocean strand to
lofty mountain, from parching desert to steaming jungle, and in each have I found
beauty unsurpassed: it may have been the odor of jasmine along the Nile, the
smile of a Nubian child, the soft touch of a Chilean fern, the iridescence of a
Nepalese sunbird, the fuzzy face of an Austrian edelweiss, the intricate structure
of a Japanese Shinto shrine, the alert stance of a tiger beetle on a jungle trail in
Malaysia, or the leap of a flying fish in the middle of the Atlantic Ocean. Each experience is a snapshot, a touchstone along the continuum of evolution through
which the eternal mystery of life unfolds.

Part of our inability to grasp the eternal mystery with our intellect is that no
two things in the universe have ever been—or ever will be—exactly the same.
Therefore, no two things can ever be equal—except in their inequality. Moreover,
all life is composed of physical relationships in ever-changing patterns and
rhythms that both affect life and are, in turn, affected by life. In this sense, life
not only is pattern seeking and pattern sensitive but also is guided by the
eternal rhythms of the universe. As well, every life form is a microcosm of the
whole—from the most simple to the most complex.

Everything in the universe is thus connected to everything else in a cosmic
web of interactive feedback loops, all entrained in self-reinforcing relationships
that continually create novel, never-ending stories of cause and effect, stories that

began with the eternal mystery of the original story, the original cause. Everything, from a microbe to a galaxy, is defined by its ever-shifting relationship to every other component of the cosmos. Thus "freedom" (perceived as the lack of constraints) is merely a continuum of fluid relativity. Hence, every change (no matter how minute or how grand) constitutes a systemic modification that produces novel outcomes. A feedback loop, in this sense, comprises a reciprocal relationship among countless bursts of energy moving through specific strands in the cosmic web that cause forever-new, compounding changes at either end of the strand, as well as every connecting strand.

How, then, is human ownership of anything possible?

Ownership assumes that we humans are somehow in control, that we can actually possess something and hold it in a condition of our choosing—a snapshot, if you will. In reality, however, we are totally incapable of owning anything because we are not in control of a single aspect of life other than our thoughts, motives, and their subsequent actions—the outcomes of which are beyond our ability to control. Therefore, all we ever do is borrow from the cosmic store of materials to form our bodies, our homes, our automobiles, the roads on which to drive them, and so on ad infinitum. Everything in every part of the universe is on indefinite loan to every other part of the universe—such is the cosmic web of never-ending stories that form the relationships of life.

In turn, individual, living organisms (that collectively form the species that collectively form the communities as they spread over the land and fill the seas) join the myriad constituents of diversity itself, such as the scales of time, space, and temperature, and the processes that shape the Earth. Together, the nonliving, physical elements and the living organisms have molded and remolded the earthscape in an ever-changing kaleidoscope. These organisms, through the exchange medium of the soil, are influenced by short-term ecological limitations even as they influence those same limitations through their life cycles. The interactions of communities and soil are controlled and influenced by the long-term dynamics that coincidentally form the three, interactive spheres of our earthscape: the atmosphere (air), the litho-hydrosphere (the rock that constitutes the restless continents and the water that surrounds them), and the biosphere (the life forms that exist within the other two spheres). We humans, however, arbitrarily delineate our seamless world into discrete ecosystems as we try to understand the fluid interactions between nonliving and living components of planet Earth. If you picture the interconnectivity of the three spheres as being analogous to the motion of a waterbed, you will see how patently impossible such divisions are because you cannot touch any part of a waterbed without affecting the whole of it.

Beauty in form is clearly visible to our senses, from the microscopic to the infinite, from the delicate design of a diatom to the violent death throes of a star. But the beauty of function is often hidden in the act of living—be it a bearded vulture riding the thermals high in the Himalayas, a polar bear wandering the Arctic sea ice in search of seals, or the "emergent properties," by means of which termites in the Australian savannah construct their twenty-foot-tall towers. Yet each of life's actions is a form of participation in a feedback loop whereby life serves life.

As the autumn of life approaches and my time on Earth wanes, I have come to understand that the biophysical principles governing our home planet and the universe function perfectly—albeit our understanding of these principles is imperfect, and our acceptance of their limitations is unwilling. And our unwillingness to accept what is, often through informed denial, causes the pain and suffering I witness in my travels—virtually all of which is human caused. Yet, despite our all-too-often inhumane treatment of one another and our environment, the collective, functional beauty of nature's biophysical principles lies in their flawless impartiality, the absolute fairness whereby every living being is treated during its life—whether we accept this truth or not.

Nevertheless, wherever my sojourn on Earth leads, my field of view is graced by the splendor of nature's patterns in form and function and my relationship with them. And it is my sense of wonder that I would share with you, as a reminder that all the horrific ugliness unleashed by the severely dysfunctional among us cannot erase nature's ineffable beauty or your place within it—if you will but choose to remain consciously focused on the eternal mystery of life. Yet, hidden within the splendor I find in living, there lurks an abiding question that must be addressed within the first part of this century if life is to be more than simply a matter of survival: Are the lifestyles we have chosen sustainable on Earth?

Be forewarned that answering this question requires a renewed sense of commitment to personal growth and social justice, a commitment that causes us as individuals and as a society to act now for the simultaneous benefit of both present and future generations. Personal growth is vital to the answer because the level of consciousness that causes a problem in the first place is not the level of consciousness that can fix it, as will become abundantly clear in the discussion of the Aswan High Dam in Chapter 5.

Our personal and social reticence to deal openly and honestly with this question calls to mind a salient paragraph from a speech Winston Churchill delivered to the British Parliament on May 2, 1935, as he saw with clear foreboding the onrushing threat of Nazi Germany to Europe and the British people:

When the situation was manageable it was neglected, and now that it is thoroughly out of hand we apply too late the remedies which then might have effected a cure. There is nothing new in the story. . . . It falls into that long, dismal catalogue of the fruitlessness of experience and the confirmed unteachability of mankind. Want of foresight, unwillingness to act when action would be simple and effective, lack of clear thinking, confusion of counsel until the emergency comes, until self-preservation strikes its jarring gong—these are the features which constitute the endless repetition of history.[1]

Consider that civilizations have evolved by similar steps: growth of intelligence through discoveries and inventions, advancement through the ideas of government, family, and property, all based on a slow accumulation of experimental knowledge. As such, civilizations have much in common, and their evolutionary stages are connected with one another in a natural sequence of cultural development.

The arts of subsistence and the achievements of technology can be used to distinguish the periods of human progress. People lived by gathering fruits and nuts; learned to hunt, fish, and use fire; invented the spear and atlatl and then the bow and arrow. They developed the art of making pottery, learned to domesticate animals and cultivate plants, began using adobe and stone in building houses, and learned to smelt iron and use it in tools. Finally, what we call civilization began with the invention of a written language, culminating in all the wonders of the modern era.

Each civilization has also been marked by its birth, maturation, and demise, the latter brought about by uncontrolled population growth that outstripped the source of available energy, be it loss of topsoil, deforestation, or the continued despoliation of its water source. But in olden times the survivors could move on to less populated, more fertile areas as their civilizations collapsed. Today there is nowhere left on Earth to go.

Yet, having learned little or nothing from history, as Churchill pointed out, our society is currently destroying the very environment from which it sprang and on which it relies for continuance. Surely, society as we know it cannot be the final evolutionary stage for human existence. But what lies beyond our current notion of society? What is the next frontier for "civilized" people to conquer? Is it outer space, as so often stated? No, it is inner space, the conquest of oneself, which many assert is life's most arduous task. As the Buddha said, "Though he should conquer a thousand men in the battlefield a thousand times, yet he, indeed, who would conquer himself is the noblest victor."[2]

In the material world, self-conquest means bringing one's thoughts and behaviors in line with the immutable biophysical laws governing the world in which one lives, such as the law of cause and effect. In the spiritual realm, self-conquest means disciplining one's thoughts and behaviors in accord with the highest spiritual and social truths handed down throughout the ages, such as: love your neighbor as yourself, and treat others as you want them to treat you.

The outcome of self-conquest is social-environmental sustainability, which must be the next cultural stage toward which we struggle. This is the frontier beyond self-centeredness and its stepchild, destructive conflict, which destroys human dignity, degrades our global ecosystem's productive capacity, and forecloses options for all generations.

To fulfill our obligation as environmental trustees for the children we bring into the world requires fundamental changes in our social consciousness and cultural norms, changes that will demand choices different from those we have heretofore made, which means thinking and acting anew. But "a great many people," as American psychologist William James observed, "think they are thinking when they are merely rearranging their prejudices."[3]

Where, for instance, is there an unequivocal voice among national and international leaders that speaks for the children who must inherit the consequences of our decisions and actions? Where is there an unequivocal voice among national and international leaders that speaks for protecting the productive capacity of the global ecosystem—our bequest to all generations? Without such a singular voice of courage and unconditional commitment in each nation, we, the adults of the world, are condemning every generation, including those of our children and grandchildren, to pay a progressively awful price for our petty psychological immaturities as we bicker among ourselves about who will do what rather than accept the sometimes-bitter pill of our adult responsibilities.

In my estimation, social insanity can be defined as doing the same thing over and over, despite the lessons of world history, while each time expecting new and dramatically different results. This is a simple summation of the way in which Western industrial society navigated the twentieth century—a century that was permeated by a deadly grapple between society's immediate wants and demands and what the environment can sustainably produce.

And what about the twenty-first century? Will we finally accept our responsibilities as guardians of planet Earth, the biological living trust, for the beneficiaries, the children of today, tomorrow, and beyond? Or will it too be a century of lethal, economic struggle among the polarized positions of the supremely dysfunctional among us? Are they—once again—to be allowed to determine the legacy we, as a society, as a nation, bequeath those who follow us? The choice is ours, the adults of the world. How shall we choose?

If we choose wisely in our care of Earth as a biological, living trust, we must understand some of the inviolate biophysical principles that govern this magnificent planet. We owe that much to the children who follow us because our decisions become their consequences, and we give them no voice in either the quality or the sustainability of their inheritance. To bestow the gift of wise decisions, however, requires understanding some basic principles of life, which are normally omitted from the discourse of sustainability. I have done my best within these pages to partially amend that oversight.

To this end, Chapter 1 is a discussion of what we are taught and what we are not taught about the biophysical dynamics of our home planet, how we affect them, and how they, in turn, affect us. Understanding our reciprocal relationship with our environment is critical because how we treat the Earth determines not only how the Earth will respond but also the rigidity of the social-environmental constraints we are bestowing on future generations. Chapter 2 is a verbal portrait of our ever-changing landscape patterns. Chapter 3 examines how species enrich our lives and the world. Some of the never-ending stories of cause, effect, and change are recounted in Chapter 4, and Chapter 5 points out how local acts affect the whole world. Chapter 6 is a short course on how to repair the ecosystems we have damaged in our living and in our pursuit of monetary wealth, and Chapter 7 both raises the question of where we go from here and provides vital direction. Finally, to help readers understand the message within these covers, a glossary of terms and an appendix of common and scientific names are included.

ACKNOWLEDGMENTS

My lovely wife, Zane, and our loving, older cat, Zoe, granted me their patience during the long hours that I plied the seemingly endless literature and the keys of my computer. Their understanding made writing this book a daily joy. In addition, Zane did a wonderful job of proofreading the galleys.

I am deeply grateful to Doreen Valentine of Rutgers University Press for her brilliant organizational editing and for once again instructing me in good writing, while simultaneously improving the clarity of the text for you, the reader. Copyeditor Pamela Fischer then took the manuscript and guided me through the process of repairing the editorial potholes I had inadvertently built into the text; along the way she taught me about the finer points of editorial repair. Beyond that, the Rutgers team took over and, with their usual excellence, produced the book you are holding.

Finally, Global Forest Science, a worldwide research foundation, graciously offered financial support for the publication of this book. On behalf of all generations, "Thank you."

Earth in Our Care

Introduction

The ultimate test of human conscience may be the willingness to sacrifice something today for future generations whose words of thanks will not be heard.–Senator Gaylord Nelson

Although planet Earth reveals its secrets slowly, we now have far more knowledge of the world in which we live than did our forbearers. Therefore, we not only have greater opportunities than they did but also are confronted with greater responsibilities because we are now part of an interconnected global society, whether or not we fully understand the idea or even like it. Just as their decisions set the stage for our reality, our decisions will determine the options of tomorrow and write the history of yesterday.

If humanity is to survive this century and beyond with any semblance of dignity and well-being, we must both understand and accept that we have a single ecosystem composed of three inseparably interactive spheres: the atmosphere (air), the litho-hydrosphere (the rock that constitutes the restless continents and the water that surrounds them), and the biosphere (all life sandwiched in the middle). And because this magnificent, living system—planet Earth—simultaneously produces, nourishes, and maintains all life, we would be wise to honor it and care for it. If we do not, if we cause too much damage to any one of these interdependent spheres, we will be the authors of our own demise—and that of all the world's children.

Here, it must be understood that every system in the universe, both living and nonliving, is governed by variability. No system is controlled by the averages. I say this because everything in the universe is defined by its ever-shifting relationship to everything else, which means that "freedom"—perceived as a lack of constraints—is always relative, never absolute. Every change, no matter how minute, constitutes a systemic modification that produces novel outcomes. Here is one of life's abiding paradoxes: change is a constant process, which honors the Buddhist notion of impermanence—a biophysical reality that forever precludes the existence of an independent variable or a constant value of any kind.

Put a little differently, nothing can exist as a separate reality that is independent of anything else. Whatever is created, therefore, is the introduction of a unique, never-ending story of cause and effect in a finite world, as eloquently pointed out by conservationist David Brower: "There is but one Ocean, though its coves have many names; a single sea of atmosphere with no coves at all; the miracle of soil, alive and giving life, lying thin on the only Earth, for which there is no spare."[1]

The invariable process of ever-shifting relationships totally negates the possibility of anything being reversible. Although we may have reams of data and use them in our bid to return something to an earlier condition (termed *restoration*), we cannot do so because the eternal now is all we have or ever will have. We can, however, physically revisit a given place and do our best to emulate—but only emulate—what we perceive a prior condition to have been like in some former time. Whatever we create will be original and immediately entrained in the perpetual novelty of change.

By way of analog, your grandmother's rocking chair is missing part of its back and has a broken rocker. The chair is now yours, and because it reminds you of your grandmother, whom you loved, you want it restored to its original condition. You therefore take it to a repair shop, where the chair is fixed but with different wood and modern finishing materials by someone who did not build it in the first place. Although the chair now functions as it is meant to, it has not been returned to its original condition and never can be. Moreover, it has again entered the conveyer of change, the continuum of time.

The concept of interdependence (and its antithesis: absolute freedom) relates to our human condition, which itself is an abiding paradox in that we are each self-aware and seemingly able to act in accord with our own dictates, yet are bound by our character and held prisoner by our fears. Ultimately, all we humans do—ever—is practice relationships with energy of one kind or another, energy that is always pulsing, never even in its flow. In so doing, we experience ourselves experiencing the practice of relationships with ourselves (emotions), with one another (friendship, hatred, prejudice), with nature (responding to the weather, enjoying a rainbow, feeling a drop of water, gliding over the surface of snow), with an idea (contemplating the meaning of an abstract thought), with time (past memories, present circumstances, future hopes and fears), with some piece of technological gadgetry (a computer or cell phone), and so on. Experiencing ourselves practicing relationships is at once the essence of life and the limitation to absolute freedom.

By way of illustration, consider that a river's water, which appears free flowing, is in fact constrained in its movement by the ever-changing configuration of its channel and the volume of water it can hold at any one time. Another way of looking at freedom is to consider the invisible tug of gravity, which always leads this precious liquid downhill to the lowest possible level. By the same token, the still water of a lake is also held in place by gravity. Nevertheless, the water of a lake journeys to the sea just as surely as that of a river, but, unlike the journey of river water, which is visible in its flowing, that of a lake is converted to a gas through the invisible medium of evaporation. Once aloft, it rides the currents of air until it condenses, falls as precipitation, and begins again its journey to the sea—a journey it may eventually complete. And speaking of water, changes in the way we use land (such as the mass conversion of forests to tree farms, row-crop agriculture, live-stock pastures, housing developments, and shopping malls) are increasing both surface runoff and evaporation, thereby influencing how the Earth functions as an ecosystem.

The significant increase in the observed flow of rivers worldwide in the twentieth century was due mainly to climate change and widespread deforestation. The elimination of trees, and thus their evaporative transpiration, caused more water to run over the surface of the soil than could be accommodated by infiltration. It is clear from historic data that changes in land use play a critically important role in controlling the amount of regional runoff, particularly in the tropics, where the alteration of land has been the most pronounced. On average, changing land uses accounted for 50 percent of the increased flow of rivers over the last century.[2]

With respect to evaporation, deforestation is as large a driving force as irrigation in causing changes in the hydrological cycle. Deforestation has decreased the global flow of water vapor from land by 4 percent, a decrease that is quantitatively as large as the increased evaporative flow caused by irrigation. Although the net change in global evaporation is close to zero, the different spatial distributions of deforestation and irrigation lead to major transformations in the regional patterns of evaporation.

Although widespread deforestation in sub-Saharan Africa would decrease the evaporative flow from the land's surface, it would increase the over-the-surface flow of water, whereas intensified agricultural production in the Asian monsoon region would substantially increase irrigation and thus evaporation from the fields. Furthermore, significant modification in the flow of water vapor in the lands around the basin of the Indian Ocean would increase the risk of altering how the Asian monsoon system behaves. Thus, the need to increase food production in one region may affect the ability to increase food production in another.[3]

What lesson might this sense of relationship hold for us as human beings? Although each of us may indeed be a whole as an individual, we are nonetheless part of a greater whole, which is part of yet an even greater whole, ad infinitum. This condition makes us at once whole and only partial.

We are each unique in our personal history and our perception of the world based on that history, and yet we are simultaneously incomplete as an inseparable entity of the human family in time, space, and developmental history, all archived in our DNA. Holding the duality of our individual and collective selves in our consciousness instills in us an enduring need to express both our individuality and our universality—our separateness from and our union with others and the environment. In turn, we reconcile this union when we look deeply enough within ourselves as an interactive part of the universe.

Although we tend to think of freedom as the relaxation of some external constraint impinging on us (on our God-given free will), freedom is an inner state of consciousness and is totally dependent on ourselves. In this sense, our free will is the beginning (the cause) of a never-ending story in which everything is an eternal effect that becomes the cause of another effect—and so on and so on forever.

Our true freedom is proportional to our truthfulness and to the extent we align ourselves with the inner law of our being or metaphysical order (spirituality), as well as with the biophysical principles that govern nature. These principles were enshrined for hunter-gatherers within the sacred rituals they practiced in atonement for their violent use of the land and its creatures—the violence of killing, which neither they could nor we can avoid as a condition of life.

The upshot is that we are, at all levels of our being, both self-organizing through our thoughts and interdependent through our actions, a tension making it clear that the conscious integration (unity) of this pair of opposites is the indispensable means by which anything creative in our lives will materialize. For example, a person who studies a discipline, say an artist or a writer, must agree to be constrained by the technical aspects of the discipline—form and color for an artist and grammar for a writer—in order to gain freedom of expression within the discipline. Of late, however, "the laws of Congress and the laws of physics have grown increasingly divergent, and the laws of physics are not likely to yield," admonishes author Bill McKibben.[4]

To have a sustainable world, one that nurtures us, we must agree to abide within the constraints of the biophysical principles that govern Earth and our inseparable place within its indivisible spheres. We are not, after all, masters of nature—despite all the religious incantations. We are, instead, simply one creature among the many and thus an infinitesimal blip in the continuum of change. If, therefore, we are willing to reside within the constraints of biophysical integrity, we can live our lives with a sense of dignity and well-being—a worthy legacy to bequeath all who come after us.

To achieve this end, we must balance our focus on the differences we are so carefully taught to observe among the myriad components and relationships we call "nature" and the functional commonalities of the biophysical principles that govern those components and relationships within and among the three spheres. Only then will we begin to truly comprehend and appreciate our role as guardians of the biological living trust placed in our care on behalf of the beneficiaries—the children of the planet Earth.

1

Of Ignorance and Knowledge

If we could see all the never-ending stories occurring in the eternal now, we would be experiencing the unifying principle of the universe: The Eternal Mystery, the Original Storyteller, the Author of The First Never-ending Story–out of which all the others are born, wherein all the others are contained.–Chris Maser

Although ignorance is thought of as the lack of knowledge, there is more to it than that. Our sense of the world and our place in it is couched in terms of what we are sure we know and what we think we know. Our universities and laboratories are filled with searching minds, and our libraries are bulging with the fruits of our exploding knowledge, yet where is there an accounting of our ignorance?

Ignorance Is Simply a Lack of Knowledge

Ignorance is not okay in our fast-moving world. We are chastised from the time we are infants until the time we die for not knowing an answer someone else thinks we should know. If we do not know the correct answer, we may be labeled as stupid, which is not the same as being ignorant about something. Being stupid is usually thought of as being mentally slow to grasp an idea, but being ignorant is simply not knowing the acceptable answer to a particular question another person is interested in.

Without ignorance, knowledge could not exist because all knowledge is born of a question, albeit often an unconscious one. In turn, all questions reside in the domain of ignorance. Like every paradox, ignorance and knowledge are two halves of the same dynamic—consciousness.

However, society's preoccupation with building a shining tower of knowledge blinds us to the ever-present, dull luster of ignorance underlying the foundation of the tower, from which all questions must arise and over which the tower must stand. Although acquiring knowledge may reduce a sense of ignorance, the greatest danger for society is the delusion of omnipotence that accompanies the certainty of its knowledge—not the scope of its ignorance, which is infinite and thus grounds for humility. Nevertheless, the search for knowledge in the material world is a continual pursuit, but the quest does not mean that a thoroughly schooled person is necessarily an educated person or that an educated person is necessarily a wise person. And despite our most fervent wishes, knowledge is no guarantee of wisdom. In fact, teachers, in common with us all, are often blinded by the certainty of their knowledge and thus unable to comprehend the vast landscape of their ignorance. The only thing worse than not knowing is being unaware that you don't know.

"Man is much more afraid of the Light than he is of the Dark," observed astrologer Alan Oken, "and will always shield his eyes against a truth, which is brought to him prematurely. He will throw stones at it or even crucify it in order to remain in the comfortable shadow of his ignorance."[1] That is the unspoken motive behind the notion "ignorance is bliss," a notion through which we absolve ourselves of responsibility by pleading ignorance. People choose ignorance because knowledge is deemed too painful, too uncomfortable, or too limiting in the array of choices it allows.

Knowledge Is Some Version of the Truth

In the confrontation between a stream and a rock, the stream always wins
. . . not through strength, but through persistence.–Anonymous

Because knowledge is always relative, all we can navigate is an ever-shifting version of the truth through the accrual of knowledge. Therefore, we must learn to accept our ignorance, trust our intuition, and doubt our knowledge.

The realities we accept as obvious, neutral, objective, and simply the way the world works are actually structures of power, which we create as we think and live. They are created by our rendition of history and by our understanding of our society, our world, and ourselves within it. Moreover, our intellectual fabrications are always partial with respect to the whole.

Over the years I labored as a research scientist, I came to appreciate how much—and yet how very, very little—we humans understand about the three spheres of which we are an incontrovertible part. There is so much for us to learn about ourselves as individuals, as a species, and about the Earth we influence in our living, that I firmly believe the complexities of life and its living are permanently beyond our comprehension. The salient point, therefore, is not our knowledge but rather our ignorance, which is all that can be proven. The validity of our knowledge rests, albeit tenuously, before the jury of tomorrow, the day after that, and the day after that, ad infinitum. After all, it's wisdom that's sacred—not knowledge.

This is not to say that knowledge is unimportant. To the contrary, it is critical because a society held in ignorance is powerless to govern itself, as Thomas Jefferson so eloquently stated: "If a nation expects to be ignorant and free, it expects what never was and never will be." Ignorance, in this sense, is a lack of understanding the relationships necessary for a people to govern themselves in a dignified, sustainable manner, something Carl Rowan, author and journalist, understood well. "The library," he said, "is the temple of learning, and learning has liberated more people than all the wars in human history."[2]

Granted, an informed populace is a knowledgeable one, but what is it we humans study in order to become "knowledgeable"? Although the answer seems obvious, few people appear to grasp its simplicity. We study nature and nature's bio-physical principles. If this is not clear, think about it this way: we study the only thing we can study—ourselves in relation to life—through a variety of scientific endeavors. This kind of self-study includes the reciprocity of our relationship with our environment and one another as we are more roughly jostled together in an increasingly cluttered dimension of space and with the illusion of accelerating time.

In other words, if a nation does not have a populace well educated in the art of civil living (one that is willing to work together, exercise personal responsibil-ity, and accept accountability for both the short- and long-term consequences of their decisions and actions), military power can easily replace civil liberty. Hermann Goering put it bluntly at the Nuremberg trials:

> Of course the people don't want war. But after all, it's the leaders of the country who determine the policy, and it's always a simple matter to drag the people along whether it's a democracy, a fascist dictatorship, or a parliament, or a communist dictatorship. Voice or no voice, the people can always be brought to the bidding of the leaders. That is easy. All you have to do is tell them they are being attacked, and denounce the paci-fists for lack of patriotism, and exposing the country to greater danger. It works the same way in any country.[3]

All you need do to validate this point for yourself is look around the world today and take inventory of the suffering of those who daily struggle to survive under military-style, dictatorial governments and "would-be" governments. But it's not just religious fanatics or military dictators who use fear to control citizens; it's also monied interests that foster fear of weakening markets due to higher prices and loss of jobs as a means of avoiding expensive, mandatory emission controls that would begin to clean the air, water, and soil. Author and publisher Satish Kumar states the problem nicely:

> Livelihood is about quality of life; living standard is about quantity of material possessions.
>
> Education aimed solely at raising living standards relates to concepts of employment, jobs and careers based on individualism and personal suc-cess. Education for livelihood is just the opposite. It is about relationships, mutuality, reciprocity, community, coherence, wholeness, and ecology.
>
> Most schools and universities are dominated by materialist and con-sumerist goals. They have taken on the mission of literacy instead of meaning, information instead of transformation, and training instead of learning. Modern-day educators have become servants of the economy

and they are oblivious to the catastrophic consequences for the people and the planet.

"Education as usual" is no longer an option.[4]

In the end, however, it is through language that we accrue knowledge, a subject on which professor and author David Orr has some thoughts: "Unable to defend the integrity of words, we cannot defend the Earth or anything else. . . . The integrity of our common language, however, depends a great deal on the cultivation of discerning intelligence in the public, and that requires better education than we now have."[5] To this, Robert Lackey, of the U.S. Environmental Protection Agency, adds: "The meaning of words matters greatly and arguments over their precise meaning are often surrogates for debates over values."[6]

Language

Circumstance does not make the man; it reveals him to himself.–British philosopher James Allen, *As a Man Thinketh*

This revelation can take place only through language, ideas, and freedom of speech.

As part of that "better education," we need to protect one aspect of our culture that we normally neglect: language. Perhaps one of the greatest feats of humanity is the evolution of language, especially written language—those silent, ritualistic marks with their encoded meaning that not only made culture possible but also archive its history.

As I mull over the probable events that led to our modern, human languages, it occurs to me that all words are the names of things, be it a touchable entity (a flower, animal, or tool—each a noun); a quantifiable time (a second, an hour, today, yesterday, tomorrow, next year—each a noun); an action (do, run, sit, speak—each a verb); or a description of the qualities of something else (pretty, ugly, hairy, large, small, fast, slow—each an adjective), in time (now, earlier, later—each an adverb) and as a degree (very, exceedingly, little, much—each an adverb or an adjective). Put differently, words define the mental boundaries of our perceptions. A child points to something, hears the utterance of sound from an adult in response to the gesture, and, lo, the rudiments of meaning are born. With repetition, a boundary of meaning (or definition) is established.

Hence, words define the mental limitations of our perceptions. When we speak, therefore, we are transferring our sense of discernment into the meaning attached to names of things, time, actions, and qualifiers, all of which are in some way concrete. With the invention of each new word (each new name), we humans are endeavoring to simultaneously explore, define, and refine the boundaries of meaning attached to our perceptions of the world around us—boundaries encompassed in the names by which we recognize what we see. In essence, we are attempting to express our perceptions by assigning definable limits of meaning to the various categories of words we use, which is like trying to fence a portion of the sky in order to own a particular constellation of the stars.

Although many people believe words carry meaning in much the same way as a person transports an armful of wood or a pail of water from one place to another, words never carry precisely the same meaning from the mind of the sender to that of the receiver. In this sense, language, in its fullest experience, is so much more than mute scratches on paper, repetitive configurations on computer screens, or even the utterance of predetermined sounds.

Words are vehicles of perceived meaning. They may or may not supply emotional meaning as well. The nature of the response is determined by the receiver's past experiences surrounding the word and the feeling(s) it evokes. Therefore, the lack of a common experience or frame of reference is probably the greatest single barrier to mutual understanding.

Feelings grant a word meaning, which is not in the word itself, but rather originates in both the sender's mind and the receiver's mind based on personal experience. Because a common frame of reference is basic to communication, words are meaningless in and of themselves. Meaning is engendered when words are somehow linked to one or more experiences shared by the sender and the receiver, albeit the experiences may be interpreted differently. Words are thus merely symbolic representations that correspond to anything people apply the symbol to—objects, experiences, or feelings.

As words accumulate, we consciously construct sounds to form new words, such as "thingamajigs," "doohickeys," and "whatchamacallits," based on utility or aesthetics or both. Think for a moment about the invention of a word and the subsequent conveyance of its meaning through the ever-shifting sands of time. How, for example, did a word like *floccinaucinihilipilification* come into being? What does it mean?

With twenty-nine letters, it is the longest non-technical word in the original edition of the *Oxford English Dictionary*, which dates the word's first use in literature to 1741 in William Shenstone's *Works in Prose and Verse*: "I loved him for nothing so much as his flocci-nauci-nihili-pili-fication of money."

Although the dictionary gives no specifics on its derivation, the word was supposedly invented as a joke by a student at Eton College who constructed the word by combining and adding suffixes to four words in a Latin textbook connoting "nothing" or "worthless."

floccus,—i, a wisp or piece of wool, used idiomatically as *flocci non facio* ("I don't give a hoot")

naucum,—i, a trifle

nihil,—is, nothing; something valueless (lit. "not even a thread" from *ni+hilum*)

pilus—i, a hair; a bit or a whit; something small and insignificant[7]

It is often spelled with hyphens, and has even spawned *floccinaucical* (inconsiderable or trifling) and *floccinaucity* (the essence or quality of being of small importance). The dictionary appears to have overlooked *floccinaucinihilipilificatious*, however, which has one letter more than the nominal form, and means "small" or "insignificant." When the common English nominal suffix *ness* is then added to the above adjective, a thirty-four-letter noun *floccinaucinihilipilificatiousness* is formed, which means "smallness" or "insignificance."[8]

To take another example, how did the words *atmosphere, litho-hydrosphere*, and *biosphere* come into being? How many millions of sounds did it take over the millennia of human evolution to reach the point where such words could even be conceived? How can we define their borders when they encompass every interdependent relationship embodied in the past, present, and future of planet Earth?

There are two caveats to the use of words, however wonderful they may sound, both embodied in Arab proverbs. The first cautions that "while the word is yet unspoken, you are master of it; when once it is spoken, it is master of you." The second stipulates that "each word I utter must pass through four gates before I say it. At the first gate, the keeper asks: 'Is this true?' At the second gate, the keeper asks: 'Is it necessary?' At the third gate, the keeper asks: 'Is it kind?' At the fourth gate, the keeper asks: 'Is this something you want to be remembered for?' If you doubt the truth of the first proverb or the wisdom of the second, notice that words set things in motion, and motion is nothing more or less than a continuous stream of cause and effect—a never-ending story, as it were—for which your word was responsible.

The wise use of words is therefore critically important to human survival because we are creatures who must share life's experiences with one another in order to know we exist and have value. The greatest poverty in the world of humanity is not being wanted and so being denied the heart and soul of human existence—love and compassion, which translate into recognition as a human being.

Reality, however, is beyond language because words are merely metaphors through which I attempt to transmit feelings and understanding by sending a verbal or written message. You, in turn, must receive, translate, and comprehend what you think I mean. The challenge is that I cannot express verbally how I feel or what I really mean; therefore, you can receive only an approximation of what I intend, after which you must translate what you think I indicated based on your understanding of the words I have used. Further, your understanding of the words is based on your experiences in life, which are different from mine—even if we're identical twins.

The relative independence with which cultures evolve creates their uniqueness both within themselves and within the reciprocity they experience with one another and their immediate environments. Each culture, and each community within that culture, affects its environment in a specific way and is accordingly affected by the environment in a particular way. So it is that distinct cultures in their living create in the collective varied, culturally designed landscapes, which in some measure are reflected in the myths they hold and the languages they speak.

Therefore, saying what we mean and meaning what we say to the very best of our ability are critical to our survival as a species. To this end, science is both a discipline and a language that helps us define boundaries of meaning surrounding the words we use while navigating the three spheres of our magnificent planet.

Science

Science does not deal with the whole of nature; other human faculties must be activated.–Klaus K. Klostermaier, German author and Hindu scholar

Another part of the "better education" Orr spoke of will be to require scientists to understand and accept that although they may attempt to be objective, they will never be successful because, being a part of nature, they must participate with nature in order to study nature. Moreover, we humans are subjective creatures and thus incapable of impartiality, if for no other reason than every question we ask is subjective.

In 2004, eight people met at Le Chateau De Taureene, built in the twelfth century by the Knights Templar near the village of Aups in France. One of their conclusions was: "The paradox today is that science, in its obsession with a materialistic and mechanistic view of nature, has itself become more like a religion [with the corresponding certainty that its point of view represents the truth] than the neutral pursuit of knowledge and understanding."[9] Nevertheless, science—despite its inherent subjectivity—is designed to help us humans understand the biophysical principles governing the three spheres and how they relate to us as an indivisible, functioning part of the natural world.

That notwithstanding, it's impossible to accurately represent nature through science because scientific knowledge is not only a socially negotiated, rigid construct but also a product of the personal lens through which a scientist peers. Scientists may attempt to detach themselves from nature to become objective, but they are never successful if for no other reason than the unavoidable subjectivity of the questions they ask to guide their inquiry. In addition, every person is a part of nature and must participate with nature in order to study nature.

We scientists see but dimly what is before us—first because we cannot detach ourselves from nature and second because our perceptions, colored as they are by our personal lens, are all we can judge as fact. We may intellectually polish our lenses, but appearance—not reality—is all we can ever hope to see, and so it is appearance to which we often unknowingly direct our questions. Knowledge is a collective outer experience of humanity's, and society's, subjective judgments about things, the truth of which cannot be known and therefore is explainable only in the illusions of its appearance. Thus, the actual objects of our inquiries, the formulations of our questions and definitions, and the mythic structures of our scientific theories and facts are social constructs.

Even history, which we tend to accept as fact, is merely someone's perception, and thus interpretation, of what actually happened. Ask ten soldiers what happened on the battlefield, and you will get ten different answers. The British historian Arnold Toynbee recognized this reality when he wrote, "My view of history is itself a tiny piece of history; and this mainly other people's history and not my own."[10]

The irony of scientific research is that nothing can be proven—only disproved. Therefore, we can never "know" anything. We can know only through intuition—the knowing beyond knowledge, which is inadmissible as evidence in modern science. Whatever truth is, it can only be intuited and approached, never caught and pinned down.

Knowledge, therefore, which is external to a person, is not knowable, and intuition, which is internal to a person, is not knowledge and therefore not subject to disproof. Intuition is inner sight—individualized, inner knowing that is beyond knowledge—for which proof is unnecessary and explanation impossible.

Intuition, a mode of knowing widely accepted during ancient times, has been clouded with ambiguity and controversy since the advent of the reductionistic, mechanical mindset, which swept Western industrialized society more that a century ago. For some, intuition is merely a meaningless after-thought of unconscious processes, but for others it's a harbinger of the deepest truths.

Intuition, an instantaneous, direct grasping of reality, is the source of our deepest truth, that sense of unquestionable knowing, of which even John Stuart Mill, a pillar of the empirical method, said, "The truths known by intuition are the original premises from which all others are inferred." Even so, our contemporary scientific theories, facts, and practices—including the scientific method itself—are largely, but not totally, expressions of today's social, political, and economic interests. In other words, they are expressions of cultural themes and metaphors and personal biases, as well as of personal and professional negotiations for the power to control, albeit minutely, the scientific knowledge of the world.

The problem is that we confuse the limiting nature of scientific method with the nature of ordinary observation and experience. Whereas the scientific method tends to be reductionistic and mechanical, life experiences are holograms made of complex threads of past experiences, present perceptions, suggestions of future opportunities, and portends of future disasters. Life is thus a moment-by-moment kaleidoscope of perceived realities, which may or may not have anything consciously in common with reality as reality. To understand what I mean, let's consider commonalities, cycles, and feedback loops, which in concert create the "commons."

Commonalities

As a boy in European boarding schools during the early 1950s, I noticed that some of the lizards were similar to those at home in Oregon, and some of the bird songs also sounded the same. A decade later, while working in Egypt, I found a skink, which is a type of lizard, that was similar to the western skink at home—the young of both had bright, blue tails. I also discovered that the desert gerbil was in many ways similar to Ord's kangaroo rat of the Great Basin in the western United States.

In 1967, while working in Nepal, I heard a great green barbet (a bird), which sounded so much like the western tanager in the coniferous forests of Oregon that, for a moment, I forgot where I was. Moreover, the jungle cat, Himalayan weasel, and the Nepalese golden jackal are similar in habits to the long-tailed weasel, bobcat, and coyote of North America.

Wherever I have been, from Oregon to Alaska and Canada, from Egypt to Nepal, Japan, Eastern Europe, Malaysia, and Chile, I have seen the similarities of nature. And yet, as a scientist, I was trained to focus on nature's differences, which I suppose should come as no surprise because American culture is primarily a divisive one that tends to focus on our differences as a collective people rather than on our similarities. In some respects, we are still a nation of settlers from many different cultures, staking our own claims, and competing for resources through the "money chase."

Today, as I look around a world at war in one way or another, from an all-out conflict in Iraq, Afghanistan, and some African nations, to the local insurgency of a drug cartel, or humanity versus nature, I find the outcome to be the same—an exaggeration of differences and the attendant sense of insecurity we humans let these perceived dissimilarities breed. I see an increasingly terrified global society because some of its factions focus so narrowly on the perceived, irreconcilable differences of right versus wrong, of us versus them fostered by the various religious beliefs. Yet I look at the same world and see the seemingly forgotten similarities—the commonalties of life: the basic human needs for love, trust, respect, and dignity, as well as the biophysical principles of nature that govern the three spheres.

Consider, for instance, that the citizens of New York in the United States and the citizens of Nairobi in Kenya are almost poles apart, geographically speaking. Given that, one would expect striking differences between them, and there are. But as much as they differ, they also have much in common. If, however, we observe only the differences, we see but a little of what makes us all human. The larger part comprises the commonalities to which we are often blinded by our choice of focus. Therefore, all we are judging is our perception of an appearance, and our observation is always a partial view, despite our knowledge and best intentions, as exemplified by the distribution of mosquitoes.

If we compare ten individuals of the same kind of mosquito found in the city of Kuala Lumpur, which is near the equator on the west coast of the Malaysian Peninsula, with ten individuals of the same kind of mosquitoes found in the city of Cayenne, French Guiana, which is near the equator on the east coast of South America, we would probably classify the two groups as distinctly different species—a classical, clear-cut "either/or" categorization, based on their extreme divergence in appearance. But if we were now to collect mosquitoes from around the world along the equator, what would we find when they were compared? That would depend on how we compared them.

If we compared each group of ten mosquitoes with their nearest neighbor (ten from location A with ten from location B), we would find an astonishing degree of similarity between the two samples. If we then compared ten from location B with ten from location C, and ten from location C with ten from location D, we would find not only continual similarities but also gradual differences. Should we then compare samples of mosquitoes from location A with location D, we would begin to see increased divergence in characteristics.

So, instead of the neighboring samples of mosquitoes being unequivocally this or that because of clearly distinguishable characteristics, they form a continuum of gradually changing characteristics from A to Z. Despite how similar or dissimilar any two samples of mosquitoes may appear, the same examination will demonstrate the shared elements of design, such as the basic structure and function of their wings, antennae, mouthparts, body, and so on. When viewed as a continuum, the mosquitoes form a circle of similarities that reveal the commonalties of design elements necessary if they are all to be aerodynamically capable of flight and sucking blood, among other shared characteristics that make a mosquito a mosquito. Ultimately, the inviolate, biophysical principles that govern all processes simultaneously bind together the universal commonalities while allowing the novelty of differences, such as those found in large populations.

Populations of animals tend to become increasingly generalized when they are released from competition with other species, a process that can either increase the variation among individuals or broaden how an individual uses the habitat. In practice, ecological generalists, which use a wide diversity of resources, are heterogeneous collections of relatively specialized individuals, meaning they exhibit stronger behavioral specialization in how they use the habitat as individuals within the generalized behavioral patterns of the species as a whole. Hence, populations of generalists overall may tend to be more ecologically variable than populations of specialists.[11] As well, the Earth's crust is composed of spatially heterogeneous abiotic conditions that provide a greater diversity of potential niches for plants and animals than do homogeneous landscapes.

For example, in one study the richness and diversity of trees and shrubs was significantly greater in sites with high geomorphological heterogeneity than in sites exhibiting little change in either the terrain or the types and conditions of the soil, such as texture, mineralogy, and organic matter. Variations in aspect and the soil's patterns of drainage were especially important predictors of biotic diversity. Hence, geomorphological heterogeneity plays a major role in determining species richness. Because biotic and abiotic diversity are intricately linked at the scale of the landscape, conservation of geomorphological heterogeneity may have significant implications for long-term strategies in caring for our natural environment, to the benefit of maintaining biodiversity in all its forms.[12] With respect to biodiversity, professor David Pearson, from the University of Arizona, Tempe, puts it nicely:

- Ecosystem processes: Biodiversity [which includes both genetic and functional diversity] underpins the processes that make life possible. Healthy ecosystems are necessary for maintaining and regulating atmospheric quality, climate, fresh water, marine productivity, soil formation, the cycling of nutrients, and waste disposal.

• Ethics: No species and no single generation has the right to sequester Earth's resources solely for its own benefit.

• Aesthetics and culture: Biodiversity is essential to nature's beauty and tranquility. Many countries place a high value on native plants and animals. These contribute to a sense of cultural identity, spiritual enrichment, and recreation. Biodiversity is essential to the development of cultures.

• Economics: Plants and animals attract tourists and provide food, medicines, energy, and building materials. Biodiversity is a reservoir of resources that remains relatively untapped.[13]

The ultimate commonality in today's world, however, may well be the variability of our changing climate.

Climate has been dynamic throughout the various scales of geological time, and it will continue to be the main driver of our planet's story of novelty within and among the three spheres. Because climate is still the primary mechanism through which the distribution of species and ecosystem processes is controlled, new twenty-first-century climatic conditions may promote the formation of heretofore-unseen associations of species and other ecological surprises. By the same token, the disappearance of some existing climatic conditions will increase the risk of extinction for species with narrow geographic distributions or climatic tolerances, and it will disrupt other communities.

Novel climates are projected to develop primarily in the tropics and subtropics, whereas disappearing climates will be concentrated in tropical montane regions and the pole-ward portions of continents. As well, some extant climates will disappear, and new ones will appear. Consequently, species with limited abilities to disperse will experience the loss of existing climate and the occurrence of novel ones.[14] As the climate is altered, so are the biophysical cycles.

Cycles

Life is composed of rhythms ("routines" in the human sense) that follow the cycles of the universe, from the minute to the infinite. Each cycle is a curvilinear spiral, which, like the coils of a spring, approximates—only approximates—its beginning, but always at a higher level in the spiral because of the irreversibility of change.

We humans most commonly experience the nature of cycles in our pilgrimage through the days, months, and years of our lives wherein certain events are repetitive—day and night, the waxing and waning of the moon, and the march of the seasons, all marking the circular passage we perceive as time within the curvature of space. In addition to the visible manifestation of these repetitive cycles, nature's biophysical processes are cyclical in various scales of time and space.

Some cycles revolve frequently enough to be well known in a person's lifetime. Others are completed only in the memory of several generations—hence the notion of the invisible present. Still others are so vast that their motion can only be assumed. In reality, however, even they are not completely aloof because we are kept in touch with them by our interrelatedness and interdependence. Regarding cycles, farmer and author Wendell Berry said, "It is only in the processes of the natural world, and in analogous and related processes of human culture, that the new may grow usefully old, and the old be made new."[15]

When thinking about landscapes, I am often reminded of the fires, both large and small, that over the millennia shaped the great forests I knew as a youth. With that memory comes the realization that no forest (or any other biotic community) has either a single state of equilibrium or a single deterministic pattern of recovery. Those fires were a selective force that killed or wounded susceptible plants and affected the environment of others. Through the pen of evolution, the plants may have authored their own fate, thereby influencing the environment they inhabited as well as that which they passed to their offspring.

But nature's cycles are not perfect circles, as they so often are depicted in the scientific literature and textbooks. Rather, they are a coming together in time and space at a specific point, where one "end" of a cycle approximates—but only approximates—its "beginning" in a particular place. Between its beginning and its ending, a cycle can have any configuration of cosmic happenstance.

In this sense, nature's ecological cycles can be likened to a coiled spring insofar as every coil approximates the curvature of its neighbor but always on a different spatial level (temporal level in nature), thus never touching. The size and relative flexibility of a spring determines how closely one coil approaches another. But, regardless of the size and flexibility of the spring, its coils are forever reaching onward.

With respect to nature's ecological cycles, they are forever reaching toward the novelty of the next level in the creative process and so are perpetually embracing the uncertainty of future conditions. In thinking about the great forests I used to know, and those parts through which I can still hike, I am awed by all the factors that must come together to create a particular place as I perceived it or remember it, not just the events themselves but also the cycles in which the events are embedded.

A forest is the collective outcome of interdependent processes in relation to time, completing its cycle only in the memory of several human generations. We do not seem to understand this time frame, however, or we ignore it, because all our models—economic, managerial, and even ecological—tend to be short-term and linear. Our models are simple not only because we chose them to be so, based on some immediate interest, but also because we do not have the capability to construct them in any other way.

Thus, although our models can predict only in a straight line in the very short term, the cyclical nature of the forest touches that line for only the briefest moment in the millennial life of the soil, the womb from which the forest grows. Yet, it is in this instant, with grossly incomplete, shortsighted knowledge, and unquestioning faith in that knowledge, that we predict the sustained yield of all our management into the unforeseeable future.

Feedback Loops

One area of our inadequate knowledge is an understanding of the feedback loops, all of which are self-reinforcing. To fully appreciate exactly what a feedback loop is, let's visit the whistling-thorn acacias of Kenya's savanna in Africa. Unlike many acacias, the whistling-thorn does not deter herbivores through the production of toxic compounds. Instead, it recruits colonies of ants as bodyguards against hungry herbivores eager to chomp its leaves, such as giraffes and elephants. At the slightest movement of a branch the ants, which live only in these acacias, swarm out and deliver painful stings to munching giraffes, elephants, or other browsers.

The whistling-thorn acacia is a fair employer, however. In addition to having regular thorns, it also has modified pairs of thorns, which are joined at the base by a hollow, bulbous swelling (called a *domatia*) that is up to a little more than an inch in diameter. These thorns provide excellent nesting sites for the ants. In addition, special glands at the tips of their leaves produce a sweet secretion for the ants to eat.

Savage competition for the whistling-thorn exists among the four species of ants that attend to it. When the branches of one tree form a bridge to another, the ants invade their neighbors and battle violently until one colony wins control of the tree, after which the colony may grow to be one hundred thousand strong. The black-headed ant, which is the least warlike, comes out very badly in these battles, losing more of its population than any of the other three species.

To defend their trees against invasion, black-headed ants actively chew off all horizontal shoots, which causes the trees to grow tall and skinny and thereby avoid contact with trees that host enemy colonies. Pruning also causes the tree to allocate more energy to new shoots, healthier leaves, and larger nectaries, but unfortunately the ants also prune off all flower buds so the tree is effectively sterilized. Perhaps the tree trades reproduction for increased vigor and protection from browsing animals. As it turns out, however, the black-headed ant's relationship with its acacia is more parasitic than mutually beneficial.

In comparison, the mimosa ant is not only the most antagonistic but also the most cooperative partner with its acacia. These ants rely heavily on the swollen domatia for shelter and are formidable protectors in return. But with no herbivores around to browse on the leaves, the ant's services are not required, and the partnership begins to sour at both ends. The tree begins to evict the ants by shrinking its pro-ant services—namely, reducing the output of its nectaries. With less food and smaller homes, the ants are twice as likely to farm sap-sucking scale insects, whose waste fluid is a sugary liquid called honeydew, which the ants drink, but to make it, the scale insects must suck the juices of the tree. Consequently, the ants are less likely to marshal a defense against such marauding browsers as giraffes and elephants.

Conversely, Sjöstedt's ants actually seem to benefit from a tree's reduced investment in maintaining the aggressive mimosa ants. Less common than mimosa ants, Sjöstedt's ants take a more relaxed attitude toward the partnership, one that could even be viewed as parasitic because it defends the tree less aggressively and ignores the swollen domatia. Instead, it occupies boreholes excavated by beetle larvae.

Because Sjöstedt's ants are dependent on these beetle-created holes, they facilitate the beetle's ability to feed on the trees. The ants don't get upset with the suffering of their competitor, however. When the acacias reduce their provisions, Sjöstedt's ants simply more than double the members of their colony.

Penzig's ant, which is the only species that does not eat the nectar produced by its host acacia, actively destroys the nectar glands in order to make a tree less appealing to the other species. Consequently, the mutualistic feedback loops between whistling-thorn acacias and resident ants break down in various ways in the absence of large herbivores, and the acacias become less healthy as a result. Large herbivores are therefore critical components in the never-ending stories of these dynamic systems. For want of a giraffe or elephant to munch on the trees, the protective ants diminish and leave the whistling-thorn acacia in dire straits.[16]

Every system in the universe is governed by self-reinforcing feedback loops, which are little understood and thus virtually ignored in "managing" ecosystems for profit. Nevertheless, when our monetary expectations fall short of the predicted yield, we call on science.

Science, however, is ultimately a discipline of disproof, whereby something can be proven only by its actual occurrence. Therefore, proof comes after the fact—not before it. Furthermore, science and technology have no sensitivity, make no judgments, and have no conscience. It is neither scientific endeavors nor technological advances that affect our collective environment, but rather the thoughts and values of the people who use the technology, who influence our overall respect for, or abuse of, the three spheres that in concert form the "commons" in which we live.

The Commons

The commons is that part of the world and universe that is every person's birthright. There are two kinds of commons. Some are gifts of nature, such as clean air, pure water, fertile soil, a rainbow, the northern lights, a beautiful sunset, or a tree growing in the middle of a village; others are the collective product of human creativity, such as the town well from which everyone draws water.

Scattered throughout various parts of the world there still exists a tree in the middle of the square around which village life revolves. In this quaint meeting place neighbors form bonds with one another, children play games, women visit about the affairs of life, and men discuss work and politics. Here, old and young mingle in a way that bridges the generations in the flow and ebb of village life, and children still experience an unstructured and noncompetitive setting in which their parents are close at hand. As such, a village commons is far more than simply a public space around a tree. It's the center in which the life of true community blossoms because it has the scale of a human face.

The commons is the "hidden economy, everywhere present but rarely noticed," writes author Jonathan Rowe.[17] It provides the basic ecological and social-support systems of life and well-being. It's the vast realm of our shared heritage, which we typically use free of toll or price. Air, water, and soil; sunlight and warmth; wind and stars; mountains and oceans; languages and cultures; knowledge and wisdom; peace and quiet; sharing and community; joy and sorrow; and the genetic building blocks of life—these are all aspects of the commons.

The commons has an intrinsic quality of just being there, without formal rules of conduct. People are free to breathe the air, drink the water, and share life's experiences without a contract, without paying a royalty, without needing to ask permission. It is simply waiting to be discovered and used.

If a good swimming hole exists, people will find it. If a good view exists along a trail, hikers will stop and enjoy it. There is no need to advertise a commons; it will be found.

A commons engages people in the wholeness of themselves and in community. It fosters the most genuine of human emotions and stimulates interpersonal relationships in order to share the experience, which enhances its enjoyment and archives its memory. To protect the global commons for all generations requires that we begin immediately to question the biophysical sustainability of our current decisions and their outcomes.

Consequently, if we are going to ask intelligent questions about the future of the Earth and our place in the scheme of things, we must understand and accept that most of the questions we ask deal with cultural values; such questions cannot be answered through scientific investigation. Nevertheless, scientific investigation can help elucidate the outcome of decisions based on those values, and it must be so employed. It's also imperative that we are free of opinions based on "acceptable" interpretations of scientific knowledge. In addition, we would be wise to consider the gift of Zen and approach life with a "beginner's mind"—one simply open to the wonders, mysteries, and innovative possibilities of a multiplicity of realities.

Although a beginner's mind is the ideal, society today is confronted with the specter of global warming and is thus beset with the need to know and the fear of knowing. This fear is more generalized than at any time in my life since World War II, in part because the world appears to be aflame with violence of all kinds, a perception heightened by the media's instantaneous and selective coverage. In addition, increasingly bold corruption slithers through the halls of government, while the thick cloak of secrecy hides collusion that erodes public confidence every time a person in a position of civic leadership does not accept responsibility for misconduct.

Nevertheless, the guild of politicians is writing, editing, rewriting, and reediting the next scene of the social-environmental play, and the next, and the next—more often than not with scant regard for the effects set in motion by the script. And the thematic undercurrent is "informed denial," which hints that negative tradeoffs embodied in the script are illusionary. Thus, we continue to direct our impromptu social drama—with little or no idea of the long-term consequences of what we are staging.

Everything Has a Tradeoff

No person, institution, or nation has the right to participate in activities that contribute to large-scale, irreversible changes of the Earth's biogeochemical cycles or undermine the integrity, stability, and beauty of the Earth's ecologies–the consequences of which would fall on succeeding generations as an irrevocable form of remote tyranny.–Professor David W. Orr, "Saving Future Generations from Global Warming," *Chronicle of Higher Education*, April 21, 2000

Informed denial is indeed a tragic script because every line has within it the seed of its opposite in the form of a tradeoff. It's axiomatic that whatever we do to the three spheres of nature, we do not only to ourselves but also to our descendants—all of them.

The Tradeoff of a Personal Decision

The following discussion on the environmental effects of marriage and divorce is based on a study by Jianguo Liu and Eunice Yu, which was published in a 2007 issue of the *Proceedings of the National Academy of Sciences*.[18] Neither the study (which is based on observable, quantifiable data) nor my use of it to illustrate the tradeoff of a personal decision and the environmental consequences one might not think about, is in any way connected to a moral judgment.

When people get married, or otherwise agree to live together, they consolidate their life requirements, including the space they occupy. By the same token, when people get divorced, they disperse not only their physical presence but also have a greater impact on the environment from which they glean the materials they need to live. And the rate of divorce is increasing throughout the world, even in the non-industrialized countries and in nations with strict religious or cultural taboos regarding this practice.

To examine this issue, let's consider the United States, where, according to data gathered by the Census Bureau, a single, divorced person headed nearly 15 percent (almost sixteen million) of the households in the year 2000. This separation of families aggravated urban sprawl by requiring six million more dwellings, which increased the number of rooms requiring heating and cooling by nearly thirty-six million, which, in turn, affected forests because home construction necessitates cutting more trees.

In addition, removing those trees can negatively affect the biophysical integrity of the water catchments from which they are taken, and that, in turn, can affect the quality of the water and its availability. Now add the fact that, in 2005 alone, households headed by a divorced person increased the annual use of water by more than 627 billion gallons at a cost of $3.7 billion—a 56 percent increase in cost over the homes occupied by married couples. Moreover, the demand for electricity, also in 2005, rose by nearly 73.5 billion kilowatt-hours, which amounts to approximately 2 percent of the nation's total use, at a cost of nearly $7 billion (46 percent more than that paid by married couples)—some of which came from hydropower. And here is an obvious highlight that is seldom considered as a tradeoff of separation: a refrigerator uses roughly the same amount of energy whether it occupies a home of four, two, or even one person(s).

In addition, to accommodate the six million extra homes, municipalities are obligated to produce and maintain other infrastructure, such as paved roads, streets, and sidewalks, which creates an impervious coating over the surface of the land. This impervious layer prevents both rain and melting snow from infiltrating the soil, where it can be stored, can be further purified, and can recharge ground-water and existing wells and nearby streams. Instead, the water collects on the surface of the roads and streets, where it mixes with pollutants that accumulate on the pavement.

Because paved roads and streets are lined with curbs and gutters, the now-polluted water is channeled into a storm drain. In addition, each house has an impervious roof, which forces precipitation into gutters. Water from the gutters is often discharged onto the street, where it also flows into the storm drain, from whence it's conducted either directly into a sewage treatment plant or into a ditch, stream, or river.

In any event, the water is not usable by the local people. Furthermore, the storm water adds to the cost of running the treatment plant, where it must be detoxified to prevent it from polluting the watercourses on its way into the ocean.

The effect of roads and houses, both of which eliminate the ground infiltration of water, is cumulative. Enough roads over time can alter the soil-water cycle in a given community. Although the quality and quantity of water is an ecological variable, most economists and land developers treat it as an economic constant in their calculations—or they discount it altogether.

In addition to the water runoff caused by streets, they are salted today as well. For most of human history, salt was a precious commodity, usually thought of as a flavoring for food. Today, however, almost twenty million tons of rock salt are spread on paved surfaces in the United States every winter to deice sidewalks and roads, thereby making them safer for people and vehicles. Although the use of salt on roads has skyrocketed since the 1940s, the annual application is forgotten once it dissolves and washes into the soil and thus eventually into streams. If the impervious surfaces of roads, parking lots, and sidewalks continue to expand, along with the current practice of deicing with rock salt, many surface waters in the northeastern United States will be unfit for human consumption within this century. Even if we were to ban rock salting today, it will require decades, if not centuries, to return groundwater to its natural freshness (zero salinity).

Once in streams, salt can kill plants and fish, contaminate groundwater, and generally make regional supplies of municipal water unfit for human consumption. In fact, salting roads has already contaminated at least twenty-three springs in the greater Toronto area of Ontario, Canada.[19]

Still, even if water were a constant, a variable is introduced with the construction of a single road, to say nothing of the multiple logging roads that would be needed to transport the timber to build those extra six million homes—and that was just one year's census data. In addition, exploitive logging and tree-farm management alter the water regime, which affects how the forest grows. Thus, a self-reinforcing feedback loop of ecological degradation in a water catchment is created, altering the soil-water regime, which in turn alters the sustainability of the forest, which in turn affects the soil-water regime, and so on. Eventually, the negative effects are felt in all communities that are dependent on a given water-catchment or drainage basin for their potable water.

In the final analysis, we must remember that only so much water is available, and, with a change in the global climate, that amount may become even more variable and unpredictable than it already is. And additional water cannot be found in the courtroom, no matter who holds custody papers for the children. It behooves us, therefore, to consider how we care directly for each other and thus indirectly for the sustainability of water catchments—lest the wells go dry.

Besides, the bourgeoning population, coupled with the increased ecological footprint per person that the extra houses produce, is the most relevant anthropogenic factor affecting the loss of biodiversity in the United States today.[20] A study of the exurban lands in Boulder County, Colorado, corroborates this point of view.

Traditionally, undeveloped lands have been subdivided into a grid of parcels ranging from five to forty acres. From an ecological perspective, this pattern of dispersed development maximizes the influence of each home on the land. Clustered housing, however, is designed to maximize open space, and is assumed to benefit indigenous plant and wildlife communities. In the study, four indicators were used to assess this idea: (1) densities of songbirds, (2) nest density and survival of ground-nesting birds, (3) presence of mammals, and (4) percentage cover and proportion of native and nonnative species of plants.

As it turns out, clustered and dispersed housing developments did not differ on the majority of variables examined. Both types of housing developments had significantly higher densities of non-indigenous and human-commensal species (which also tend to be generalists in their behavioral patterns), and far lower densities of indigenous and human-sensitive species. (Commensal is the relationship between two different species from which one derives food and other benefits without negatively affecting the other, such as house sparrows benefiting from their association with humans in town.) Conversely, the undeveloped areas in the study had significantly higher densities of indigenous and human-sensitive species.[21]

Here, the positive benefits of marriage counseling might be a real incentive to reduce one's ecological footprint for the sake of one's children, their children, and beyond, because getting back together can work to a child's advantage while reversing the negative, environmental effects of separation. If a personal choice can have these kinds of impacts on the environment, what effects might a commercial decision have?

The Tradeoff of a Commercial Decision

In 2006, a virulent outbreak of food poisoning associated with fresh spinach from a single farm in California swept the nation, despite existing rules that limit how close to livestock, composting manure, surface waters, and noncrop vegetation farmers may plant their crops. Moreover, other regulations require the frequent testing of irrigation water for contaminants and strict practices of sanitation for farm workers and the use of tools.

The outbreak was caused by a specific bacterial strain of *Escherichia coli*, which is transmitted through animal feces. In addition, antibiotics given livestock to promote their growth escape into the soil and water, where they become incorporated into the tissues of plants, such as spinach and lettuce.[22]

As a consequence, new restrictions—the Leafy Greens Marketing Agreement— were added to the already-existing ones to protect consumers. Although protecting the health of the public is necessary, these new constraints have negative environmental consequences that will cause far-reaching problems across the entire country.

The new parameters encourage the elimination of fencerows and hedgerows created to control soil erosion because they might harbor wild animals that carry disease. However, killing unwanted vegetation with herbicides or plowing the edges of fields to eliminate fencerows and hedgerows can reduce crop yields by removing nearby nesting and rearing habitat for pollinators, such as bees and flower flies, as well as the nectar corridors they rely on when not servicing crops. In addition, many species of insectivorous birds nest in such well vegetated, tree-lined fencerows and help to control insects that are deleterious to the farm crops.

Nevertheless, as a consequence of the new regulations, some farmers are now clearing all vegetation, thereby creating bare-ground buffers around their fields, activities that will increase soil erosion. Others are putting out poisons to kill rodents—poisons that not only can kill non-target wildlife and poison the soil where they die but also can get into the groundwater, ditches, streams, rivers, estuaries, and oceans. In any case, nothing is foolproof: slugs can carry the same strain of *E. coli* found in the spinach.[23]

Although there is much people can do to prevent this kind of outbreak, it takes vision, sincere caring about other people, and personal commitment, as well as political will, to address the issues before they become problems. And doing so takes a higher level of consciousness than is evident in today's market-driven agribusiness.

The Collective Tradeoff of Social Decisions

Although the previous tradeoffs can be traced to definite decisions and the consequences they set in motion, the overall complexity of social actions and their long-term environmental effects are unimaginable. The intricacies of the self-reinforcing feedback loops forming the three spheres are not only influenced by the collective behavior of individual humans but also return to affect people on an individual basis. The effects of our choices range from the most complex and integrated, such as the direct pollution of the air, which indirectly pollutes the soil and the waters of the world, to something as relatively simple as an increase in kidney stones. A case in point is the likely northward expansion of the present-day, southeastern U.S. *kidney-stone belt* (Alabama, Tennessee, South Carolina, North Carolina, Mississippi, Louisiana, Georgia, Florida, and Arkansas), so called because these states have higher incidences of kidney stones than the rest of the country.

An unanticipated result of global warming is the percentage of the population that will ultimately suffer from kidney stones—technically termed *nephrolithiasis*. Kidney stones, which are formed from dissolved minerals that concentrate in low volumes of urine, result from either drinking too little fluid or losing too much through perspiration due to high temperatures. Either event directly boosts the probability of developing kidney-stone disease. Although perhaps not the most profound effect of global warming, it could be the most painful.

Nephrolithiasis is projected to grow in the high-risk zones from 40 percent of the population in 2000 to 56 percent by 2050, and to 70 percent by 2095. In fact, by the year 2050, a large chunk of Illinois will fall within America's kidney-stone belt, with the Chicago area alone seeing up to one hundred thousand extra cases each year. Overall, approximately 2.2 million additional climate-related cases of kidney stones are predicted to occur each year by 2050. Nationwide, the estimated cost increase associated with this rise in nephrolithiasis would be $0.9–1.3 billion annually (at the dollar value in 2000), which represents a 25 percent increase over current expenditures.[24]

In the end, we are all affected by the outcomes of one another's choices and subsequent actions, be they political agendas, commercial endeavors, personal relationships, or individual responses to local conditions. It behooves us all, therefore, to become increasingly conscious of the environmental impacts we cause through our life decisions because the negative outcomes will surely come home to roost for each of us in some unforeseen and unpleasant way. If we each accept personal responsibility for the consequences of our actions, we will find that, collectively, we can heal the three spheres of the Earth to everyone's benefit, even as we have injured them to everyone's detriment.

2

Our Ever-Changing Landscape Patterns

There are single-minded, idealistic people who perceive us humans merely as a blight on the face of the Earth, a species whose sole purpose appears to be the despoliation of the planet. We, however, have a right to be here by the very fact that we exist as an inseparable part of the global ecosystem. In this sense, what we do is natural, despite the fact that our actions are often shortsighted, unwise, and destructive.

Nevertheless, I submit that people's destructive behavior is born not of intentional malice but rather of their unconscious, yet palpable fear of living in a world governed by the uncertainties of perpetual change and its unpredictable novelty. When people feel their very survival is threatened, either individually or collectively, they retreat to their assumptions and don fear's armor in an irrational and ultimately impossible attempt to control the universal process of continual change.

Early Landscapes

In the beginning, humanoids had relatively little impact on the landscapes in which they lived. Their impact increased, however, as they evolved languages, cultures, and societies because they could then participate with nature in more intensely organized ways. Today, the human species has essentially altered the landscapes of the entire world through airborne and waterborne pollution and through the unbridled exploitation of nature's bounty.

Human society changes the dynamics and the design of every landscape with which it interacts, and it has been doing so for thousands of years. The history of England is illustrative.

About 450,000 years ago, the northern hemisphere was locked in a sheet of ice over half a mile thick that covered Scandinavia, most of Britain, and much of the North Sea. During this time, the water carried by Europe's northward-flowing rivers collected in an immense glacial lake along the ice sheet's southern boundary, in what today is northern Europe. The lake was contained by a tenuous link—the

Weald-Artois chalk ridge, which had connected southeast England and northwest France for millions of years.

Eventually, as the ice age began to end and the world warmed up, the lake spilled over the isthmus. The spillage grew rapidly, turning into a torrent, and then a megaflood, one of the largest scientists have ever identified. Although megafloods, which involve sudden discharges of exceptionally large volumes of water, are rare, they can significantly affect the evolution of landscapes and continental-scale drainage patterns, as well as change regional climate.

In a matter of days, the water sliced through the chalk ridge and down to bedrock, thereby creating the Strait of Dover, which, just over twenty miles wide, is the narrowest part of today's English Channel. However, the early humanlike creature, known scientifically as *erect man*, was already living on the newly formed island. The megaflood not only isolated erect man from the continent and influenced later human travel to and from the island but also reorganized the large-scale palaeodrainage patterns of northwestern Europe. Erect man was eventually replaced by the more modern Paleolithic humans (Old Stone Age, earlier than twelve thousand years ago).

As the Pleistocene epoch drew to a close between twelve thousand and ten thousand years ago, the ice withdrew, although not in a single smooth recession. During this time, groups of Paleolithic humans occupied the warmer places in the south of what today is England, where they seem to have had an ecological impact with their selective dependence on wild horses and reindeer for food and raw materials.

As the climate ameliorated, the trees in southern Europe and the Caucasus, which had survived the glaciation, gradually spread again to the once-glaciated areas until a self-sustaining mixed deciduous forest was established about eight thousand years ago. (The Caucasus is the region between the Black Sea and the Caspian Sea, which covers some 154,250 square miles.)

Then, about seven thousand years ago, present-day England separated from the mainland. By this time, Mesolithic hunter-gatherers had replaced the Paleolithic humans on the newly formed island, where they remained until the coming of agriculture about five thousand years ago. (Middle Stone Age, between twelve thousand and five thousand years ago.)

The earliest cultural landscapes of the area—those purposely manipulated with fire—were formed in the middle to late Mesolithic period, between seven thousand and five thousand years ago. As far as we know, that first cultural landscape came from the conversion of a mixed deciduous forest into a mosaic of high forests, open-canopy woodlands, and grassy clearings with fringes of scrub and bracken fern, patches of wet sedge, and bogs of peat. Among these habitats, groups of late Mesolithic peoples, without knowledge of crop-based agriculture, moved about gathering food.

One of the great turning points in Western European history, the Neolithic period, or New Stone Age, began with the introduction of agriculture from Asia around five thousand years ago. For Neolithic people, the arrival of agriculture was not gradual. Instead, the complement of farming tradition and myth probably came as a fully developed package from the East, even if accessory hunting persisted. The earliest agriculture in Western Europe was a mosaic of small clearings, which were abandoned as the fertility of the soil became exhausted or the weeds became too bothersome. As new clearings were made, abandoned ones reverted to forest.[1]

And so began, in that far distant time, the humble tinkering with the environment whereby humans attempted to gain greater control of their destiny; the cumulative effects of such tinkering through the millennia have become the human-caused patterns across the landscapes of today and tomorrow.

Economics and Ecology

Two commonly held views about nature lead to a systematic misunderstanding and faulty measurement of the biophysical processes. The economic view conceives of the natural environment as a warehouse of resources available for human exploitation and sees nature's services as being provided by mechanistic ecological processes. The economic perspective fails to recognize the sharp distinctions between ecological and economic processes by positing that issues of environmental sustainability can be successfully addressed by "economizing ecology and ecologizing the economy."[2] This mixing and matching obfuscates the driving force of today's environmental crisis.

Both ecology (which represents nature) and economy (which represents humanity) have the same Greek root: *oikos*, "house." Ecology is the knowledge or understanding of the house. Economy is the management of the house. And it's the *same* house, a house we humans have divided at our peril. At issue here is "whether the environment is part of the economy *or* the economy is part of the environment."[3]

Economic activities that destroy habitats and impair the free services performed by ecosystems will create costs to humanity over the long term that will undoubtedly exceed in great measure the perceived short-term economic profits. Yet, because most of these services, and the benefits they provide, are not traded in economic markets, they carry no visible price tags to alert society to their relative value, changes in their supply, or deterioration of the underlying ecological systems that generate them.

These ecological costs are usually hidden from traditional economic accounting but are nevertheless real and are borne by society at large. Tragically, a short-term economic focus in current decisions concerning nature's ecological services often sets in motion great costs that are bequeathed by myopic adults not only to their own children but also to all the children of the future. Unfortunately, history teaches that humanity finds the real value of something taken for granted only when that something is lost—thus, in hindsight. The upshot is that

social-environmental sustainability demands that we frame all economic decisions in generational time scales because what may appear to be a good short-
term economic decision (the benefits of which we reap) can simultaneously be a
bad long-term ecological decision and so a bad long-term economic decision, the
cost of which future generations must bear.

Human Alteration of Landscapes

Human alteration of landscapes and their respective ecosystems cause large
changes in the immediate environment, as well as biogeographically.
Fragmentation of ecosystems is perhaps the most common worldwide alteration
of landscapes, one that generally results in a mosaic of habitat "islands" of different sizes surrounded by and within a matrix of agricultural and urban development. The ongoing incursion of this cultural matrix into undeveloped habitats
causes fluctuations in the landscapes' receptivity to solar radiation, as well as the
movement of wind, water, and nutrients. These changes have variable influences
on the biota within and among remnant areas and between the edge of a habitat
island and the surrounding matrix. With respect to birds, as an example, the rate
of nest predation is nearly twice as high along abrupt, permanent edges of all types
as it is along more gradual edges, where plant succession is allowed to occur.[4]

Consequences of the above alterations vary based on the length of isolation
and the degree of connectivity within and among remnants and their locations
within the overall landscape pattern. These influences are modified by the size,
shape, and position of the islands and are driven primarily by factors arising in the
surrounding landscape, such as the conversion of intact habitat through residential and commercial development.[5] Because of the many factors that influence the
alteration of ecosystems, all major areas of the world differ from one another in
their landscape patterns.

Arctic

In addition to global climate change, Arctic ecosystems are experiencing large-
scale developments to extract resources. These activities are in turn producing
anthropogenic disturbances, which will interact with natural disturbance regimes
and are bound to have dramatic effects on local and regional patterns of vegetation and the migration of plants.

Disturbance is important not only because it produces patches of partially or
totally denuded ground that permit propagules to become established but also
because it may open areas to the effects of erosion. Even disturbances of relatively
low intensity and small scale have immediate and persistent effects on the Arctic
soils and vegetation. The disturbed patches support new vegetation in relatively
stable conditions on all but the wettest sites. Where slopes are minimal, such disturbances can expand over large areas in as little as four years. The effects result in
an artificial mosaic of habitat patches that, although varying greatly in quality and

quantity, constitute areas wherein terrestrial herbivores can feed and nest.[6] Although the process creates places that support plant and animal life, there is a dramatically negative side to the story.

Examination of fifty-five paleolimnological records (a study of the archeological and geological properties of circumpolar Arctic lakes) reveals vast changes among species, as well as ecological reorganizations among algae and invertebrate communities, since approximately 1850. The remoteness of these sites, coupled with the ecological characteristics of taxa, indicate that climate warming through lengthening of the summer growing season and related limnological changes were the primary factors effecting changes. The widespread distribution and similar characteristics of these changes point toward a lost opportunity to study Arctic ecosystems unaffected by human influences.[7]

A record of varves in Lower Murray Lake, northern Ellesmere Island, Nunavut, Canada, extends back over the last millennium. (A *varve* is the couplets of coarse/fine layers of deposition frequently found in the sediments of glacial lakes, much like the annual growth rings in a tree.) As with other studies of lakes from the High Arctic, varve thickness is a good proxy for the summer temperatures of Lower Murray Lake, where those temperatures in recent decades were among the warmest of the last millennium. The last time such conditions were comparable was in the early twelfth and late thirteenth centuries.[8] And change continues to take place, especially in Arctic ponds.

Characteristic of most Arctic regions are the many shallow ponds that dot the landscape, where they have been permanent hotspots of biodiversity and reproduction for microorganisms, plants, and animals over the millennia in this otherwise extreme terrestrial environment. These shallow ponds are now drying out completely during the polar summer because of their relatively low volume of water and high surface areas. Their growing disappearance is therefore linked to increased evaporation and less precipitation, presumably associated with climatic warming. The final ecological threshold for these aquatic ecosystems has now been crossed—complete desiccation.[9]

Temperate Forests

Forestry is another factor that needs to be taken into account when dealing with human alterations of the landscape matrix around the world. Here, the challenge is that both commercial forestry and agroforestry have—and are—introducing alien species of trees into natural and semi-natural ecosystems, where they often become invasive. The magnitude of the problem has increased significantly over the past few decades, with a rapid increase in afforestation, which is the process of establishing a forest on land that is not a forest or has not been a forest for a long time by planting trees or their seeds; the consequential changes in land use are significant. The species causing the greatest problems are generally those that are widely distributed and have the longest histories of intensive planting.

Pine trees are especially problematic in the southern hemisphere, where at least nineteen species are invasive over large areas, and some of the species cause major problems. One notable problem species is the Monterey pine, also known as radiata pine, which is a native to three limited areas of Santa Cruz, Monterey, and San Luis Obispo counties, in coastal California. The most invasive species of pines have a predictable set of attributes: low seed mass, short juvenile period, and short intervals between large episodic crops of seeds.

Pine invasions, such as the plantations of radiata pine I saw in southern Chile, have severely affected extensive areas of grassland and scrubland in the southern hemisphere by reducing structural diversity, increasing biomass, disrupting pre-vailing vegetational dynamics, and changing patterns of nutrient cycling. These inevitable negative consequences of intensive forestry with exotic species are spilling over into areas set aside for the conservation or the production of water or both.[10] However, these introductions are often dependent on clear-cutting forests to gain areas for planting, and such cutting also fragments the forested landscape. Before dealing with specifically structured edges, I will point out a few commonalities.

The effect of edges on species within a given landscape is influenced by the overall amount of habitat fragmentation, the configuration of the forest (the amount of edge in the landscape), the elevation, and the edge contrast based on the structure of the opposing habitats. *Habitat fragmentation* is the breaking up of once-extensive landscape-scale plant communities, such as a forest or a grassland, into smaller, isolated remnants of the former habitat surrounded by new types of habitat, such as agricultural fields or housing developments.

With respect to species behavior, the commonalities are that: (1) forest fragmentation can be described in similar terms across most regions of North America, (2) widespread species may exhibit clear and similar negative responses to habitat fragmentation, (3) the sensitivity of local species to fragmentation varies geographically and may be lower in regions with greater overall forest cover, and (4) the results from single-species or local studies cannot be extrapolated to other species or regions, including the patterns of seed dispersal: by animals and by wind.[11]

Tropical forests are not alone in having animals disperse the seed of trees. There is a large contingent of scrub jays around my hometown, and they are exceedingly efficient and effective in planting hundreds, if not thousands, of acorns and filberts every autumn—most of which I sometimes think get planted in my garden. Then, in spring, when the acorns and filberts begin to grow, the jays dig out and eat the cotyledons, thereby leaving the now-rooted oak or filberts viably planted and intact. In fact, if I did not routinely pull them out, I would have a forest in my garden. (A *cotyledon*, or seed leaf, is a significant part of the embryo within the seed of a plant, which, upon germination, may become the embryonic, first leaves of the seedling.)

Scrub jays are not the only birds that plant seeds in my garden. American robins also make annual deposits of cotoneaster, holly, and hawthorn via their feces, which they scatter at will during their winter visitations. In addition, members of the neighborhood population of indigenous western gray squirrels seem to delight in planting the huge seeds of black walnut wherever they can—seemingly to test my vigilance each and every autumn. Of course, they return each spring when the walnuts sprout to dine on the cotyledons, while studiously leaving the already-rooted seedling in place.

In the drier forests of ponderosa pine, which is common throughout the intermountain West, the mantled ground squirrel is a prodigious planter of pine seeds. I have often watched one of these squirrels collect pine seeds in autumn, bury them, and almost instantly forget where they are secreted. Going back to some of the same locations in late spring, I have been greeted by little clumps of bright-green pine seedlings. The fact that the seeds were buried, in turn, explains why, as a boy in the 1940s and early 1950s, prior to the extensive logging of today and the cumulative effects of fire suppression, I saw so many tight clumps of three, four, five, and sometimes even six old ponderosa pine trees growing together almost as a single tree out of the same spot. Other seeds within these forests, however, were blown about by the wind.

The height from which a seed is released and the velocity with which it is blown are the primary determinants of long-distance dispersal of seeds by wind. Yet potential determinants at the ecosystem level, such as the seasonal dynamics of foliage density characterizing many deciduous forests, have received much less attention.

Sparser canopies (deciduous forests in winter after leaf fall) are characterized by more organized, vertical eddy motions, which promote long-distance transport by lifting seeds to higher elevations, where winds are stronger. Yet, sparser canopies are also characterized by reduced wind speed aloft. Nevertheless, a vertical eddy motion more than compensates for the reduced wind speed with respect to the favorable long-distance dispersal of seeds, which may account for the tendency of many temperate tree species to release their seeds either in early spring or in late autumn, when the canopy is relatively thin.

The role of seed dispersal in maintaining genetic connectivity among forest fragments has largely been ignored, however, because the flow of genetic material through the distribution of pollen is expected to predominate. That notwithstanding, the long-distance dispersal of seeds among the forest remnants across a chronically fragmented landscape is commonly mediated by the way genetic material is disseminated. The relative importance of seed-mediated gene flow may have been underemphasized in fragmented systems, where diagnosing the response of forest trees to anthropogenic disturbances requires a genetic assessment after seedlings are established.

For example, the European beech dominates forests over large regions of Europe, where habitat fragmentation has led to genetic bottlenecks and the disruption of its breeding system. In turn, this disruption is leading to significantly elevated levels of inbreeding, population divergence, and reduced genetic diversity within populations. Thus, in contrast with the conclusions of previous studies, a fractured landscape has a negative genetic impact, even on this widespread, wind-pollinated tree.

The circumstances of the European beech illuminate the critical necessity of genetic assessments because, although habitat fragmentation poses a serious threat to plants through genetic changes associated with increased isolation and reduced population size, the longevity of trees, combined with effective dissemination of seed or pollen, can enhance their resistance to the genetic effects of fragmentation—but not always.[12] Moreover, anthropogenic fragmentation is not the same dynamic as the disturbance regimes with which the trees have evolved to cope.

Habitat structure and vertebrate wildlife adapted to various disturbance regimes have common denominators, as exemplified by fire in British Columbia, Canada. Those species that breed in early successional stages tend to multiply with the increased size of a fire, whereas those breeding late in succession tend to decrease. Cavity nesters decrease, however, as the size of a fire increases because fewer snags survive the flames. But species using downed wood as habitat increase as the interval between fires lengthens and downed wood accumulates. Ideally, the practice of forestry would be designed to maintain biodiversity by emulating such natural disturbance patterns as they differ across forest types, in part by retaining an ecologically viable degree of connectivity among unmanipulated, roadless patches of habitat.

The effects of early forest fragmentation on both invertebrates and vertebrates, whether from natural or human-caused disturbance, is poorly understood, partly because most studies of fragmentation have taken place in agricultural or suburban landscapes, long after the onset of fragmentation. Incipient forest fragmentation, however, may affect populations differently from later stages of fragmentation, when the geometry of the landscape has reached a more stable configuration.

Nevertheless, densities of several forest-dwelling species of birds, and presumably mammals, can increase within a remnant of forest soon after the beginning of fragmentation as a result of displaced individuals packing into the remaining habitat. Along with higher densities in the new forest remnant, pairing success in one species, the ovenbird, for example, is lower in fragments than in continuous forest, possibly because of behavioral dysfunction as a result of high densities. Thus, the denser the population of ovenbirds, the lower their reproductive rate is. However, the duration and extent of increased densities following onset of fragmentation depend on many factors, such as the sensitivity of a species to an edge, the encroachment of individuals displaced by habitat loss, the duration and rate of fragmentation and the subsequent habitat loss, and the proximity of a forested habitat to the disturbance.[13]

Although fragmentation is usually thought to consist of a single, vertical edge between one type of habitat and another, this view is overly simplistic. First, the edge effect is not only complex to begin with but also changes as the edge matures because of the self-reinforcing feedback loops of the organisms involved. Second, the impoverishment of biodiversity, as a consequence of habitat fragmentation and its resultant edges, has profound effects on how an ecosystem functions by altering its ecological processes, such as the interactions among trophic levels. (A *trophic level* is a stage in the food chain. It reflects the number of times energy has been transferred from one organism to another through feeding; for example, a grass is eaten by a mouse, which is then eaten by a weasel, which is then eaten by an owl.) For example, herbivory by leaf-mining insects and their overall rates of parasitism decrease as woodland remnants become smaller.

The size of a remnant, the ratio of its interior habitat to its edge, and the location one might examine all determine the intensity of both the trophic processes of herbivory and parasitism in insect-plant food webs composed of hundreds of species; these processes take place between a remnant's interior and its edge. Moreover, the magnitude habitat fragmentation can be so pervasive that it affects trophic processes of highly complex food webs, thereby altering how an ecosystem functions.[14]

Third, a factor that adds to the overall complexity of edges is that they come in several dimensions, which increase their overall complexity. These dimensions include the edge of a road through a forested area, the horizontal edge within a forested area wherein a vegetative layer has been removed or altered, edges between successional stages within a forest, edges between forests and different habitats.

In forests, both paved and unpaved roads add to habitat fragmentation more than clear-cuts by dissecting large patches of quality habitat into smaller pieces and by converting forest interior habitat into edges. For example, in a 60,490-acre section of the Medicine Bow–Routt National Forest in southeastern Wyoming, the road network produced edge habitat that was 1.54 to 1.98 times greater than that created by cumulative clear-cuts. In addition to their extensive nature, roads are major physical disturbances for small creatures in otherwise continuous habitat.

Roads create long, narrow edges that extend well beyond their surface or adjoining sides. Furthermore, the soil compaction of roadbeds may persist more than forty years after a road is abandoned and embedded in seemingly continuous forest. In addition, roads have direct effects on species, such as mortality from construction and being hit or run over. They also have indirect effects: modifying animal behavior, disrupting the physical environment, altering the chemical environment, spreading exotic species, and changing how humans use the environment. The ecological effects of roads may be just as severe as the effects of other edges created by the alteration or loss of habitats, both of which have important implications for the function and diversity of forest ecosystems.

To illustrate, in the southern Appalachian Mountains of North Carolina, the abundance of an assemblage of woodland salamanders near roads has been significantly reduced, and those along the edges of roads are predominantly large individuals. Further, the road-effect zone for these salamanders extended 115 feet on either side of a relatively narrow, low-use forest road within the Nantahala National Forest. Moreover, the number of salamanders was significantly lower on old, abandoned logging roads than on adjacent sites upslope from the roads. These results indicate that forest roads and abandoned logging roads have negative effects on forest-dependent species, such lungless salamanders.

Such data exemplify a problem created by current and past land-use activities, especially the building of roads in forested regions. As with other taxa, the effect of roads on salamanders, such as the red-backed salamander, reaches well beyond their boundary, and the abandonment or the decommissioning of roads does not reverse detrimental effects but rather has significant repercussions for generations to come. Once again, the amount of quality-forested habitat is reduced, this time between 28.6 percent and 36.9 percent of the area with roads.[15]

The horizontal edges in a forest begin with the ground cover of herbaceous plants, which may or may not include grasses and sedges. The next layer is generally composed of low shrubs, above which is a tall-shrub layer, then a layer of small trees. Thereafter, a layer of taller trees may or may not exist before reaching the upper layer of the canopy.

In the past, low-intensity ground fires have affected two or more layers to various degrees. Today, the harvesting of non-forest products, such as moss and ferns, affects these lowest layers, whereas thinning from below may affect all layers under the canopy, especially if thinning is severe enough to allow the interior microhabitats, such as fallen trees, to dry to an appreciable extent, which can negatively affect such animals as salamanders.[16]

In addition to these generalized changes, many western forests of ponderosa pine and mixed conifers of the interior United States have undergone dramatic change since the westward expansion of European Americans.

Historically, these forests consisted of widely spaced, fire-tolerant trees with an understory of grasses. Since the early twentieth century, however, they have developed into dense stands of more fire-sensitive and disease-susceptible species. These alterations have been attributed primarily to active suppression of fires (which formerly reduced tree recruitment, especially of fire-sensitive, shade-tolerant species) and selective logging of larger, more fire-tolerant trees.

There is, however, a third factor, livestock grazing, which, in its own way, is as important as fire and logging.

Long-term grazing by domestic livestock, especially cattle, has created a "horizontal edge" along the surface of the forest floor through the alteration of the plant-community structure and thus ecosystem processes. Grazing causes several problems: it reduces the density and diversity of understory grasses and sedges, which otherwise smother conifer seedlings, thereby helping to maintain open stands of trees; it reduces the abundance of fine fuels, which formerly carried the many low-intensity fires that helped to keep the forests open; and it reduces the cover of herbaceous plants and litter, which allows the disturbance and compaction of soils, the reduction of water infiltration, and increased soil erosion. Grazing by domestic livestock has thus contributed to increasingly dense western forests and to changes in the composition of tree species and forest structure.[17] White-tailed deer have a similar effect in the northeastern and upper midwestern United States.

Browsing by white-tailed deer can profoundly affect the abundance and population structure of woody and herbaceous species of plants. Prior to European settlement, forests in northern Wisconsin contained relatively sparse populations of deer, but populations started increasing in the late nineteenth century with the onset of logging. Both habitat fragmentation as a result of continual, scattered timber harvests and the purposeful creation of wildlife openings to improve deer forage have encouraged and maintained the currently high densities of deer throughout much of the Northeast.

Because deer wander widely, the effects of high population densities penetrate deeply into remaining stands of old and mature forest, thereby greatly modifying the composition of their plant species, such as preventing the regeneration of the once common woody species: Canada yew, eastern hemlock, and white cedar. And because deer progressively removed the habitat below the browse line, the numbers and diversity of insects, which are primary consumers of the understory vegetation, are affected. Thus, the abundant horizontal edge inside the forest, coupled with the vertical edge between older forest and early succession engendered by the large numbers of deer, represents a significant, mobile threat to these late-successional plant communities and the wildlife that depends on them.[18]

Edges between successional stages are vertical in structure and are either soft or sharp depending on how close the successional stages are to one another in their development. For example, an old forest juxtaposed to a mature forest has a relatively soft edge, one that is somewhat indistinct, as opposed to the sharp edge between an old forest and a recent clear-cut. To understand what I mean, let's quantify these edges.

Suppose each of the six successional stages (grass-forb, shrub-seedling, pole-sapling, young forest, mature forest, and old forest) were given a value from 1 through 6—with a clear-cut being 1 and the old forest being 6. Now, if the value of the clear-cut (1) is subtracted from that of the old forest (6), then the edge contrast is 5, a sharp edge. But if a mature forest (with a value of 5) is subtracted from an old forest (with a value of 6), the edge contrast is 1, a soft edge. Therefore a 1 is the softest edge and a 5 is the sharpest edge, with some variation in between and effects that correspond to the amount of contrast.

The pendulous, fruticose lichen called witch's hair that grows on Norway spruce in northwestern Sweden, among other places, exemplifies a sharp edge. There, sharp edges between mature Norway spruce forest and large clear-cuts (82 to 643 acres) were studied to determine their effects on witch's hair lichen. Both the distance from an edge and the age of the edge significantly affected the length of witch's hair.

The maximum effects of an edge extended 25 to 2,110 feet into the forest at moderately exposed sites, where physical damage from strong winds was the major factor in reducing the lichen's abundance. Such pendulous lichens are prone to the fragmentation of their vegetative portions, particularly at newly formed edges. At older edges, however, witch's hair had recovered to a distance of 66 to 98 feet inside the edge, probably as a result of increased growth in response to more sunlight reaching the lichens.[19]

At ground level, amphibians share several biological characteristics that may cause them to be sensitive to abrupt transitions in microhabitat and its associated microclimate, which occur across forest edges. Although salamanders are generally more sensitive to even-aged harvesting and the associated effects of sharp edges than are frogs, there are habitat generalists and specialists within every forest. Some structural microhabitat variables are potentially limiting to amphibians near forest edges, including the percentage of canopy closure; the amount and distribution of forest litter on the ground; the number, size, and distribution of stumps, snags, and their root channels. Clearly, the impacts from intensive forest practices extend beyond the boundaries of harvested stands.[20]

With respect to interior-dwelling forest birds, the primary demographic parameter of a population's dynamics is its reproductive capacity, which varies with the distance an individual's territory is from a forest's edge. As our landscapes become increasingly fragmented, that proportion of forest habitat near edges grows geometrically, which reduces the overall fecundity of interior-dwelling birds. Impaired reproduction by some forest birds in a fragmented landscape can, by itself, disrupt a population and thereby cause the types of declines and shifts in distribution observed in the fragmented forests of southern Wisconsin. Here, without the immigration of reproductive recruits from other areas, populations of interior-dwelling birds can become locally extinct in a severely fragmented landscape.[21]

The problems faced by many species of Neotropical, migrant songbirds are different from those experienced by interior-dwelling species. The relative threat of predation to nesting songbirds changes with forest fragmentation. Historically, interior forest-dwelling birds have been subjected primarily to the activity of small predators, whereas predation by large animals tends to increase as forests become increasingly fragmented—a process against which songbirds have no defense. It seems, therefore, that predation on the breeding grounds in North America plays a larger role in the decline of migratory songbirds than does the deforestation of their wintering grounds in the tropics.[22]

Although bats historically used the basal hollows in old coast redwood trees of California as roosts, they now frequently use basal hollows in remnant fragments of old-growth redwoods that are twelve acres or less. The use of such small fragments may be due, in part, to the proximity of remnants to stream courses. In addition, bats, such as little brown bats, big brown bats, and silver-haired bats, which typically feed in open areas along forested edges or over cleared areas may

be attracted to the hollow-bearing trees in remnants, which are potentially important habitat for a variety of bats. These trees do not necessarily replace a hollow
one with more contiguous forest, where such interior-dwelling bats as the hairy-
winged bat traditionally feed.[23]

With respect to small, ground-dwelling mammals, habitat fragmentation
would probably have the least effect on the deer mouse or the Oregon creeping
vole because the mouse is highly adaptable and exploratory and occupies a wide
range of habitats, whereas the vole shifts its entire diet, depending on whether it
is living in a deep forest or in the grass-forb stage of a clear-cut. The California red-
backed vole is at the other end of the spectrum, however, because it has a specialized diet of truffles—the underground fruiting bodies of mycorrhizal fungi, which
quit fruiting when the forest is clear-cut. Hence, the forest/clear-cut edge has a
critical affect on the vole's food. As the forested portion of the edge dries out
because of increased sunlight and drying winds, the fungi stop fruiting and the
voles are increasingly confined to that area of the forest remnant that still functions fully as a forest—thus, the larger the remnant, the better.[24]

A study of the invasion of old-growth forest islands by exotic plants in rural central Indiana was carried out using paired sample sites along warm (south and west)
edges and cool (north and east) edges in each of seven forest remnants. The fragments generally appeared to resist the invasion exotic plants. Where it did occur,
however, the species richness and frequency of the aliens dropped sharply as one
progressed inward from forest edges, and the interiors were relatively free of exotics.
The main limiting factor is probably the low availability of light in forest interiors,
although little disturbance of indigenous plant communities may also be important.

Despite the fact that aliens were more diverse and more frequent on warm
edges than on cool ones, their invasion is discouraged by the maturation of a remnant's edge as it develops a dense wall of bordering vegetation that reduces interior light and the penetration of wind. The most successful invaders are mostly
escaped ornamentals or other species not commonly found in adjoining fields. In
agriculturally dominated landscapes, however, vast areas of cropland surrounding
these forest islands may slow the spread of aliens at the landscape scale.[25] Here,
you might wonder how temperate forests differ from tropical forests.

Tropical Forests

The spatial distributions of 36 to 51 percent of the tree species in three diverse
Neotropical, primary forests in Colombia (La Planada), Ecuador (Yasuni), and
Panama (Barro Colorado Island) show strong associations with the distribution of
soil nutrients. These results indicate that the availability of belowground
resources plays an important role in the assembly of communities of tropical trees
at local scales. With respect to climate change, the response of tropical forests will
depend on individual species' nutritional strategies.

A group of tropical trees, which form the forest canopy, and ferns, which live
in the understory, rely on a common pool of inorganic nitrogen rather than

specializing in the extraction of different nitrogen pools. Moreover, these tropical species abruptly changed their dominant source of nitrogen in unison in response to a climate-driven change in precipitation. This threshold response indicates a coherent strategy among topical species to exploit the most available form of nitrogen in soils. Such apparent community-wide flexibility indicates that diverse species within tropical forests can physiologically track changes in the cycling of nitrogen brought about by shifts in the climate.

This kind of flexibility is critical when it comes to changing land use in the tropics, where vast areas of habitat have been damaged and biodiversity has been reduced. Yet the effects of dwindling biodiversity are rarely considered in terms of their intimate relationships to ecosystem processes. Data from studies in Puerto Rico, Southern China, Dominica, and Nicaragua indicate that functional diversity—not simply species richness—is important in maintaining integrity despite the fluxes of nutrients and energy. High species richness may, nevertheless, increase the resilience of tropical ecosystems following disturbance by increasing the number of alternative pathways along which resources can flow.[26] Nevertheless, Brazil's Amazon forest remained largely intact until the Transamazon Highway inaugurated the "modern" era of deforestation in 1970.

Although it accelerated relatively slowly, the rate of deforestation in Amazonia has exhibited a rapid upward trend since 1991. Amazon forests are cut for various reasons, but the predominant motive is to create pastures for raising domestic cattle, an activity that has created large and medium-sized ranches, which account for roughly 70 percent of all deforestation. Additionally, the impact of increasing livestock production extends well beyond the forest.

The cycles of nitrogen and carbon are closely connected to livestock's role in current land use and pending changes in land use, which includes grazing land and cropland dedicated to the production of feed crops, such as corn and fodder. If we consider emissions along the entire commodity chain, livestock currently contribute about 18 percent to the effect of global warming. Livestock also contribute about 9 percent of total carbon dioxide emissions, and they are responsible for a whopping 37 percent of the methane and 65 percent of the nitrous oxide. Moreover, livestock will substantially increase over the coming decades because pastureland is currently at its maximum expanse in most regions. Hence, the future expansion of the livestock industry will be increasingly crop-based, and thus will greatly increase the livestock-based pollution emitted along the entire commodity chain—from the ecological integrity of the soil to that of the world's water and air.[27]

Here a lesson from Australia might be applicable, namely the ecological integrity of the topsoil as demonstrated by something as small as the biodiversity of ants.

Although Western Australia is a relatively unpopulated region, agricultural clearing, rangeland grazing, urbanization, road construction, and mining have modified considerable areas of native vegetation. Not only is the diversity of ants reduced by each of these land uses but also the community composition is changed. Road construction has the greatest long-term effect on the alpha diversity of ants, followed by agricultural clearing, mining, urbanization, and rangeland grazing. Clearing land for agriculture, followed by livestock grazing, is responsible for the greatest loss of integrity of Australian ants.

All the same, cattle ranching is only one source of income that makes deforestation profitable. Logging is another.

Both the degradation of Amazonian forests from logging and the ground fires facilitated by logging translate into fragmentation and the formation of vast areas of vertical edges between forest remnants and cleared areas. This degradation contributes not only to the loss of forest cover but also to the loss of biodiversity, reduced water cycling (and rainfall), and increased global warming.[28]

The growing loss of biodiversity and ecosystem processes and services caused by destruction of the rain forest and the intensification of agriculture is a prime concern for both science and society. Potentially, tropical ecosystems could produce satisfying economic gains with limited ecological losses, as demonstrated in cacao agroforestry in Sulawesi, Indonesia. Expansion of cacao cultivation by 230 percent since the 1980s has been triggered by economic markets as well as by rarely considered cultural factors.

The transformation of near-primary forest to cacao agroforestry had little effect on overall species richness, but it did reduce biomass of plants and storage of carbon by 75 percent. The richness of forest-using species was also reduced by 60 percent. In contrast, when the intensity of cacao agroforestry was increased, coupled with a reduction in shade-tree cover from 80 to 40 percent, only minor quantitative changes in biodiversity were detectable. Moreover, high levels of ecosystem functioning were maintained while the farmers' net income was doubled. Because unshaded systems further increase a farmer's net income by 40 percent, current economic incentives and cultural preferences for yet more intensification of agricultural practices put shaded systems at risk. Nevertheless, low-shade agroforestry provides the best available compromise for satisfying a farmer's economic gains while limiting ecological losses.[29] But this kind of artificial stability does not work everywhere.

A twenty-two-year study of ecosystem decay in fragments of Amazonian forests highlights the fact that edge effects play a critical role in the dynamics of habitat fragmentation. Moreover, the resulting landscape matrix has a major influence on the connectivity of remnants and how they function as islands of habitat because many Amazonian species avoid even small clearings, those less than three feet wide. The effects of fragmentation are highly eclectic in that they alter species richness and abundance, forest dynamics, the trophic structure of communities, and a variety of ecosystem processes and services. In addition, forest fragmentation interacts synergistically with such events as hunting, fires, logging, and the invasion of alien species, thereby posing an even greater collective threat to the rain forest biota and the self-reinforcing feedback loops they create and perpetuate.[30]

Obligatory mutualists provide an example of self-reinforcing feedback loops. Fig trees and their pollinating wasps, which are principally groups of tropical organisms, are *obligatory mutualists*, a form of symbiosis wherein the survival of one partner requires the survival of both. The fig/wasp feedback loop works in the following way. Fig wasps began to pollinate and co-evolve with figs ninety million years ago, even before continental drift separated what we today think of as the Old and New World groups of figs, of which there are over 750 recognized species. As for the fig wasps, they form a complex of cryptic species that evolved separately for more than 1.5 million years. A *cryptic-species complex* is a group of species that satisfies the biological definition of species—that is, they are reproductively isolated from each other but are virtually indistinguishable on a morphological basis. The only way to tell them apart is through DNA sequencing.

Some cryptic species of fig wasps are actually *sibling species*, which means they not only shared the same ancestor but also probably evolved within a single species of host fig or very closely related species. However, genetically identical wasps may also be found in fig hosts of two different species, a finding that suggests new associations are formed now and then.

Each species of fig has one or more species of small wasp, the female of which pushes her way into the fig while it's still green and hard. As she squeezes herself inside, her wings are torn off. Once within the fig, she is confronted with three kinds of flowers: male, short female, and long female.

To fulfill her life's purpose, she must make her way past the immature male flowers, which, as yet, have no pollen. She then moves further down into the hollow, where she dusts the female flowers with the pollen she brought from the fig in which she grew up. A female wasp can only reach the ovaries of the short female flowers with her ovipositors, in which she lays her eggs and dies soon after. Meanwhile, the long flowers become the fig's seeds.

The eggs laid, the fig tree chemically detects their presence and surrounds them with plant tissue. The eggs hatch and the fig provides the rapidly developing larvae with enough food to grow and restart the cycle. While the larvae grow inside the fig, the male flowers mature.

The male wasps are born first, but look nothing like a wasp because they have neither eyes nor wings; yet they soon detect the baby females and mate with them. Before their brief lives are forfeited, the males perform one other duty: they enlarge the original entrance, which allows the females to exit with their wings intact.

For their part, the female wasps are dusted with pollen as they squeeze past the male flowers to exit the fig. Once outside, they carry the pollen to another green fig, which they, in turn, will pollinate on their way to laying their eggs. When the wasps have left or died, the figs become bright in color, soft, and succulent as they ripen.

In return for providing a home and nourishment for the wasps, fig trees get their flowers pollinated. Without their tiny symbiotic wasps, the figs would not ripen, and the tree would eventually become extinct. Nevertheless, the population of wasps can be maintained only if figs are produced year-round, and because individual fig trees flower synchronously, the wasps that pollinate them have to locate a new individual host tree at each generation.

Once the figs have ripened, the tree fulfills its role of providing a staple diet for many fruit-eating birds and mammals, which disperse the figs' seeds, complete with fecal fertilizer. Without the figs, which are rich in carbohydrates and calcium, many forest animals could not survive. Therefore, the ability of the wasps to find new host trees is important in many tropical ecosystems because figs are a fundamental resource for frugivores, such as parrots, squirrels, and fruit-eating bats, which in turn are essential to plants, besides the fig trees, because the bats disperse their seeds.[31]

Flying foxes, which are large, Old World, fruit-eating bats, are extraordinarily important, both ecologically and economically, throughout forests of the tropics from Nepal to Australia, where nearly two hundred species perform the essential services of pollinating and dispersing seeds. Although these bats are frequently misunderstood, intensely persecuted, and therefore exceptionally vulnerable to extinction, at least 289 species of plants rely to varying degrees on large populations of flying foxes for propagation. These plants, in addition to their many ecological services, produce some 448 economically valuable products.[32]

Forest fragmentation is a greater threat to communities of vertebrates and their feedback loops than to communities of tropical trees because individual trees are typically long-lived and require small areas for survival. Nevertheless, fragmentation of Amazonian forests provokes surprisingly rapid and profound alterations in the composition of tree communities.

A two-decade study of exceptionally diverse tree communities in forty plots of 2.5 acres in both fragmented and intact Amazonian forests of 267 genera and 1,162 tree species showed that abrupt shifts in floristic composition were driven by sharply accelerated mortality and recruitment of trees within 328 feet of a fragment's edges. This shift caused a rapid turnover of species, coupled with declines in populations or local extinctions of many large-seeded, slow-growing taxa of old trees. Even among old trees, species composition in fragments is being substantially restructured because trees under the canopy that are obligate crossbreeders and rely on animals to disperse their seeds are facing a strong disadvantage in the community assembly.

However, there not only is a striking increase in a smaller set of disturbance-adapted species that are physically dispersed by such mechanisms as wind but also a significant shift in the distribution of tree sizes. These changes in tree communities will have a wide impact on forest architecture, plant-animal feedback loops, and the ability of the forest to store carbon.[33]

In addition to fragmentation, the loss (through poaching or otherwise) of those species of vertebrates that disperse the seeds of tropical trees has a significant effect on the species composition of the remnant plant community. To examine this phenomenon in different successional stages, we'll travel to the North Negros Forest Reserve, a forest fragment that is one of the last wet, tropical rain forest ecosystems in the biogeographic region of the central Philippine Islands. Because 80 percent of the trees in this forest community are zoochorous (depend on animals to disperse their seeds) the question is: What would happen to forest regeneration if the fruit-eating birds and mammals were hunted to extinction?

To answer this question, nineteen species of birds, fruit bats, and other mammals were observed in three habitats: early succession, mid-succession, and late succession. As well, species of trees were identified that are potentially at risk because their seeds are dispersed by animals. The study clearly indicated that early-successional tree species were visited by a wide spectrum of frugivores, whereas hornbills and fruit pigeons primarily visited mid- and late-successional trees. Late-successional tree species were the most specialized with respect to which vertebrates dispersed their seeds and could therefore be susceptible to extinction.[34]

In another study, however, poachers reduced the abundance of herbivorous mammals that, in turn, altered seed dispersal, seed predation, and seedling recruitment for two palm trees (la palma real and palma negra) in central Panama. Mammals, other than bats, were the only seed-dispersal agents, and rodents and beetles were the only seed predators. The large, durable endocarps are easily located on the forest floor and bear characteristic scars when a rodent or beetle eats the enclosed seed. (An *endocarp* is the hard inner layer that forms the pit or stone of a cherry, peach, or olive.)

In protected areas, 85 to 99 percent of the seeds were carried away from beneath trees of the same species, whereas 3 to 40 percent were so dispersed in areas where poaching was most intense. The percentage of the dispersed seed destroyed by beetles ranged from none up to 10 percent in protected areas, but from 30 to 50 percent in sites where poaching occurred. Conversely, the proportion of dispersed seeds destroyed by rodents ranged from 85 to 99 percent in protected sites but ranged only from 4 to 50 percent, depending on the level of hunting, in unprotected locales. Consequently, the densities of seedlings were also directly dependent on the intensity of poaching.[35]

Even in the dry deciduous forest of western Madagascar, mammals are critical to the dispersal of seeds. Aside from possibly the bush pig, only one vertebrate species of the dry forest, the brown lemur, ingests seeds greater than one-half-inch long and passes them unscathed through its digestive tract.

In one study, fewer lemur-dispersed tree species regenerated in forest fragments without the brown lemur than would be expected, based on the presence of mature trees whose seeds are typically dispersed by lemurs. Primary forests with the brown lemur, however, had a full complement of trees, including those whose seeds can also be dispersed by other vertebrates. Thus, the brown lemur is a critical component in the dry deciduous forest of western Madagascar if a complete set of tree species is to be maintained in the primary forest.[36]

As far as birds are concerned, small islands of tropical forest in the eastern Usambara Mountains have lost understory species of birds following forest fragmentation. The Usambara Mountains are part of a twenty-five-million-year-old necklace of mountains called the Eastern Arc, which is located northwest of Dar es Salaam, Tanzania. Here, as is the general pattern elsewhere, forest-dependent understory species vary greatly in their vulnerability to habitat fragmentation. The relatively rare species and the forest-interior dwellers are the most adversely affected by forest fragmentation. Whereas habitat fragmentation clearly affects populations of birds typically found within continuous forest, some effects, such as the absence of a species, are obvious, but others may be subtle and easily overlooked.

The wedge-billed woodcreeper and the white-crowned manakin, both captured during a study of forest fragments north of Manaus, Brazil, had a significantly slower growth rate of the outer right rectrix (quill feather of the tail) than did birds of the same species captured in continuous forest. Based on recapture data, wedge-billed woodcreepers probably grew their feathers where they were first captured, whereas white-crowned manakins were highly mobile and rarely recaptured. Nevertheless, data from this study indicate that birds in poorer physiological condition were more likely to be caught in fragments than in unbroken forest. Therefore, habitat fragmentation may have subtle yet important effects on species inhabiting altered landscapes.[37]

Prairies

The region of southeastern Washington and adjacent Idaho, which is composed of rolling hills on deep soils, is known as the Palouse prairie. The forests of northern Idaho bound its eastern border, and the Snake River forms its southern boundary. To the north and west, areas of flat terrain and shallow soils, where the deep soils were scoured away by ice or water during past glaciations, bound the Palouse. Some scientists, however, have a more inclusive view of what constitutes the prairie; they consider areas to the west and south, and even parts of northwestern Montana, to be part of the Palouse prairie ecosystem.

In the mid-1800s, the typical vegetation throughout the Palouse consisted of perennial bunchgrasses, which grew in tufts or clumps, accompanied by many different kinds of flowering forbs, or wildflowers, that, when seen together in spring and early summer, gave the appearance of a lush meadow. The principal bunchgrasses were Idaho fescue, bluebunch wheatgrass, and prairie Junegrass. Short shrubs, especially snowberry and wild rose, were common, and mosses and lichens added an important but inconspicuous feature. This type of vegetation grows in environments with almost enough precipitation to support trees.

The distinctive ecosystem is more than just plants. Although mule deer and Rocky Mountain elk fed on the Palouse prairie throughout the Holocene, American bison were abundant there only during cool and moist periods. (The *Holocene epoch* is the geological period that began approximately 11,550 years before the present and continued to about three hundred before the present, or to the 1700s.) In addition, archaeological data indicate that human hunters took deer, elk, and bison more frequently than small-mammal prey throughout the last ten thousand years and that the decrease in ungulates was due both to hunting and to changes in climate. Elk have been abundant only since the twentieth century because they were transplanted into areas where predators had been exterminated.

Small mammals, such as ground squirrels, gophers, and voles, were common prey for North American badgers, golden eagles, hawks, and owls. Hummingbirds pollinated some of the more brightly colored flowers, and Brewer's sparrow nested in the thickets of shrubs. In the spring, huge flocks of sharp-tailed grouse gathered on strutting grounds, where male birds danced to attract females.

Insects were important, but less obvious, members of the Palouse prairie, where they fulfilled such important roles as pollinating flowers and dispersing seeds. Below ground, the soil also teemed with life. Fungi, bacteria, algae, and invertebrates recycled dead plants through such processes as breaking down the massive underground root systems of those that died. The activities of these organisms affected and were affected by soil fertility and texture in complex ways.

Areas with especially deep, moist soils provided habitat for the endemic giant Palouse earthworms, which reached up to three feet in length! These white, lily-scented denizens of the prairie's fertile soils play important roles in soil health by transferring plant debris from the surface deeper into the soil, which they enrich; they also aid the penetration of air and water by digging its tunnels.

Prior to European invasion, the Palouse people inhabited the Palouse River drainage, although people from the Nez Perce tribe spent much of their time in the southern part of the prairie, while members of the Coeur d'Alene and Spokane tribes used its northern fringes. The Cayuse people used the area to the southwest. These patterns were fluid, however, with much overlap among tribes as they followed the annual cycle of hunting and gathering.

The cycle began with gathering camas roots at low elevations in spring. Low-lying, wet swales supported dense stands of camas, which is a member of the lily family. Camas provided many important things to the people, including large quantities of a nutritious food, a staple that could be stored for long periods, an economically valuable resource controlled by the women, and a valuable item for trade. In fact, camas plants were so dense in the wet meadows that early explorers mistook the masses of blue flowers for water. In 1806, Meriwether Lewis wrote in his journal that the camas resembled "a lake of fine clear water, so complete is this deseption [sic] that on first Sight I could have sworn it was water." As the seasons progressed, however, plants and animals were harvested at progressively higher elevations.

The prairie has been dramatically altered since the mid-1800s by agriculture, and virtually all of it has been invaded to some extent by aggressive alien species, weeds that were either deliberately or accidentally introduced from the Old World.

The Palouse prairie is now considered one of the rarest ecosystems in the United States, in part because native bunchgrasses do not have adaptations that make it easy for them to recolonize a site after it has been cultivated; in other words, the grasses are not in a good position to cope with the change. As a result, the native grasses have become less abundant, as have many species of flowering forbs that are—or were—characteristic of the Palouse.

At about the time that the Palouse was prepared for growing crops, plants adapted to cultivation arrived from their native Eurasia, which included many species of exotic grasses and wildflowers. These species evolved in agricultural regions and consequently have developed the ability to germinate on areas of bare soil that are exposed after cultivation. For instance, annuals cheatgrass (or downy brome), medusahead, and wild oats; broad-leafed plants, such as ox-eye daisy, Canada thistle, field bindweed (morning glory), and common teasel became abundant at the expense of the plants native to the prairie. Now, any disturbance that results in areas of bare soil provides sites that can be readily colonized by these invaders. So much for the plants. What about the effect of habitat alteration on the native birds?

The historical and current distribution of prairie vegetation on different soil types in eastern Washington was evaluated, as were bird communities at seventy-eight sites from 1991 to 1993. The abundance of avian species was then compared among soil types and the conditions of available habitats.

The pattern through which the shrubsteppe ecosystem was converted to agriculture has resulted in a disproportionate loss of deep-soil communities. Eight species of birds reflected strong relationships with soil type and three with the condition of the habitat. These associations probably resulted from the influence of soil type and the historical use of the vegetation in these communities.

The Brewer's sparrow and sage sparrow are most abundant in deep, loamy soils, whereas the loggerhead shrike is most abundant in deep, sandy soils. Sage sparrows occur most frequently in areas with intact fragments of shrubsteppe habitat, a finding that indicates a negative relationship with fragmentation.

Thus, it seems clear that fragmentation of the shrubsteppe ecosystem and the pattern of its conversion to agriculture among the soil types have had detrimental effects on numerous species. For example, the available landscape for species with an affinity for deep loamy-soil communities has changed considerably more than the overall area loss of the shrubsteppe ecosystem would indicate. To maintain the remaining diversity of avian species (as well as other species), it will be necessary to protect the extant shrubsteppe communities on deep soils and reduce further fragmentation of this ecosystem.[38]

Yet, as incongruous as it may seem in light of the foregoing discussion, because of the world's burgeoning human population, one of today's most critical actions is to protect prime agricultural land as farmland instead of paving it over with superhighways and parking lots or covering it with housing developments, shopping malls, and industrial complexes.

Agricultural Lands

The integration of peasant farms and natural ecosystems into agro-ecosystems forms a continuum where plant gathering and crop production are actively practiced. Many of these traditional agro-ecosystems are still found throughout the non-industrialized countries, where they constitute major in situ repositories of germplasm for both crop plants and wild plants. (*Germplasm* is a collection of genetic material for an organism. For nonarboreal plants, germplasm may be stored as a collection of seeds; for trees, the germplasm may be maintained by growing them in a nursery.)

Domesticated plants have evolved, in part, under the influence of farming practices shaped by various cultures and thus are directly dependent on the care given them by particular groups of humans. Because genetic material is archived more effectively as living rather than nonliving systems, maintaining traditional agro-ecosystems is a realistic strategy for in situ protection of the genetic properties of both crop and wild plants. In addition, it would be wise to link the ethnobotanical knowledge of the people with their self-sufficiency. (*Ethnobotany* is an inquiry into the complex relationship between plants and people and how a given culture uses plants.) Protecting traditional agro-ecosystems can be achieved only in conjunction with the maintenance of the local people within their culture, which, in turn, protects cultural diversity and thus is a snapshot of the human family.[39] Over time, however, the ecological effectiveness of small, diversified family farms has been replaced by the economic efficiency of larger and larger monocultural farms, but in this case larger is not necessarily better, as exemplified in the Ghibe Valley of Ethiopia.

Successful control of tsetse fly–transmitted sleeping sickness in the Ghibe Valley appears to have accelerated the conversion of wooded grassland into farmland; in turn, wildlife habitat has been affected. To assess the influence of this expanded agricultural land use, the species richness of and composition of bird communities were used as indicators of environmental impacts.

At the height of the growing season, the number of bird species and associated vegetative complexity were greater in the oxen-plowed fields of small farms and riparian woodlands than in wooded grasslands or in the tractor-plowed fields of large farms. Species composition differed greatly among the types of land use, with many species found in only a single habitat type. Although this result implies that converting land from wooded grasslands to small farms may not adversely affect the numbers of bird species in this region, it will alter the composition of their communities within different habitats.

All the same, moderate land use, which creates a mosaic of small-farm fields in the Ghibe Valley, increases habitat heterogeneity and bird-species richness, as opposed to land use in the large, tractor-plowed fields of intensive-use farms. Therefore, tsetse control may indirectly maintain species richness of birds in the valley by encouraging the differential spread of these small-scale, heterogeneous farms in place of large-scale, homogeneous farms. But if the small farms significantly exceed their current number, the bird-species richness and community composition may be negatively affected.[40] Thus, once again, it is relationships and not numbers that convey relative stability to ecosystems. In a sense, the farms of north-central Florida in the United States are a counterpart to those of the Ghibe Valley with respect to their attractiveness for birds.

A two-year analysis was conducted on paired organic and conventional farms, during which the diversity, distribution, and insect-foraging activity of native birds on farms were assessed. The results indicate that farms supported 82 to 96 percent of the land birds known to breed in the region; species richness and abundance varied significantly with the habitat matrix in which the fields are situated, as well as with the kind of border habitat surrounding the fields (but not with the year or type of farm management); the highest numbers of birds were associated with mixed crops, field borders, and the adjacent habitat, which was composed of forest and hedge; and the ten species identified as functional insectivores (because they are most likely to contribute to the control of unwanted, crop-damaging insects) were attracted significantly more to mixed crops than to monocultural crops.[41]

I, however, watched helplessly as the small, protected fields of the family farms around my hometown of Corvallis, Oregon, increasingly gave way (from the 1940s through the 1960s) to larger and larger homogeneous fields of corporate-style farms, as fencerows were cleared to maximize the amount of tillable soil, to squeeze the last penny out of every field. Gone are the fencerows with their rich, fallow strips of grasses and herbs, of shrubs and trees, which interlaced the valley in such beautiful patterns of flower and leaf with the changing seasons. Gone are the burrowing owls from the quiet, secluded fields I once knew. Gone is the liquid melody of the western meadowlark I so often heard as a boy. Gone is the fencerow trill of the rufous-sided towhee. Gone are the song sparrows, Bewick's wrens, yellow warblers, and MacGillivary's warblers. Gone are the wood-rat nests, the squirrels, and the rabbits. Although these species may still occur along the edge of the valley and in isolated patches of habitat, they vanished with the fencerows from the agricultural fields of the valley floor.

Today, compared with the time of my youth, the valley's floor offers little in the way of habitat other than a great, depersonalized, open expanse of silent, naked fields in winter and a monotonous sameness under the sun of summer. And even now, forest, fen, and native grassland are giving way to agricultural fields.

Despite widespread recognition of the importance of forests, the extent and rate of deforestation, even in temperate regions, remains poorly understood. The conversion of forestland in Saskatchewan, Canada, was assessed for the entire boreal transition zone, which amounts to 19,246 square miles, to determine whether the density of roads, rural developments, arable land, landownership, or some combination thereof, influenced the distribution and rate of change in forest cover.

Forestland in the boreal transition zone was reduced from 6,900 square miles in 1966 to 5,200 square miles in 1994, an overall conversion of 73 percent of the boreal transition zone to agriculture in the province of Saskatchewan since European settlement. The annual conversion rate was 0.89 percent over those twenty-eight years—approximately three times the world average.

The factors that influenced the conversion of forests to agriculture in Saskatchewan's past are still present today. These factors include lands that are privately owned, have soils with high agriculture potential, have a good network of roads, and are in the southern portions of the boreal transition zone. In addition, the areas of remaining forest tend to be clustered. Collectively, this set of circumstances indicates that deforestation is more likely to occur on privately owned land than on land controlled by the provincial government. Nevertheless, no programs are in place to slow or halt deforestation, despite dramatic alteration of forested areas in the boreal transition zone and despite their importance to a wide variety of forest-dwelling wildlife, as well as for the capture and storage of water, which will be affected by the progressive changes taking place in the global climate.[42]

Primarily subsistence or smallholder farmers in non-industrialized countries will feel some of the most important impacts of global climate change. Their vulnerability comes both from being predominantly located in the tropics and from various socioeconomic, demographic, and policy trends that limit their capacity to adapt to changing conditions. The necessary adaptations are particularly difficult because of the intrinsic characteristics of these systems, primarily their extreme complexity, their location specificity within a given region, as well as the integration of existing biophysical constraints and cultural strategies. Although adapting agriculture to our changing climate may be most difficult in the tropical regions of the world, it will not be easy for people to adjust in other areas either, but it is possible given the extreme adaptability of the human family through time.

Many options are available for small adaptations in existing agricultural systems, often variations of existing responses to the variability of current weather patterns. Implementation of these options is likely to produce substantial benefits for some cropping systems (which simply means how crops are grown) under moderate changes in the climate. There are, however, limitations to their effectiveness under more severe alterations in the climate.

As climatic conditions become more intense and sustained, systemic changes in the allocation of resources must to be considered, such as the conscious diversification of crops and production systems as well as of the type of livelihood they engender. Achieving this level of adaptation will necessitate the integration of risks involved in the variability of the climate and in the volatility of the market, as well as the shifting policies around such things as sustainable development.

Dealing with the many barriers to effective adaptation will require a comprehensive rethinking of current approaches to agriculture, including providing educational opportunities for farmers to help them deal with new, weather-related risks to their chosen crops. In addition, new market strategies must be established to facilitate effective responses to the ongoing adaptations in the fields.

Science also has to adapt. Multidisciplinary problems require multidisciplinary solutions; scientific disciplines must focus on dissolving traditional proprietary boundaries and foster integration and synthesis, as well as strengthen the interface with decision makers. Such adaptability can flourish only if scientists let go of the illusion that any scientific endeavor is objective and accept that all science is subjective simply because all questions are subjective. A crucial component of this approach is the implementation of interdisciplinary frameworks through which potential adaptations can be assessed for their relevance, robustness, and ease of use by all necessary and interested parties.[43] Part of this robustness and ease of use must include an interface with urban areas that protects farmlands, forestlands, and open spaces for the infiltration and storage of water—protection that is now missing.

Urban Areas

Suburban, exurban, and rural development in the United States consumes nearly 2,471,054 acres of land per year and thus is a leading threat to biodiversity.[44] To gain a sense of what this means, housing data derived from the 2000 U.S. Census were integrated with a 100-foot resolution into the U.S. Geological Survey's land-cover classification.

The number of housing units in the Midwest grew by 146 percent between 1940 and 2000. Growth in the number of houses was particularly strong at the fringe of metropolitan areas (suburban sprawl) and in rural areas rich in such amenities as lakes and forests (rural sprawl). Medium-density housing (four to thirty-two housing units per six-tenths of a square mile) increased the most in area. The growth of suburban housing was especially high during the decades following World War II, whereas rural sprawl was highest in the 1970s and 1990s.

Pervasive sprawl is cause for concern when it comes to the continuing fragmentation of our landscapes and their ability to respond to the growing changes in global climate. Suburban sprawl has major environmental tradeoffs in comparatively small areas because the large number of housing units concentrates the required inflow of resources from all over the world at a tremendous cost of energy in the form of fossil fuels. In contrast, rural sprawl affects larger areas but with less intensity because of the lower density of housing. Nevertheless, the environmental effects per house are higher because people are continually altering the commons by creating smaller and smaller, isolated, more discrete, product-oriented packages; in these circumstances economic value takes precedence over the intergenerational quality of life.[45]

In Britain, for example, human activities exert the dominant influence on the commonness or rarity of plants, which can be detected at the regional scale. In Western Europe, the identity of increasing and decreasing plants appears to depend on the density of human populations, which, in itself, is a crude measure of the human impact on landscapes. Although intensive agriculture is conventionally regarded as the greatest threat to British wildlife, urbanization may be at least as significant a danger—in the United States as well, where urbanization has a generally negative affect on such organisms as frogs and toads.[46]

In one study, woodpecker diversity was related to several socioeconomic indices in twenty central European countries, where the basic physiogeographic conditions are similar. The diversity of woodpeckers was low in highly industrialized countries with a long history of intensive land use. Conversely, diversity of woodpeckers was much higher, with no species lost, in less-industrialized countries. The negative correlation between the degree of urbanization and the diversity of woodpeckers is seen as a causal link between present, unrestrained biological impoverishment and degradation of the neotechnological landscape and the decline of biodiversity. The relative importance of particular species of woodpeckers for the level of woodpecker diversity shows that species depending on naturally dynamic temperate forests are particularly sensitive to anthropogenic changes.[47] And so is marine life in the oceans of the world.

Oceans

The structural size of phytoplankton assemblages strongly influences the transfer of energy throughout the food web and the carbon cycling of the ocean. Because the prevailing climate affects stratification in the ocean, a global, abiotic influence has been responsible for dramatic evolutionary changes in the dynamics of marine planktonic communities over the past sixty-five million years. Large-scale transitions among alternative states in ecosystems, such as the periods of rapid climate and oceanographic change that took place over the past twenty thousand years, are known as *regime shifts*.

Major disruptions in the benthic communities during this period started with Heinrich Event I, first described by marine geologist Hartman Heinrich, which took place from about seventeen thousand to sixteen thousand years ago. During this period, which occurred after climatic warming had begun, armadas of icebergs broke off from glaciers and traversed the North Atlantic, where masses of rock frozen in their bottoms (dubbed *ice-rafted debris*) dropped onto the sea floor as the ice melted. Loss of latent heat from the ocean's water, as the icebergs melted, created near-glacial sea-surface temperatures and kept much of the subpolar North Atlantic cold, despite increasing warming in the Northern hemisphere. The cold-ocean conditions influenced the immediately adjacent terrestrial climates and may also have affected remote regions.

This event caused prodigious amounts of fresh water to be added to the North Atlantic, where it may well have altered the density-driven thermohaline circulation patterns of the ocean that often coincide with fluctuations in the global climate. The thermohaline circulation is composed of surface and subsurface currents, which form a closed loop by sinking in some regions only to return again to the surface.

Although other periods of severe disruption occurred, the transition from the last glacial period, which ended around sixteen thousand years ago, to the present interglacial period was punctured by a brief, but intense, return to cold conditions around eleven thousand years ago. First identified and studied in northern Europe, the Younger Dryas Event, as it is known, is now understood to have been experienced throughout the entire North Atlantic region to the West Coast of Canada; as well, evidence of this event has been found in the southern hemisphere. This episode is a prime example of the dramatic and rapid oscillations in climate.

The latest disturbance occurred during the beginning of the Younger Dryas Event, between 13,100 and 12,200 years ago. This disruption caused the largest collapse in the deep-sea benthic community of ostracodes in the northwestern Atlantic, which exhibited a faunal turnover of up to 50 percent, not an unusual event during major, climatically driven, oceanographic changes. What is more, the collapse was characterized by an abrupt, two-step decrease both in the upper North Atlantic deep-water assemblage and in species richness.

The ostracode fauna at this site did not recover fully until eight thousand years ago, with the establishment of Labrador-Sea-Water ventilation. But ecologically opportunistic slope-dwelling species prospered during this community collapse. Other abrupt community collapses during the past twenty thousand years generally correspond to millennial climatic events, which demonstrate that deep-sea ecosystems are not immune to the effects of rapid changes in climate.

And today changes in the thermal, physical, and optical properties of the snow/sea-ice system (such as global warming) and the feedback loops between various temporal and spatial scales affect the accumulation of microalgae on the bottom of the sea ice. The maximum biomass of ice algae has been observed under the intermediate cover of snow. Under a thin cover of snow, the biomass of algae declines steadily, coincident with seasonal warming and desalination of the ice cover. A deep cover of snow, however, negatively affects the biomass of algae.

These results suggest that a thin covering of snow is associated with a warming that causes the sloughing of algae, whereas under deep snow algae are limited by the amount of light but are thermally insulated from the warming atmosphere. Both these circumstances highlight the importance of snow cover on the sea-ice system operating below. Consequently, in the face of potential global warming, shifts in snow depth would result in decreased biomasses of ice algae.[48]

In addition to shifts in overall climate, ocean temperatures affect marine organisms, such as dinoflagellates, which are one-celled algae with long, whiplike structures called flagella that let them turn, maneuver, and spin about in the water. Roughly 90 percent of these algae dwell in the ocean, and about half of them employ photosynthesis to obtain energy. Moreover, the median size of dinoflagellate cysts (the resting stage, which fossilizes) is controlled by the thermal gradient between the surface waters and the deep waters, and thus the magnitude and frequency of available nutrients may have acted as a selective factor in the macroevolution of cell size in the plankton. (In evolutionary biology today, *macroevolution* refers to any evolutionary change at or above the level of species.)

Speaking of deep water, microbial activity in the ocean sediment is critical in determining whether particulate organic carbon is recycled or buried. Diverse consortiums of anaerobic microorganisms (primarily bacteria that are able to survive and grow in environments without oxygen) break down organic compounds and thereby mediate the mineralization of organic matter in the anoxic sediments, which have no dissolved oxygen. Concurrently, small changes in temperature affect the efficiency with which organic matter is recycled in these marine sediments.

Taken all together, the pelagic ecosystem of the ocean is by far the largest on Earth. Although its assemblages may be as rich locally as many terrestrial ecosystems, its global diversity is low at both a species and an ecosystem level. There are, however, latitudinal trends in the diversity of pelagic species similar to those in many terrestrial taxa. Nevertheless, the zones with the greatest species richness occur at the boundaries between different types of oceanic water, where dissimilar faunas are mixed together, but the geographical locations of these boundaries are highly flexible and shift hundreds of miles with the seasons.

Meanwhile, on the surface, marine phytoplankton are experiencing competition, predation, infection, and aggregation across distances that range from fractions of an inch to inches. The consequences of these relatively minute interactions, however, influence global processes, such as climate change and the productivity of fisheries.

In active turbulence, patches of phytoplankton, on the order of four-tenths of an inch, have repeatable asymmetry and are regularly spaced over distances of inches to more than a hundred feet. The regularity and hierarchical nature of the patches means that phytoplankton in mixed ocean water are distributed in a dynamic, yet definite seascape topography, in which groups of patches coalesce between intermittent, turbulent eddies.

These patches may link large-scale processes and microscale interactions, thereby behaving like fundamental components of marine ecosystems that influence efficient grazing, species richness, initiation of aggregation, and subsequent carbon flux. *Carbon flux* is an abbreviated phrase referring to the net difference between the sequestration of carbon dioxide through photosynthesis and the respiration of carbon dioxide by such organisms as plants and microbes.

Moreover, ocean water is typically resource-poor; therefore, bacteria could gain significant advantages in growth if they could exploit the ephemeral nutrient patches that originate from numerous small sources. As it turns out, the rapid chemotactic response of the marine bacterium (*Pseudoalteromonas haloplanktis*— no common name) substantially enhances its ability to exploit nutrient patches before they dissipate. Therefore, marine bacteria that possess strategies for chemotactic swimming in patchy nutrient seascapes can exert a strong influence on the turnover rates of carbon by triggering the formation of microscale hot spots of bacterial productivity. (Chemotaxis is a movement in which bodily cells, bacteria, and other single-cell organisms direct their movements according to certain chemicals in their environment, as when bacteria find food by swimming toward the highest concentration of food molecules or when they flee from poisons.)

Phytoplankton accounts for 50 percent of the primary production worldwide; this diverse group is therefore a major component in the global-carbon cycle. A fraction of an inch of seawater may contain tens to hundreds of species from very different taxonomic groups. Yet, despite their obvious importance, we have limited knowledge concerning the functional role of phytoplankton diversity, or of microbial diversity in general for that matter. In addition to conceptual problems regarding speciation within and among unicellular organisms whose reproduction is largely asexual, it is generally unknown whether or how microbial diversity relates to the function of an ecosystem in ways similar to those observed in higher organisms.

The tight coupling between microbial diversity and how an ecosystem functions implies that factors impairing this diversity are likely to alter the effectiveness of ecosystem processes. For example, pollution of marine phytoplankton, especially from toxic substances, often manifests itself through a loss of biodiversity. This phenomenon has similar effects in microbial communities in soils, where toxic compounds not only reduce the diversity of natural communities but also lower their functioning and make them increasingly susceptible to further stress. In addition, globalized economics is today driving a lot of anthropogenic change—among it, increasing pollution—in the world's oceans, such as the Black Sea.

Once described as healthy, the Black Sea ecosystem was dominated by various marine predators. By the late twentieth century, however, it had experienced dramatic anthropogenic impacts, such as overfishing, eutrophication through human activities, and invasions by exotic species. *Eutrophication* is the process whereby chemicals, typically compounds containing nitrogen or phosphorus, are introduced by humans into an aquatic system, where they act as excess nutrients that stimulate excessive growth of plants such as algae.

As a result of human influences, two major shifts took place: a depletion of marine predators and an outburst of the alien comb jellyfish. Overfishing triggered both regime shifts, which resulted in system-wide upheaval in the marine food web.

A similar situation of overfishing and pollution has occurred in the Bohai Sea in China since the late 1980s. Together with the influence of the Yellow River cut-off, which is a shut-off valve to stop the water's flow, the Bohai ecosystem experienced a dramatic change in community structure between the 1980s and the 1990s. This shift was not only over the geographical regions but also at both the species and family taxonomic levels.

For the sake of the world's oceans and their importance to the global commons, wiser approaches than we now employ to the caretaking of our oceans and to controlling the sustainability of fisheries are vital. Even if better care is taken to repair the marine fisheries, a prolonged warming of the oceans will surely alter today's options. For example, changes in the environmental features of China's

Yellow Sea during the last twenty-five years of the twentieth century included increases in the water temperature that are consistent with the recent global warming in northern China and the adjacent seas, such as the Bohai and the East China. The reduction of dissolved oxygen is probably attributable to the increase in temperature and the decrease in primary production in these regions. On the one hand, the increase in dissolved inorganic nitrogen is attributable mainly to precipitation and partly to the discharge of fresh water from the Chang Jiang (Yangtze) River basin. On the other hand, decreases in the concentration of phosphorus and silicon are due to their declining concentrations in seawater that flows to the Yellow Sea from the Bohai. As a result, the ratio of nitrogen to phosphorus is greatly increased in the water of the Yellow Sea.

Moreover, some responses of the Yellow Sea ecosystems to changes in physical variables and chemical biogenic elements include strengthening nutrient limitation, decreasing chlorophyll *a* (the most common type of chlorophyll), succession of dominant phytoplankton species from diatoms to nondiatoms, as well as changes in fish-community structure and species diversity.[49]

Seagrasses, which are any one of four submerged, marine, flowering plants, are sometimes termed ecosystem engineers because they create their own habitat by slowing ocean currents; they thereby increase sedimentation, which not only gives seagrasses more nutrient-rich substrate in which to grow but also augments their roots and rhizomes in stabilizing the seabed. Their importance to associated species is due mainly to their three-dimensional structure in the water column, which provides both shelter and vegetated corridors between and among different patches of habitat, such as coral reefs and mangrove islands. Like land plants, seagrasses require sunlight for photosynthesis and thus are limited in their distribution by the clarity of the water in which they grow. In addition, they produce oxygen.[50]

Seagrass serves as habitat for the settlement of larval recruits and is thus likely associated with the increased abundance of lobsters found in isolated habitats connected by corridors of seagrass. In one study, immigration and emigration of juvenile lobsters were three to four times higher on islands connected by seagrass than on islands surrounded by bare rubble or sand. Rubble fields functioned as barriers to the sea-floor dispersal of all but adult lobsters. Hence, the effects of insularity on a population of lobsters could be lessened by surrounding habitats if they have important functional roles as areas of larval settlement, foraging grounds, or corridors of relative safety through otherwise hostile territory.

Conversely, vegetated corridors can facilitate the access of such predators as the blue crabs to beds of oysters, on which they prey. Accordingly, the spatial proximity of one habitat to another can strongly influence both the population and community dynamics of both. Understanding the trade-off effects of landscape characteristics in estuarine habitats could be useful in predicting the consequences of habitat fragmentation in marine ecosystems, especially where the conservation or repair (or both) of a system is required for the sake of biodiversity and its associated services.[51]

Marine reserves, where fishing is excluded, have been argued to be an effective means of repairing complex reef fisheries while protecting populations of species vulnerable to overfishing. The argument rests on predictions of increases in abundance and size of fishes after the elimination of anthropogenic mortality; in turn, these increases lead to greater production of eggs per area of reef and greater pelagic dispersal to fishing grounds. These concepts proved out in the responses of fish populations to areas closed to fishing in a small Caribbean reserve surrounding the island of Saba in the Netherlands Antilles. But this strategy is likely to work as intended only if networks of stepping-stone reserves are established within a relatively short distance of one another. Within these reserves, ocean currents form the corridors between and among larger areas of protected habitat throughout the fish community's areas of reproduction, larval transport and settlement, and feeding grounds for adults. Planning marine reserves will require some serious forethought, however, because the jet stream drives the ocean currents. A shift in the jet stream caused by global warming will affect the location of the various currents and will thus have a potential impact on existing and future networks of marine reserves.[52]

Yet, despite the best planning, there are some marine creatures that reserves may not be able to help survive. Maintaining a particular biophysical service may necessitate returning the threatened population of an ecologically pivotal species to near its former abundance, but it is often difficult—and in some cases nearly impossible—to estimate the historic size of a species' population, once it has been heavily exploited. To illustrate, the gray whales in the Eastern Pacific, which play a fundamental role in their Arctic feeding grounds, are widely thought to have once again achieved their prewhaling abundance. At previous levels, gray whales may have seasonally resuspended 24,720,266,705 billion cubic feet of sediment while feeding, as much as twelve Yukon Rivers, and thus provided food to a million sea birds.

Although recent spikes in their mortality might signal that the population has reached a long-term carrying capacity, an alternative explanation for this decline is that it is due to shifting climatic conditions in their Arctic feeding grounds. Using a genetic approach to estimate the prewhaling abundance of gray whales, researchers determined that a population of 76,000–118,000 individuals was the norm, approximately three to five times more numerous than today's reputed average population of 22,000 individuals. Amalgamating data suggest that an average of 96,000 individuals was probably distributed between the Eastern and currently endangered Western Pacific populations, which means that the Eastern population is at most at 28 to 56 percent of its historical abundance and thus should be deemed depleted. Therefore, human-caused mortality in this population necessitates a reduction from 417 to 208 individuals killed per year.[53]

A potentially significant loss of ecosystem services may have resulted from a decline of 96,000 gray whales to the current population. It must therefore be noted that the loss of a single species—be it a plant, animal, or otherwise—would impoverish, rather than enrich, the world.

3

How Species Enrich Our
Lives and the World

Since the first living cell or cells came into being, nothing living has ever again been alone on planet Earth because life has kept life company. Although this statement is true in the abstract sense on the physical plane, it is not true in the psychological realm. Here, the paradox is that even though we are compelled to share our life's experiences with one another to know we exist and have value, we are forever well and truly alone with each and every thought, each and every experience in our own never-ending stories, from birth to death.

We Are Truly Alone in Life

A baby comes into the world with its own experience of the womb and the birth process, something the baby can never share, even with the mother. The mother, for her part, has her own experience of the birthing process, which she can never share with her own child, albeit they both coexisted with each other for nine months of their respective lives in perhaps the closest association two human beings can have. When we die, even surrounded by family and friends, we pass out of life, as we know it, without being able to share the experience. Thus, we are born and we die alone—the only person in the world who will ever experience the experience, like a soldier alone in his foxhole.

I use the metaphor of a foxhole to describe one of the most severe tests we face along the spiritual path because it has always seemed to me that the men—and now women—who have been hailed as heroes and have received recognition and medals during war have been those whose actions have often been spectacular and were witnessed by someone else. For the most part, such actions were instinctive and occurred in the space of seconds in which there was little or no time to think about consequences to oneself.

I do not mean by this statement to in any way detract from what these men and women have done. I, myself, have committed some deeds that were relatively dangerous without the time to consider the possible consequences. So I have some sense about the speed of instinctive action, even when other-centered.

But now consider a man in a foxhole, a man sitting wet, hungry, cold, and cramped in a muddy hole on a moonless night with a steady rain falling. Sleep is impossible even if he could quell his fear of the unknown creeping about in the black night of his imagination. This man has all the time in the world to consider what might happen as alone and afraid he faces the unknown—that time when the enemy has no face.

There is no one to support him, no one to see his bravery as he constantly strains his eyes and his ears for some slight hint of danger, all the while choking back his fear. And there is no hero's welcome, no special recognition, no medal of honor when the war is over. There is only the private knowledge that he did his duty to the best of his ability in a strange place, under difficult circumstances, and that he faced his test—his fear—totally alone.

This is the test of the foxhole, the test no one sees, the test only you know about. This is the test of your courage to keep on keeping on when all about you people are oblivious to your struggle, and others seem to get all the attention and the awards. This is your personal, hidden, silent test of the agony of doubt in the material world.

Even if we could verbally share an experience with someone who had been through a similar experience, we would still be alone with our own rendition of it because all we can share are metaphors of feelings and emotions through a chosen combination of words available in the language we are speaking. Furthermore, our ability to share the meaning of the metaphors we choose depends on how conversant the person with whom we are visiting is with the language. We cannot, however, share the feelings, emotions, or thoughts themselves because they cannot be expressed through language. Even two people in the midst of a deeply intimate, sexual union have vastly different, private experiences, which neither can accurately portray to the other.

If the notion of being alone is expanded into the arena of life, it soon becomes apparent that we are alone with each thought we have, each question we ask, each decision we make, each rainbow or flower we observe, each bird's song we hear or symphony we listen to, and each emotion we experience. We are alone—totally alone—within the psychological world of our own making, regardless of how extroverted or introverted we are. Be it a world of exceeding beauty or terrific horror, we are the sole creator of the life we experience, and we live it alone both as creator of our thoughts and as prisoner of our thinking. With the first hominid-animal bond, however, the conscious notion of love—as our earliest ancestors understood it—was extended beyond the human dimension and language to fill the world.

Keeping Us Company

Somewhere in the everlasting twilight of the past, a child may have noticed a mouse sharing the family cave and put food out for it. When the mouse responded, the child put more out. The mouse then began to "expect" to find food in a certain place each evening, and the child delighted in the mouse's eating the food. Over time, a bond of trust was established, a friendship developed, and the idea of a pet came into being. The taming of the first animal became a thread in the ever-expanding story of humanity, a thread that forever changed how humanity treated the animals with which it shared the planet as well as the planet itself. In fact, animals, as pets, have become so much a part of modern human life that a whole industry has grown up around it—one that often threatens the animals whose company we seek, such as parrots.

The poaching of nestling Neotropical parrots for the pet trade is thought to contribute to the decline of many species, its effects have been poorly demonstrated. Therefore, rates of mortality were calculated for nest poaching of Neotropical parrots in twenty-three studies; they evaluated 4,024 nesting attempts of twenty-one species of parrots in fourteen countries. The average rate at which nestlings were poached was 30 percent across all studies.

Aside from the obvious theft of nestlings, tropical birds are relatively long-lived and produce few offspring, which develop slowly, including in the egg, and mature relatively late in life. Additionally, tropical birds have a reduced rate of basal metabolism that, when coupled with their life history and a slow pace of life, compounds the severity of stealing nestlings.

This study shows that the legal and illegal parrot trades are related, despite assertions to the contrary made by some people with interests in aviculture. Moreover, the study indicates that stealing baby parrots for economic gain is both a widespread and biologically significant source of infant mortality in these Neotropical birds.[1]

Unlike the trade in parrots, however, the commercial capture of reef fish for the aquarium trade has begun to change, at least in some parts of the world, from destructive methods, such as the use of cyanide and dynamite, to less-destructive methods, such as fishing with hand-nets.

The Banggai cardinalfish, for example, has become popular in the aquarium trade since its rediscovery in Indonesia in 1995, and thousands have been exported, primarily to North America, Japan, and Europe. A paternal mouth-brooder that lives in groups of two to two hundred individuals in the proximity of sea urchins, this fish has limited abilities to disperse because it lacks a pelagic larval phase. Furthermore, it is believed to be endemic to the Banggai archipelago off the east coast of Sulawesi, Indonesia.

To assess the effects of the aquarium trade on wild populations of the Banggai cardinalfish, a field study was conducted to quantify density, age distribution (the numerical ratio of juveniles to adults), and quality of habitat, as indicated by sea-urchin density at eight sites in the Banggai archipelago. Although the effects of fishing on the density of fish were marginally detrimental, it had a significantly negative outcome on group size in both cardinalfish and sea urchins. The detrimental consequences of the aquarium trade on wild populations of reef fish are thus demonstrated despite the widespread use of relatively nondestructive meth-ods of fishing.² As it turns out, wild animals in our midst are sometimes in danger of being loved into domestication.

The long-term consequences of how we humans treat wild animals are seldom part of our consciousness. Here, the salient questions are How do we, as individuals and a society, deal with nature's long-term, rational impartiality and our short-term, subjective emotional inclination when it comes to allowing each living being to fulfill its appointed role in the evolution of our planet? How do we know what that role is—or can we even make such a determination?

We must honor the integrity of animals as part of the sacred evolution of nature. But we must consciously decide when our notion of helping animals actu-ally interferes with the integrity of their naturalness, when our attempts to rescue animals interfere with the evolution of nature's processes.

Humans often rush to the aid of wildlife in such cases as a whale trapped in a pocket of open water as the ice begins to freeze the polar sea, deer and North American elk starving during severe winters, and feral horses dying of thirst dur-ing severe drought. People are likely to mobilize in a supreme, costly effort to res-cue individual animals from imminent death wrought by nature's impartial evolutionary processes, while often doing little or nothing to prevent the extinc-tion of entire species through exploitation or the destruction of habitat.

Why? Is it because an individual animal is a concrete entity that tugs at our emotions, while a species is an intellectual abstraction? What is our "need" to res-cue? Although I have no succinct answer, I believe most efforts to rescue animals in distress are not the benevolent acts we would like to think they are, but rather are a way to reduce our discomfort of participating in the impartial, often harsh ways of nature, ways we do not understand in our emotional selves and therefore want to control at any cost. This is not to say that some efforts to rescue individual animals are not motivated by true compassion. However, I wonder how many efforts to rescue individual animals are carried out because of our unconscious guilt at the terrible devastation we have wrought on our environment and the habitats of these selfsame animals.

Deer and elk, for example, periodically starve in great numbers when they overpopulate their habitat and thus their food supply. To rescue them under these circumstances may relieve our immediate stress at being out of control but will only prolong their overpopulation, which we may have largely caused in the first place by eliminating their predators and being opposed to hunting as a way to control their numbers within viable limits. Overpopulation of such animals as deer has profound ecological implications besides the ultimate fate of starvation.

The concentration of deer in the forests of eastern North America affects the abundance and population size of forest birds by structuring the vegetation under the forest canopy. A nine-year study of bird populations was conducted by moni-toring the density and diversity of vegetation and birds in eight ten-acre sites in northern Virginia, where four of the sites were fenced to exclude deer. Both the density and the diversity of understory woody plants increased within the fenced areas.

The numerical response of shrubs at the sites lacking deer was predicted on the ratio of organic carbon to nitrogen in the soil. Bird populations, as a whole, increased following removal of deer, particularly those species oriented to life on the ground and in the intermediate understory. The diversity of birds did not increase significantly, however, because the original species were replaced as the understory vegetation proceeded through the succession stages. Nevertheless, succession in the understory vegetation accounted for most of the variability in the abundance and diversity of birds. Consequently, controlling the number of deer and other ungulates in an area not only helps to prevent mass starvation but also maintains a good quality of habitat for other species—not only birds.[3]

These kinds of overpopulations used to be much more easily avoided in the days before predators, such as wolves, were exterminated. Cottonwood trees, for example, are an important component of the plant communities along the riparian zones within the winter range of the northern herd of Rocky Mountain elk in Yellowstone National Park. Young cottonwoods, both narrowleaf and black, are highly palatable to ungulates; however, their recruitment declined at some locations after wolves were extirpated from the park in the 1920s. Elimination of the wolves removed the elk's main predator, which allowed them to increase virtually unchecked. Since the reintroduction of wolves, the elk population is again experiencing the normal cycle of predation, which has allowed the riparian zones to begin recovery from their long overuse by the elk.[4]

There is more to predator-prey relations than simply hunt, kill, and eat. Predators can also affect the population density of their prey through intimidation, which affects the prey's demographics at least as strongly as direct consumption. Intimidation can stimulate costly defensive strategies, such as reducing the prey's ability to acquire energy through nervousness, keeping the prey moving, and thereby lower the success of mating. Intimidation, traditionally ignored in predator–prey ecology, may actually be the dominant facet of trophic interactions.[5]

Whereas wolves chase their prey and run it to ground, cats, such as the puma or mountain lion, stalk theirs and then ambush it. Pumas are highly adaptable generalists. For the most part, resident adults confine their movements to specific areas year after year, but there is also a contingent of younger, transient adults. Resident pumas occupy fairly distinct, contiguous winter–spring and summer–autumn home ranges. Generally speaking, resident pumas use larger areas in winter than they do in summer, and males tend to travel more widely than females. The tendency of pumas to increase their movements during late winter is a result of the scarcity of food. Pumas hunt almost continuously, rarely spending more than a day in the same location. Except for the longer periods of heavy rain in spring and autumn, the activity of pumas seems largely independent of the weather.

During winter, a male ranges over a minimum area of about sixty-four square miles, while a female covers a minimum area of about thirteen square miles and a maximum of about fifty square miles. Pumas have a high degree of tolerance for one another in their areas but are, nevertheless, decidedly unsocial in that they avoid contact with another individual. This behavioral mechanism of mutual avoidance keeps them distributed without injury.

Among pumas, land tenure is based on prior rights, and home ranges are well covered. Young adults establish home ranges only as vacancies occur. The land-tenure system acts to maintain the density of breeding adults below the carrying capacity of the available supply of food. (*Carrying capacity* is the maximum number of pumas or other animals that their prey base can sustain without altering the integrity of the ecosystem that supports them.)

In the short term, a puma's home range is in a state of constant change created by the availability of prey. Over the long term, however, the conditions in certain parts of the home range are such that a cat tends to be more successful in making kills there and, as a result, spends more time in those areas. There is a definite advantage for a cat to be thoroughly familiar with its home range, especially for a female rearing her kittens.

The big cats' staple diet is deer, elk, and porcupines. A hunting cat normally zigzags back and forth through thickets, around meadows, and under the overhang of cliffs. It goes up and down small draws and back and forth across creeks. This method of travel enables it to detect prey and thus stage a successful attack. Pumas do not attempt to indiscriminately capture prey, but, being stalkers, they must find prey in a location that allows an approach close enough for a successful attack; finding prey in this way may require more than an hour of patient stalking.

Each large animal that is killed has some physical or behavioral defect with respect to its long-term survival and is the most vulnerable to predation by the cats, which ultimately cull them from the herd. In addition, predation by pumas keeps deer and elk moving, especially on the winter range.

The mere presence of a puma does not usually alarm deer or elk, but when a kill is made, the reaction is striking. Both ungulates immediately leave the area, crossing to the far side of the drainage, or even moving into a different drainage. This forced redistribution helps to prevent deer and elk from overpopulating an area and causing the type of severe damage to their habitat that the extirpation of wolves in Yellowstone National Park allowed the elk to do.[6]

On the one hand, our attempt to rescue is a form of participating only with the symptom, which likely will allow the cause to worsen. If, on the other hand, we do not interfere and allow nature's cycles to treat the symptom in relationship with the cause, the entire herd will rebuild toward the next moment of vigor and balance, which will last only until the next correction becomes necessary. In short, we, as a species, are so out of tune with nature's rhythms that we too often intervene at inappropriate moments to reduce our own angst, which, I am the first to admit, is a notion I have long debated within myself.

To me, the salient questions surrounding the issue of our human interference in the evolutionary process are How do we as individuals and a society deal with nature's long-term, "rational" impartiality and our short-term, subjective emotional inclination when it comes to allowing each living being to fulfill its appointed role in the evolution of our planet? How do we know what that role is—or can we even make such a determination?

Addressing these questions reminds me of a story about feral cats, a story I will tell from two points of view: mine and my interpretation of the point of view of my wife, Zane. My point of view in this story is one of rational logic and impartiality. I have achieved such a point of view slowly, painfully over forty some years of living with animals and studying ecosystems in a variety of places. I have worked with literally thousands of animals and have seen death well over a thousand times, including human death. My view of life is one of immediate circumstance mediated by the long-term effect of my choices, decisions, and actions.

My interpretation of Zane's point of view in this story is one of purely emotional responses to immediate circumstances in which the long term is seldom, if ever, initially considered. In fairness, I point out that Zane has not had the experiences with animals, death, ecosystems, or nature's check and balances that I have had. Nor is that her interest. There is no right or wrong in either point of view; they are only different, with different long-term outcomes, and each is a matter of personal conscience.

One day while Zane and I were living in Las Vegas, Nevada, a magnificent Siamese cat wandered into our walled backyard. Zane had seen this beautiful cat roaming the neighborhood and wondered whether it had a home. Finally, concluding that it was indeed a stray, she wanted to feed it.

I suggested that we not feed it because our house was then for sale, and we daily hoped to sell it and move. Looking ahead as best I could to the long-term consequence for the cat if we began feeding it, I asked Zane what the cat would do when we left and its food was suddenly cut off. She did not know, but that was clearly irrelevant at the moment. To her, the cat was both a stray and hungry, and that was the immediate point. With that thought, Zane exercised her prerogative of independent action and fed it.

As it turned out, the cat did have a home down the block, but before we learned that, three bona fide strays showed up. I objected to feeding them for the same reason as before. But I still found no way to articulate my concern that it not only would be unfair to the cats when the food they had learned to depend on was suddenly cut off because we had moved but also would put them in a life-threatening position as they tried to return to their extremely harsh and competitive way of life, but without their former hunting territories. Because I could not get this point across, I agreed to feed them until we left.

The three cats soon became five, then seven, ten, twelve, fifteen, and perhaps even more under the cloak of darkness. It was clear that we had a major problem because our charitable feline eatery began drawing customers from far and wide.

In time, three scrawny kittens showed up, and Zane, who loves baby animals, began trying to tame them. Thinking of nature's impartiality, I again objected, but to no avail. Zane said that she simply couldn't resist them.

I argued that they were domestic cats gone wild and would have to remain wild if they were to survive after we moved, which could be any day. But Zane would have none of it. To her, I was just being intellectual because she could neither see nor understand the long-term consequences of her immediate actions. I, for my part, could not only see but also appreciate her emotional response because I had often struggled with the same response myself over the years, too often to the long-term detriment of the animals I tried to rescue.

Zane, with time and patience, was able to tame the skittish black and white kitten that tugged most strongly at her heart. Although our two older cats were less than thrilled at the prospect of a new housemate, we adopted the kitten. Then, as often happens, we went to the local pound to find our new kitten a little friend—and came back with two. We now had five cats!

Then something shifted in Zane's understanding of the plight of our eatery cats, and she became concerned about their well-being once we were gone. She was now ready to reach beyond the emotional immediacy of the moment to see the integrity, the naturalness, of the cats as they were—wild cats. Now she was able to understand that by feeding and taming them, we would hinder their future survival because, by feeding them, we had stolen some of their wildness, some of their integrity, some of their ability to fit into their highly competitive environment after we left.

In addition, we had caused them to concentrate in unusually dense numbers in a small space, which effectively shrank with the addition of each new stray. We had displaced most of the cats from their original hunting areas. These areas, once vacated, would be taken over by other feral cats, and the ones we had fed for six months would have no place to go without a fight. They would now be the outsiders, a position from which claiming or reclaiming an already-occupied area in which to hunt would be most difficult.

We therefore began cutting back on the cats' food to force them to hunt more. Zane remained somewhat uncertain about the short-term "cruelty" of starving the cats into hunting. But when she realized that the wildness of these cats was born of necessity and that the integrity of their wildness was their survival insurance in the naturalness of their environment, she saw the whole drama in a new and different light.

She now understood that we had altered the cats' abilities to participate with nature by making such keenly honed participation unnecessary, and that, if we really cared about them, we had a moral obligation to restore that which we'd stolen—the integrity of their wildness and their naturalness. Therefore, we had to withhold food, to love them with a closed hand. Such love transcends the unruly emotions, such as guilt and sadness, and enters the realm of rational impartiality, which in nature is often the greatest love of all because it accepts the intrinsic perfection of the evolutionary process.

Zane was as right from her original point of view as she is from her present point of view, albeit they differ by 180 degrees. She had originally seen what she considered to be unnecessary hardship, and she was right when you consider the origin of these cats—the personal irresponsibility of those people who did not neuter their cats or dumped their unwanted cats alongside some road. Nevertheless, the cats were here now, and she began to see the integrity of the cats in the context of their participation with nature, the excellence of their fit into their environment.

But where in all of this, she asked, is compassion? Compassion is not stealing an animal's integrity in the first place. Compassion is the commitment to adopt the wild kitten plus two other abandoned kittens from the animal shelter. Compassion is acting from the conscious awareness of the long-term consequences we cause as a result of our short-term decisions. But, most important, compassion is the heart-centered understanding that motivates our behavior.

A woman who lived down the street from us illustrated this kind of compassion. Unbeknownst to us, as we were cutting back on the cats' food, she began feeding those same cats. The wonderful part of this story is that she planned to live there for many years and made a long-term commitment to the cats, part of which was to live-trap each cat and have it neutered. In that way, she could continue to feed and care for them, and they, in their turn, began to show the friendly, loving sides of their nature, but without multiplying their numbers.[7] This story took place in 1992.

Today, however, as urban sprawl gobbles up more and more habitat worldwide, the story is different. Wild animals are becoming domesticated with increasing frequency when they live in our midst. One example is the endangered Key deer, a subspecies of the white-tailed deer, which lives in the Florida Keys. As with the feral cats, the Key deer are given food and water by the residents; the results are an increase in the size of the groups of deer dependent on the handout of food and water and a decrease in the distance they are willing to travel from the security of the handouts. This illegal feeding has caused changes in the density, group size, and distribution of Key deer in a manner that is indicative of domestication.

Because fresh water and food are the primary selective pressures for Key deer in the wild, illegal feeding and watering may cause genetic changes to occur in the future, which could render the domesticated Key deer unfit to survive without human intervention. For those who value wildness in wildlife, domestication of wild species presents a serious psychological and ecological problem.[8] Yet, in the mists of time, just such a circumstance may have led to the domestication of animals, other than pets, in the first place.

Sharing Our Life, Labor, and Strife

Consider, for example, that a tamed dog could sound an alarm when danger was near, as well as help in the hunt for meat; in return, the dog was fed a portion of the capture. But once an individual, family, or group of people could domesticate sheep and goats and raise them as a hedge against starvation, the dog's role began to change. Now, instead of being hunters, dogs were trained to protect the growing flocks.

Of course, one might argue that a group of nomadic hunter-gatherers had less impact on the land they used than did a group of nomadic herders of domestic animals, such as sheep, goats, cattle, camels, or reindeer, because the animals, being an extension of the people, had a direct impact on the land beyond that caused by the people themselves. In addition, the nomadic herders were less at the mercy of nature than the hunter-gatherers because the herders took part of their food supply with them—their animals. In addition, domesticated animals provided much of the herders' clothing and shelter in the form of skins, as well as implements in the form of antlers, horns, and bones from the animals slaughtered for food.

With the advent of domesticating and herding animals, however, came the necessity of continually finding enough pasture on which to graze one's herd. The more people in a given vicinity who had flocks of sheep or herds of goats, and later herds of cattle or horses or both, the more inevitable it became that competition for grazing lands would find its way into culture. Such competition would be accentuated, of course, if the people viewed their animals as their wealth and thus purposefully increased their flocks or herds to numbers beyond those necessary for mere human survival.

At the same time, dualism—us versus them, as well as the haves versus the have-nots—began creeping into the human psyche, a seed that grew into one of the deepest causes of conflict. Here, the challenge is that once the world is divided into us-versus-them, people perceive the necessity of acting collectively in self-interest and self-defense, which today translates into our-national-interest versus everyone-else's-interest. This sense of dualism is the seat of humanity's increasingly fragmented view of a seamless world, a fragmentation that was augmented with the continued domestication of larger and larger animals.

The domestication of such animals as oxen, and eventually the Asian elephant, created a human-animal relationship out of which was born the ability to use animals for work. But an ox by itself could not do much, other than carry things on its back, and was, therefore, good primarily for food, hide for clothing and leather goods, horns for holding things and for the making of utensils, sinew for thread, and bones for various implements. Something was missing.

Let's suppose that someone, seeing the strength of his ox, began toying with the notion of placing a vine or piece of hide around its neck and attaching the opposite end to an object, say a large piece of wood, to see whether the ox would pull it. New technology was suddenly available, and with it a lesson from the hunter-gatherer cultures began to fade—other-centered cooperation for the good of the whole. With time, people discovered that leather rope and then two ropes, one down either side of an ox, made pulling large objects even easier.

If one ox could pull something, someone wondered, could two oxen pulling together drag something even bigger? The answer, of course, is yes, but it's difficult to make two independent oxen pull in unison. Then someone wondered whether two oxen could be made to pull in unison in order to move a heavy object. After much experimentation, a crude ox yoke was invented, which revolutionized the possibilities of putting oxen to work. For a time, the focus was on perfecting the design and craftsmanship of the yoke.

Then someone began to wonder if a still-better way could be found to move heavy objects with an ox or even a team of oxen, and the wheel was born. Again, after much experimentation, the first crude wheel was engineered. Consequently, a mechanism then had to be invented to hook wheels together so they would work in unison. Then someone had to figure out how to hitch the ox yoke to the wheels so that something could be pulled. If two wheels worked well, would four wheels not work better? It was only a matter of time before the first cart was invented.

But oxen are slow compared with a horse. If a yoke could be invented to harness the energy of oxen, could not a similar device be invented to harness the energy of a horse, giving the owner of both oxen and horses a greater array of options for work and thus more potential wealth and power? With this in mind, and after much experimentation, the first, crude horse collar was produced, tested, improved, tested, improved, tested, improved, and so on.

Each of these inventions became the means by which the owner of the technology was relieved of manual labor and ostensibly was freed from the potential dangers of life. In other words, technology, by its very conception, was continually designed to make life easier by making unnecessary as much manual labor as possible, while simultaneously making life as safe and predictable as possible. For example, it would be much easier to ensure a continual supply of wood for the fire if, rather than having to carry it in small bundles by hand, one could haul home a large portion of a fallen tree from afar with a team of oxen or horses and thus increase the effectiveness and efficiency with which one could use the environment. Using animals this way in turn allowed people to embrace a more sedentary lifestyle, but with an increasingly greater impact on the environment in both space and time.[9]

Just Being There

Crows come each winter morning to remind me, in their noisy fashion, to put the food out. I can neither touch nor hold one, for they are wild creatures, despite their insistence on my feeding them. I find great joy, however, in their mere presence, and I am gladdened by their antics. They don't have to pretend to be something they are not. Their presence is sufficient to lift my spirits during our gray, cloudy, northwest winter days. With the advent of spring, however, they are suddenly gone until the following winter.

Come spring, birds fill the air with their songs, while butterflies add flitting color to flowers in the meadow and in my garden. Again, they grace my life with unimaginable beauty in form, color, and function by simply being there.

I once heard it said that God was inordinately fond of beetles. I'd say God was inordinately fond of insects in general, as judged by their astonishing array of sizes, colors, shapes, habitats, abilities, and behaviors.

Although I marveled at the few insects I was aware of along the roadside ditch I played in as a child in the 1940s, I had no inkling of the magnitude of the insects' world. I knew about the water skippers in the ditch, for example, but had no idea that a species of water skipper also lives on the open seas. I knew about grasshoppers because I used them as bait to catch fish but had no concept of their incredible variety, as I have since discovered in Egypt, Nepal, Japan, Malaysia, and throughout much of Europe, the United States, and Canada.

Insects, even here in and around the pond of my garden, range in size from the minute aphid to the dragonfly, which is several inches long. They also range from a dull white (a termite), to the brilliant hues of the rainbow (a tiger beetle), to glorious primary colors and all combinations thereof (butterflies). Some insects are blind (various termites), whereas others have five eyes (bees)—three small, simple eyes set in a triangle between two large compound eyes.

Insects also come in every conceivable shape, including round (ladybug), oblong (click beetle), vertically flat (flea), horizontally flat (bark beetle), skinny (walking stick), fat (scarab), smooth (leaf beetle), long-snouted (weevils), frilly (tropical praying mantis), plainly visible (yellow jacket), and well-camouflaged (Saharan pebble beetle). They live in my garden below ground, inside leaves, in bushes and trees, inside dead wood, and in the water of my pond. Beyond my garden, insects occur from the Arctic to the tropics, from the jungle to the desert, from the ocean shore to above timberline in the Himalayas, in every kind of water and in every kind of soil.

The insects in my garden fly, hop, crawl, climb, swim, walk on water, and burrow. The dragonfly, for instance, can hover, as well as fly forward and backward, up and down, to this side and that. The flea, flea beetle, and grasshopper can hop. In fact, if I could jump proportionately as high as a flea, I could leap over a very tall building. Then again, if I could pick up and carry a load proportional to that of an ant, I'd be the strongest person in the world. If I could dig with the speed of the ground cricket of southern Nepal, I could, with my bare hands, out-dig anyone with a shovel.

Although people think of most insects as crawling, if I could cover the terrain with the speed and endurance of an ant, I would easily walk fifty miles in an eight-hour day without tiring. If I could leap a distance comparable to that of a grasshopper, I'd forever be the world's broad-jumping champion by a wide margin. If I could climb up a vertical surface like a housefly, I could safely walk up the side of a twenty-story building. If I could emulate a water skipper, I could literally walk on water. If I could swim with the speed of a water boatman, I'd be the fastest human swimmer on Earth. If I could maneuver in the water like a whirligig beetle, I'd be little more than a blur of motion. If I was as light and well engineered as an ant, I could fall from a ten-story building and sustain not a single injury.

Insects pollinate flowers, cut leaves, produce honey, produce acid, eat wood, eat every conceivable kind of plant and all of their parts, cultivate fungi, make silk, make wax, metamorphose, carry a cold light in their bodies that is blinked for communication at night, "herd" and "milk" other insects as though they were cows, recycle decomposing vegetation and flesh, chew up wood and spit out paper, sting, bite, suck, spit, keep blood from clotting, live totally under water at one stage of their lives and totally on land at another, and make music.

Some insects are social and construct elaborate homes in the ground; chew elaborate tunnels in wood; and even build huge, vertical colonial homes out of soil. Others make more modest homes of mud with arched entryways, whereas others create abodes of paper. Although many insects live with members of their own species, there is a kind of scarab that lives as a guest in ant colonies.

Insects are predaceous, vegetarian, fungivorous, frugivorous, parasitic, mutualistic, and cannibalistic. There are even adult insects, such as mayflies, that do not eat at all; in fact, some even lack mouthparts. In addition, insects transmit diseases among themselves and to other animals, plants, and humans. Some wasps make paper and did so long before humans learned how. If all the truly miraculous aspects of the world's insects were incorporated into me as a single person, I would be the most amazing and adaptable creature on Earth, although not always socially acceptable.

Insects have myriad abilities that we humans lack and could never emulate and thus are naturally superior to us in some ways; they should inspire us to contemplate our place in the scheme of things and gain a little humility. Then, perhaps, we would treat our home planet with more respect than we currently do. Simple-minded as insects may be by our reckoning, at the rate we are destroying our global environment, they will likely be around well after we humans have made planet Earth uninhabitable for ourselves.

However simple we may think insects to be, Ralph Waldo Emerson pointed out that "the invariable mark of wisdom is to see the miraculous in the common." Perhaps we need to look more closely at the impressive beauty in form and function of the species—all of them—that share this magnificent planet with us. Then, perhaps, we would begin to see the vast array of services our fellow travelers bestow on us, free of charge, simply by being there.

According to Kate Brauman and colleagues, "Ecosystem services, the benefits that people obtain from ecosystems, are a powerful lens through which to understand human relationships with the environment and to design environmental policy. The explicit inclusion of beneficiaries [children of all generations] makes values intrinsic to ecosystem services, whether or not those values are monetized.[10]

Environmental Services

As landscapes are altered to accommodate ever more intensive agriculture and sprawling urban development, three changes occur in humans' relationship with nature: we increasingly attempt to purify and specialize crop plants—particularly through genetic engineering; we accelerate our move toward monocultures, which decrease diversity and therefore stability; and we increasingly view plants and animals that exert any perceived negative effect on the desired economic outcome as pests to be eliminated. This perception necessitates a growing outlay of capital, time, energy, and such materials as fertilizers and pesticides. Intensified management (meaning attempted control) ensures that many normal biological processes will be seen as competition, which conflicts with production goals and will call for continued artificial simplification of ecosystems. And, yet, the most ubiquitous and irreversible environmental problem society already faces is the loss of biological diversity.

The changing global environment affects the sustainable provision of a wide variety of inherent ecosystem services. Although the availability of these services is strongly influenced by the biophysical effects of land use, as well as the changing climate, it is also moderated by the functional diversity of biological communities.

Because of the importance of nature's inherent services, usually thought of as ecosystem functions, it is worthwhile to examine one worldwide service in greater detail—pollination.[11] Indeed, wild and semi-wild pollinators service 80 percent, or 1,330 varieties, of all cultivated crops, including fruits, vegetables, coffee, and tea. Between 120,000 and 200,000 species of animals perform this service.

Bees are enormously valuable to the functioning of virtually all terrestrial ecosystems and such global industries as agriculture. Pollination by naturalized European honeybees is sixty to a hundred times more valuable economically than is the honey they produce. In fact, the value of wild blueberry bees is so great that farmers who raise blueberries refer to them as "flying $50 bills."

While more than half the honeybee colonies in the United States have been lost since the late 1950s, 25 percent have been lost since the beginning of this century. Widespread threats to honeybees (other than viruses and mites) and other pollinators are the fragmentation and outright destruction of their habitat (hollow trees for colonies in the case of "wild" honeybees), intense exposure to pesticides, a generalized loss of nectar plants to herbicides, as well as the gradual deterioration of nectar corridors that provide food to migrating pollinators.

In Germany, for instance, the people are so efficient at weeding their gardens that the nation's free-flying population of honeybees is rapidly declining, according to Werner Muehlen of the Westphalia-Lippe Agricultural Office. Bee populations have shrunk by 23 percent across Germany since the late 1990s, and wild honeybees are all but extinct in Central Europe. To save the bees, says Muehlen, "gardeners and farmers should leave at least a strip of weeds and wildflowers along the perimeter of their fields and properties to give bees a fighting chance in our increasingly pruned and . . . [sterile] world."[12]

Besides a growing lack of food sources, one fifth of all the losses of honeybees in the United States are due to exposure to pesticides. Wild pollinators, such as flower flies (hoverflies) and bumblebees, are even more vulnerable to pesticides than honeybees because, unlike hives of domestic honeybees, which can be picked up and moved prior to the application of a chemical spray, colonies of wild pollinators cannot be relocated. Because wild pollinators service at least 80 percent of the world's major crops and domesticated honeybees only 15 percent, honeybees cannot be expected to fill the gap by themselves if wild pollinators are lost.

Ironically, economic valuation of products, as measured by the Gross Domestic Product, only credits—never debits—and thus fosters many of the practices employed in modern, intensive agriculture and exploitive forestry, which actually curtail the productivity of crops by reducing pollination. An example is the high level of pesticide used on cotton crops to kill bees and other insects, which reduces the annual yield in the United States by an estimated 20 percent, or $400 million. In addition, herbicides often kill the plants that pollinators need to sustain themselves when not servicing crops. Finally, squeezing every last penny out of a piece of ground by killing as much unwanted vegetation as possible in the practice of exploitive forestry or by plowing the edges of fields to maximize the agricultural planting area can reduce yields by disturbing or removing (or both) nearby nesting and rearing habitat for pollinators, as well as nectar corridors.

Despite widespread declines in pollinators, little is known about the patterns of change in most pollinator assemblages. Studying bee and hoverfly assemblages in Britain and the Netherlands produced evidence of declines (before and after 1980) in local bee diversity in both countries; however, divergent trends were observed in hoverflies. Depending on the assemblage and location, the declines were most frequent in species with narrow habitat parameters, in species that completed their life cycles in a single year or were yearlong residents or both, and in species that were flower specialists. Concurrently, those species of plants that rely on insects for cross-pollination are themselves declining relative to other kinds of plants. Taken together, these findings strongly suggest a causal connection between local extinctions of functionally linked species of plants and their pollinators. The loss of birds is a second example of imperiled ecosystem services.

Overall, 21 percent of the species of birds are currently extinction-prone, and 6.5 percent are functionally extinct, which means their numbers are so small their contribution to ecosystem services is negligible. A quarter or more of the fruit-eating birds and those with a catholic diet are susceptible to extinction. As well, one-third or more of the species of birds that eat plants or fish or that scavenge are extinction-prone. Moreover, by 2100, between 6 and 14 percent of all bird species will likely be extinct, and between 7 and 25 percent (28 to 56 percent on oceanic islands) will be functionally extinct. Should this rate of decline pan out, critical ecosystem processes will diminish accordingly, especially decomposition, pollination, and seed dispersal.[13]

In addition, overfishing, among other human influences, is reducing the diversity and abundance of fish worldwide, thereby jeopardizing how freshwater aquatic systems function. Recycling of nutrients is an important ecosystem service influenced directly by fish, species of which vary widely in the rates at which they excrete nitrogen and phosphorus. Thus, an alteration in fish communities could affect how nutrients are recycled. To better understand the ramifications of the extinction of fish in freshwater aquatic ecosystems, data were collected on the population sizes of fish species and the rates at which they recycled nutrients in a Neotropical river and in Lake Tanganyika, Africa.

In both these species-rich ecosystems, a relatively few species dominated recycling, but contributions of individual species differed with respect to nitrogen and phosphorus. Therefore, alternative patterns of extinctions were widely divergent. For example, the loss of species targeted by fishermen would lead to faster declines in nutrient recycling than if extinctions were to take place in the order of rarity, body size, or position in the food web (the trophic hierarchy).

However, if surviving species were allowed to increase after extinctions, these compensatory responses—ecological backups—had strong moderating effects even with the disappearance of many species. These results underscore the abiding complexity of predicting the consequences of losing ecosystem services to extinctions of the service providers from species-rich animal communities. Nevertheless, the importance of the ecological services performed by exploited species in nutrient recycling indicates that overfishing could have particularly detrimental effects on how an ecosystem functions.[14]

Such declines in biodiversity have prompted concern over the unknown consequences of compromising ecosystem processes that account for nature's goods and services, on which we humans rely for the quality of our lives. However, relatively few studies have evaluated the functional ramifications of simplifying ecosystems within a realistic, biophysical context. Understanding the real-world outcomes of declining biodiversity will require addressing changes in species composition, as well as their respective contributions to and performances within nature's interactive web of life.[15]

In addition, we must understand and account for the continuance of life's biological backup systems, those multi-species relationships that act simultaneously as environmental shock absorbers and insurance policies, whereby ecosystems are largely protected against functional collapse. Another major challenge with the loss of the individual services performed by different species is the gradual dismantling of nature's self-reinforcing feedback loops, which collectively are the functional engine of every ecosystem. If diversity of species, in the form of feedback loops, is essential, how do species act in concert with an entire forest?

Biologist Louise Emmons made a critical observation with respect to this question: "For want of a squirrel, a seed was lost; for want of a seed, a tree was lost;

for want of a tree, a forest was lost; for want of a forest . . ." In other words, the tremendous biodiversity of the tropical rain forests could be lost without cutting a single tree! How? Let's consider the rain forest in Gabon, Africa, whose fascination to Emmons lies in "its stunning complexity." In this forest, says Emmons, "You can stand anywhere and be surrounded by hundreds of organisms that are all 'doing something,' going about their living in countless interactions—ants carrying leaves, birds dancing, bats singing, giant blue wasps wrestling with giant tarantulas, caterpillars pretending they are bird droppings, and so on."[16]

In Gabon, Emmons found that nine species of squirrels all live together in one forest. Each is a different size; three have specialized diets or habits, which leaves six that feed on nuts, fruits, and insects and could therefore be potential competitors for food. But a closer look reveals that three of the six species—one large, one medium, and one small—live exclusively in the canopy of the forest, where the largest one, a "giant" squirrel, feeds primarily on very large, hard nuts, while the smaller ones eat proportionally smaller fruits and nuts. The other three species— again one large, one medium, and one small—live exclusively on the ground, where they eat the same species of fruits and nuts as do their neighbors in the canopy, except that they eat the fruits and nuts once they have fallen.

The forest in Gabon is evergreen. Fruit can be found on the trees throughout the year, but any one species of tree produces fruit for only a short period each year. To support three species of squirrels, eight species of monkeys, and eight species of fruit-eating bats (and so on) in the canopy, this forest must have a wide variety of species of trees and lianas (high-climbing, usually woody vines), each of which produces fruits and nuts in its own rhythm. The varying sizes of the fruits and nuts can support different sizes of squirrels with different tastes, whereas these same fruits and nuts when they fall to the ground can feed a whole analogous array of species. Halfway around the world, high in the Amazonian rain forests, is a microcosm of the feedback loops described for the forests of Gabon, but there, the actors are ants.

Among invertebrates, ants have the primary major role in dispersing specialized seeds from thousands of plants. An outstanding example occurs in the Amazonian rain forest, where arboreal ants collect seeds of several species of epiphytes and cultivate them in nutrient-rich nests, thereby forming a multitude of conspicuous hanging gardens known as ant gardens. Both the ants and the plants they cultivate are dominant members of the lowland ecosystems, where they are obligatory mutualists: for one partner to survive, both must survive, regardless of which species they are. In this case, the ants locate and accept seeds from particular plants while rejecting others. The selection process is based on chemical compounds, which produce odors that emanate from the seeds as olfactory cues whereby the ants are directed not only in their initial choice of seeds but also in their seed-carrying behavior.

Just how rich in species of plants is a tropical rain forest? Al Gentry, of the Missouri Botanical Garden, counted the species of trees and lianas in tropical rain forests for many years. Perhaps the richest site he found was a 2.5-acre plot near Iquitos, Peru, where he counted an incredible 283 species of trees over four inches in diameter. He found 580 trees of this size in the plot, which means that there was an average of only two individual trees per species; he found an astounding fifty-eight species among the first sixty-five individual trees counted.

Moreover, pioneering species of trees, on average, have a much lower longevity than nonpioneers, whereas among old trees, emergent species have greater longevity than do those of the canopy. In essence, ages of the various tree species are consistent with radiocarbon-based studies that suggest Amazonian trees can occasionally exceed one thousand years of age.

Worldwide, tropical rain forests seem to have from about 90 to 283 species of large trees within every 2.5 acres (one hectare). Even the "poorest" tropical rain forest has an average of about five individual trees per species every 2.5 acres.

In contrast, a dry tropical forest, such as occurs in northern India, has about half as many species of trees as does a wet, tropical forest. The richest forests in the United States have about twenty species of trees over four inches in diameter, with an average of about thirty individuals per species, in each 2.5 acres of ground. Most temperate forests, however, are much poorer than this.

Clearly tropical rain forests are amazingly rich in species of trees—not just any trees, however, but those whose fruits are eaten and whose seeds are dispersed by birds and mammals, termed *zoochorous trees*. Not surprisingly, therefore, tropical rain forests are also rich in mammals and birds—not just any mammals and birds, however, but those that eat fruits and disperse their seeds. There are, for example, 126 species of mammals within a single area of forest in Gabon and 550 species of birds within a single lowland site in the Amazon basin of Peru. Furthermore, the life cycle of each species is interdependent with the life cycles of the other species. The enormous number of vertebrate animals appears to be supported by the large number of species of plants that act as sources of food all year.

If all this biodiversity is to be maintained, each individual tree must succeed in leaving offspring. Seeds and tender young seedlings are among the richest foods available to forest animals, and their succulence greatly increases their chance of being eaten by the large numbers of hungry animals searching for food around the bases of fruit- and nut-bearing trees. Similarly, such organisms as fungi, worms, and insects soon accumulate where the seeds and seedlings are concentrated and spread from one seed or seedling to another.

Under such circumstances, seeds carried away from such concentrations of hungry organisms are more likely to succeed in germinating than seeds that remain in place. Another major benefit of seeds being carried away from the parent tree is the likelihood they will fall in places with different conditions. A new condition might be a pocket of better soil on a mound created by termites or in a spot where a dead tree has created a hole in the canopy that emits sunlight.

It is certainly no accident that 80 to 95 percent of the species of trees in tropical rain forests produce seeds that are dispersed by birds and mammals.

By dispersing those seeds, the birds and mammals also maintain the rich diversity of species of trees, which not only formed their habitat in the first place but also perpetuates it. This is an ideal example of a self-reinforcing feedback loop.

Many species of trees in tropical rain forests, especially those that germinate in the dark understory, have large seeds, which carry enough stored energy to grow leaves and roots without much help from the sun. Such fruits and seeds are often so large that only proportionately sized birds and mammals can swallow or carry them. In Gabon, for example, monkeys dispersed 67 percent of the fruits eaten by animals in the area studied by Emmons.

Seed-dispersing animals, such as large birds and large monkeys, are critical in the replacement of the large trees and lianas of the forest canopy and thus help the species survive. These animals are, however, the first to disappear when humans hunt for food. These species, along with elephants, have already been hunted so heavily that they have either been drastically reduced in numbers or have been eliminated completely over vast areas of the African forest, and a similar situation exists in the tropical rain forests of Central America and South America.

For the most part, foresters have overlooked how the interdependence of plants and animals affects the biodiversity of a plant community. Elephants in the Ivory Coast, for example, disperse the seeds of thirty-seven species of trees. Of those, only seven species have alternate means of dispersal (by birds and monkeys). In one study area, elephants dispersed 83 species of trees out of a possible 201 species.

In one forest where humans had eliminated elephants a century earlier, few juvenile trees of the elephant-dispersed species were left, and the two major species had no offspring at all. One of these two species is the single most important species for the two largest squirrels that Emmons studied in Gabon—one eats the large, hard nuts in the canopy, and the other eats the same nuts once they have fallen to the ground.

Once the large species of birds and mammals are gone, the stunningly rich tropical rain forests will change and gradually lose species of trees, lianas, and other plants. Smaller seeds dispersed by wind will replace large seeds dispersed by large animals. Those species of plants whose seeds grow in the shaded understory will not survive, and the land will gradually be forested by fewer, more common species.

Frugivores are highly variable in their contribution to the removal of fruit in local populations of tropical trees. Data are lacking, however, on species-specific variation in two central aspects of seed dispersal: distance of dispersal and probability of dispersal among populations through long-distance transport.

To understand this process, the distribution of seeds from the Mahaleb cherry (a native of western Asia and parts of Europe) by small- and medium-sized passerine birds and carnivorous mammals was studied. Small passerines carried most seeds short distances into covered microhabitats (50 percent dispersed seeds less than 170 feet from the tree of origin). Mammals and medium-sized birds dispensed seeds long distances into open microhabitats (50 percent of the mammals carried seeds more than 1,625 feet, and 50 percent of medium-sized birds dispersed seeds more than 360 feet). Thus, the behavior of different frugivorous species determines the distance seeds are carried and the microhabitat into which they are deposited.

Mammals accounted for two-thirds of the seeds introduced into the population, whereas birds accounted for one-third, a finding that demonstrates that frugivores differ widely in their effects on seed-mediated gene flow. Despite the fact that highly diverse groups of mutualistic frugivores spread seeds, critical long-distance dispersal might rely on a small group of large birds and mammals. Declines in the population of these frugivorous species may seriously impair seed-mediated gene flow in fragmented landscapes by curtailing long-distance transport and thus may limit the deposition of seeds in available microsites within forest fragments.

As the forests become poorer in species of plants, the number of species of birds, mammals, and other creatures will correspondingly decline. The entire complex, interconnected, interdependent feedback loops among plants and animals will gradually simplify. The species that constitute the feedback loops will be lost forever—and the feedback loops along with them. This is how the evolutionary process works. Ecologically, it is neither good nor bad, neither right nor wrong, but those changes may make the forest less attractive and less usable by other species that rely on its products, including humans.[17]

Maintenance of feedback loops, like those discussed above, is based on ten postulates of genetic diversity: (1) genetic variation within a species is related to the size of its population; (2) genetic variation within a species is related to the size of its habitat; (3) genetic variation is related to population size within taxonomic groups; (4) widespread species have more genetic variation than restricted ones; (5) genetic variation in animals is negatively correlated with body size; (6) genetic variation is negatively correlated with the rate of chromosome evolution; (7) genetic variation across species is related to population size; (8) vertebrates have less genetic variation than invertebrates or plants; (9) island populations have less genetic variation than mainland populations; and (10) endangered species have less genetic variation than non-endangered ones. Empirical observations support these hypotheses, which means genetic variation is related to the size of a population—large size increases evolutionary potential, whereas small size reduces evolutionary potential in wildlife species. The ability of a species to resist some pathogens is also related to the potential genetic variability of its population size.[18] These postulates bring to the fore another kind of feedback loop, one that occurs among the synergisms of many simultaneous environmental changes that are posing unprecedented threats to such areas as the Amazonian forests.

The rapid fragmentation and actual loss of forests in the Amazon basin makes areas other than intact forests more prone than they used to be to periodic damage from droughts brought about by the El Niño–Southern Oscillation (ENSO); these droughts cause increased tree mortality, increased litterfall, shifts in phenology (recurring annual cycles) of plants, and other ecological changes, especially near forest edges. Moreover, self-reinforcing feedback loops in deforested areas, fragmentation of forests, fire, and change in the regional climate appear not only increasingly probable but also increasingly detrimental to human values.

The major cause of deforestation, which leads to such extensive burning, is the conversion of tropical forests to pastures for cattle. Simple harvesting of timber also causes problems, however, because once the canopy of the forest is opened, the understory environment changes drastically, and the forest can no longer sustain itself. Never in the history of humanity has so much of the world's tropical forests been disturbed in such a foreign and catastrophic way on such a large scale as it has since the late 1970s. The significance of this statement lies in the fact that tropical rain forests—one of the world's oldest ecosystems—occupy only 7 percent of the Earth's surface but are home to more than 50 percent of all species. What does such deforestation mean for the Amazonian tropical forest?

An intact rain forest creates its own internal and external climate, in which about half of all the rainfall originates from moisture given off by the forest itself. Deforestation reduces plant evapotranspiration, which in turn constrains regional rainfall and thereby increases the vulnerability of forests to fire. Once the forest is gone, drought is likely to occur, which increases the probability of fire and decreases the probability that the forest will ever return.

The environment in the deforested areas of the Amazon has been altered to such an extent that the ecological processes that maintained the tropical forest are irreversibly unraveling. Once the forest has been even partially cleared or logged, the environmental conditions change swiftly and dramatically. Removal of the trees not only alters the internal microclimate of the forest by exposing its heretofore protected, moist, shaded interior to the sun but also leaves behind large accumulations of woody material that are exposed to the sun's drying heat. Daily temperatures soar in the deforested areas by ten to fifteen degrees, which allows woody fuels to dry and become extremely flammable.

Forest fragments are especially vulnerable because they have dry, fire-prone edges, are logged frequently, and often adjoin cattle pastures, which are burned regularly. The net result is that there may be a critical deforestation threshold above which Amazonian rain forests can no longer be sustained, particularly in relatively seasonal areas of the basin. Global warming could exacerbate this problem if it promotes drier climates or more severe droughts because of the stronger El Niño–Southern Oscillations. Now it may not be a matter of if the area will burn, but rather when it will burn. The ultimate result may be a quick, dramatic change from a dense, closed-canopy forest virtually immune to fire to a weedy, flammable pasture in which fires are common and often occur repeatedly—to the exclusion of a new forest.

In addition, changes in mean climate, as well as its variability, currently contribute to the worldwide increase in the runoff of precipitation, as opposed to its infiltration and storage. Historic land-use data show that change in the use of land plays an added, important role in controlling the amount of runoff on a regional basis, particularly in the tropics, where changes in landscapes are the most pronounced. One consequence of simplifying the tropical ecosystems is a discharge of water from the land that is substantially larger than that caused solely by climate change.[19] In turn, more sediment is entering rivers, but less is making it to the sea from many major rivers because of dams, which deprive the oceans of vital, terrestrially based nutrients. As well, sediment is collecting behind dams, which decreases their water-holding capacity over time.[20]

In yet another kind of feedback loop, animals act as unwilling monitors of our human activities—the proverbial canary in the coalmine, as it were. I first became aware of this service in 1969, when I captured a number of montane voles (a kind of meadow mouse) living along an irrigation ditch that drained an agricultural field. The voles, which normally had a brownish-gray pelage, all had abnormally yellowish fur. Suspecting a chemical interaction with their fur, I took some live voles to the Department of Agriculture at the local university, but no one was willing to acknowledge that agrochemicals could be involved, and thus they refused to test the fur. Their refusal was particularly troubling to me because the voles lost the abnormal color when held in captivity and fed normal laboratory food. Those living along the ditch, however, remained yellowish.[21]

Since then, many examples have emerged of animals being adversely affected by contact with agricultural chemicals, and now gonadal abnormalities, such as those found in Florida in cane toads (also known as marine toads), are likely to reduce the reproductive success of affected individuals. This phenomenon might help explain why amphibian populations exposed to pesticides are declining worldwide or have already become extinct.

With respect to the cane toad, dramatic increases in gonadal abnormalities are associated with the intensity of agricultural activities. Moreover, secondary male sexual traits, such as skin color, forearm width, and the number of nuptial pads, are altered in toads from agricultural areas; hence, testicular function in these toads may be compromised. These abnormalities were statistically linked in a dosage-dependent manner with the percentage of an individual's exposure to intensive agricultural activities. In addition, many of these gonadal abnormalities are organizational, which means they are likely to remain throughout the toads' adult lives.[22] So what, you might ask, is the significance of this sexual alteration?

Well, let's suppose that the toads' primary function is to control certain insect pests in and around the agricultural fields. When, however, agrochemicals are applied, be they pesticides or fertilizers, they disrupt the male toads' ability to reproduce, and each application will further diminish the ecological service these amphibians perform. As the quality—and quantity—of their insect-eating service declines, more pesticides are applied to achieve the same outcome in pest control;

pesticide application further jeopardizes the male toads' reproductive abilities, and so on in a self-reinforcing feedback loop that is tantamount to participating in a circular firing squad. Fortunately, nature provides more that one species with a similar function in every ecosystem, which most people think of as "redundancies." But these species are lost when we humans simplify an ecosystem beyond some ecological threshold of biophysical diversity.

There is an additional tradeoff here: fertilizers and pesticides exact an ecological cost from the entire planet because of the pollution created during their manufacture and the added dispersal of that pollution through application of the chemical compounds. As well, there is an economic cost to the farmer, which is passed along through the marketing chain to the consumer. With the progressive uncoupling of the ecosystem and the demise of its free services, each generation in its turn will pay a higher price for the same amount of produce. These costs are yet another aspect of the self-reinforcing feedback loop that began with the inception of agriculture.

Redundancy or Backups?

There is a basic misunderstanding with respect to redundancy in ecosystems. Meaning no disrespect to Professor Shahid Naeem of Columbia University, in a paragraph he wrote he displayed an understanding of the biophysical principle, but he did not use the correct language to convey it. Moreover, the principle is illustrated with the linear certainty of an engineering example, whereas nature's biophysical principles are cyclical and forever unpredictable. That is why a clear understanding of and use of language is paramount to saying what we mean and meaning what we say to the very best of our ability. Ultimately, our survival as a species depends on precise communication. To this end, I quote the salient paragraph by Professor Naeem:

> The concept of species redundancy in ecosystem processes is troublesome because it appears to contradict the traditional emphasis in ecology on species singularity. When species richness is high, however, ecosystem processes seem clearly insensitive to considerable variation in biodiversity. Some elementary principles from reliability engineering, where engineered redundancy is a valued part of systems design, suggest that we should rethink our stance on species redundancy. For example, a central tenet of reliability engineering is that reliability always increases as redundant components are added to a system, a principle that directly supports redundant species as guarantors of reliable ecosystem functioning. I argue that we should embrace species redundancy and perceive redundancy as a critical feature of ecosystems, which must be preserved if ecosystems are to function reliably and provide us with goods and services. My argument is derived from basic principles of reliability

engineering, which demonstrate that the probability of reliable system performance is closely tied to the level of engineered redundancy in its design. Empirical demonstrations of the value of species redundancy in ecosystem reliability would provide new insights into the ecology of communities and the value of species conservation.[23]

Although Naeem did not coin the notion of ecological redundancy, he provides a good example of the challenge we too often have in verbally conveying our understanding of dynamic, multi-dimensional, multi-scaled ecosystems. This difficulty is particularly apparent in today's reductionist scientific thinking, wherein mechanistic language is often employed to explain biophysical phenomena entrained in a continuum of change and novelty, something no artificially engineered, mechanical system is capable of explaining. Yet, from an engineering standpoint, Naeem is correct in his use of the term *redundancy*, which he defines as "the installation of *duplicate* [my emphasis] electronic or mechanical components or backup systems that are designed to come into use to keep equipment working if their counterparts fail." The words in this definition convey the reductionist sense of *identical* pieces with a predictable sameness in function, whereas nature employs *similar* species that exhibit only varying degrees of complementariness in their ecosystem services.

Redundant is defined first and foremost, however, as something not being needed or no longer being needed. In fact, the synonyms of redundant are: superfluous, outmoded, disused, surplus, extra, unneeded, the same, unnecessary, uncalled-for. Within the context of an ecosystem, however, nothing is ever superfluous, outmoded, disused, surplus, extra, unneeded, the same, unnecessary, or uncalled-for.

Therefore, while Naeem has the right idea in an ecological sense, he simply chose the wrong word to convey his meaning. The word Naeem needed to use is *backup*, which is a substitute or reserve system that can be used if the initial component (such as a species) or subsystem fails to operate as required to achieve a given systemic outcome.

In this light, consider that no one knows the precise mechanisms whereby ecosystems cope with stress, but one mechanism is closely tied to the genetic selectivity of an ecosystem's species. Thus, as increasing magnitudes of stresses influence an ecosystem, the replacement of a stress-sensitive species with a functionally similar but more stress-resistant species maintains the ecosystem's overall productivity. Such replacements of species—backups—can come only from within the existing pool of biodiversity.

Backups are a fundamental part of the design of nature that we humans increasingly ignore at our own peril because backups are an essential part of daily life. For example, built-in backups give ecosystems, as well as cities, the resilience to either resist change or to bounce back after disturbance or both. In this way, a backup system effectively protects the functional continuance of a system after a major disruption, despite the fact that some might deem having a backup to be inefficient—a waste of money.

And yet, with the proven probability that our ever-increasing reliance on electrical power is making us vulnerable to blackouts, virtually every village and city is converting as much of their functional systems to computerization as possible, thereby eliminating the manual component as unnecessary duplication—redundancy—and thus an unnecessary waste of money. Here, it may be instructive to take a lesson from a forest about the value of backup systems.

Backup systems, in the biological sense, comprise the various functions of different species that act in concert as an environmental insurance policy. To maintain this insurance policy, an ecosystem needs three kinds of diversity: biological, genetic, and functional. *Biological diversity* is the richness of species in any given area. *Genetic diversity* is differences in the way species adapt to change. The most important aspect of genetic diversity is that it can act as a buffer against the variability of environmental conditions, particularly in the medium and long term. And *functional diversity* equates to the variety of biophysical processes that take place within the area. The upshot is that healthy environments act as shock absorbers in the face of catastrophic disturbance.

To better understand this concept, think of each of these kinds of diversity as the individual leg of an old-fashioned, three-legged milking stool. When so considered, it soon becomes apparent that if one leg (one kind of diversity) is lost, the stool will fall over. Fortunately, however, biological diversity passed forward through genetic diversity effectively maintains functional diversity.

This backup results in a stabilizing effect similar to having a six-legged milking stool, but with two legs of different kinds of wood in each of three locations. So, if one leg is removed, it initially makes no difference which one because the stool will remain standing. If a second leg is removed, its location is crucial because should it be removed from the same place as the first, the stool will fall. If a third leg is removed, the location is even more critical, because the loss of a third leg has now pushed the system to the limits of its stability, and it is courting ecological collapse—in terms of the value we, as a society, placed on the system in the first place. The removal of one more piece, no matter how well intentioned, will cause the system to shift dramatically, perhaps to our long-term social detriment.

When, therefore, we humans tinker willy-nilly with an ecosystem's composition and structure to suit our short-term economic desires, we risk losing species, either locally or totally, and so reduce, first, the ecosystem's biodiversity, then its genetic diversity, and finally its functional diversity in ways we might not even imagine. With decreased diversity, we lose existing choices for manipulating our environment. This loss may directly affect our long-term economic viability because the lost biodiversity can so alter an ecosystem that it is rendered incapable of producing what we once valued it for or what we, or the next generation, could potentially value it for again. Maintaining backup systems is thus a critical link in the shared relationship between ecosystems and communities—social-environmental sustainability. With respect to humans tinkering with ecosystem processes, what might be the effect of overexploiting "bushmeat" on the self-reinforcing feedback loops and their backups that are so critical to the biophysical sustainability of tropical forests?

Bushmeat

Plant-animal interactions and vegetative structure were studied in two geographically close tropical Bolivian forests, each of which experienced different intensities of hunting. It was hypothesized that a reduction of ground-dwelling mammals weighing 2.2 pounds or more in an "intensively hunted forest" would decrease the predation of seeds and the trampling of seedlings, as well as increase seedling survival and density, whereas an "occasionally hunted forest" (which held 1.7 times as many mammalian species as the intensively hunted forest) would have decreased tree-species diversity at the seedling stage in relation to the adult stage.

As predicted, predation on the seeds of the murumuru palm (which has edible nuts) was lower in the intensively hunted forest, and seedling survival in the intensively hunted forest was higher than in the occasionally hunted forest. But the intensively hunted forest displayed lower seedling densities and a higher ratio of seedling diversity to tree diversity than did the occasionally hunted forest.[24] Reduction of collared peccaries from the intensively hunted forest may explain much of the between-site differences in seed predation, trampling, and seedling survival. Yet reductions of canopy-dwelling birds and mammals that disperse seeds, such as those in the previously discussed forests of Gabon, Africa, can have a decidedly devastating effect on the diversity of plants and thus forest structure.

Hunting for bushmeat has profound negative effects on the species diversity, the standing biomass (the total weight or mass of living organisms in a defined area), and the size and structure of vertebrate assemblages in otherwise largely undisturbed Amazonian forests—and tropical forests in general. These effects are aggravated by forest fragmentation because fragments not only are more accessible to hunters but also allow, at best, a low rate of re-colonization from nonharvested refugia. Further, fragments, especially small ones, may provide a low quality of habitat and of resource base for the birds and mammals that eat fruits and seeds.

The size distribution of 5,564 fragments of topical forest—estimated from satellite images of six regions of southern and eastern Brazilian Amazonia—clearly shows that the tracts rarely exceed twenty-five acres. It thus seems patently obvious that persistent overhunting in these areas is likely to decimate the populations of most midsized to large vertebrates, not only from the fragments themselves but also from the fractured landscape. In particular, persistent hunting markedly reduces the number of large-bodied species greater than eleven pounds. This synergism has increasingly negative consequences for the life-sustaining feedback loops in Neotropical forests, which are dependent on interactive vertebrate assemblages with a full complement of avian and mammalian species. It is imperative, however, to account for long-lived individuals because they have low rates of reproduction and long generation times, which makes them more vulnerable to extinction than species with short-lived individuals, high rates of increase, and shorter generations.

Twenty-three and a half million animals are killed annually in Amazonian forests for the bushmeat trade (an equivalent of 89,224 tons) with a market value of US $190.7 million, which is a conservative estimate. This number illustrates the enormous socioeconomic value of bushmeat to the rural population of Brazilian Amazonia and highlights the staggering effect of hunters on tropical-forest vertebrate communities. Much of the illegal bushmeat hunting is clearly facilitated and intensified by the presence of logging concessions (large areas of land set aside for extracting timber), which are commonly backed by a loan from the African Development Bank in the Democratic Republic of Congo.

For example, in one such logging concession in the Sangha region of Congo, its geographic isolation, resulting transportation costs, and market demands forced commercial loggers to exploit only the most valuable timber. Here, the selective extraction destroyed an average of 6.8 percent of the canopy and thus, unlike clear-cutting, had far less impact on wildlife populations in and of itself. Nevertheless, line-transect surveys clearly showed that the abundance of primates was exceedingly low in the logged forest.

This outcome was not a direct consequence of thinning in the canopy, but rather was the result of extremely intensive market hunting, which coincides with surveying timber and the subsequent logging. Moreover, weapons and hunting camps were common, and vehicles belonging to the logging company were used daily to transport primates, duikers, and other animals. In addition, laws pertaining to wildlife in the Congo are openly violated and not enforced.

Whatever tropical forest one contemplates, the expansion of roads increases both human access and its accompanying hunting pressure. In the rain forests of southern Gabon, hunting had the greatest impact on duikers, forest buffalo, and red river hogs, which declined in abundance, and lesser effects on lowland gorillas and carnivores. In addition to hunting, the avoidance of roads further depressed the abundance of duikers, whereas sitatungas (swamp-dwelling antelopes, also called marshbucks) and forest elephants simply avoid roads. Although five species of monkeys showed little response to either roads or hunting, some rodents and pangolins (also known as scaly anteaters) increased in abundance, possibly in response to greater disturbance within the forest.

Roads had the greatest impacts on large and small ungulates, with the magnitude of road avoidance increasing with the increased pressure of local hunting. Nevertheless, even moderate hunting pressure can markedly alter the structure of mammal communities in central Africa and thus their ability to perform the ecological services that provided tropical forests with their complex, self-sustaining feedback loops.

Because oil companies have large tracts of land set aside for exploration and production (also termed *concessions*), wherein they accommodate human access by constructing roads and pipelines, these set-asides are often relatively safe areas for illegal hunting. For example, a bushmeat-trade monitoring program was established in the town of Gamba to assess the human pressure exerted on wildlife by the number of people attracted by the oil industry in the Gamba Protected Areas Complex.

The Gamba Protected Areas Complex is a 6,835-square-mile preserve on the southwest coast of Gabon, Africa, which not only supports significant habitat and species diversity but also has the country's largest reserves of onshore oil. The complex, which is a designated series of connected faunal reserves and hunting areas, was established in the 1950s to protect areas of exceptionally rich wildlife diversity. Today, two national parks frame the Gamba complex to the east and west: Moukalaba-Doudou, a 2,796-square-mile area of mountains rich in biodiversity refugia and great apes; and Loango (963 square miles), which is known for its terrestrial-marine megafauna, intricate habitat mosaics, and potential for ecotourism.

During 279 days of observations in Gamba, nineteen species of mammals, four species of bird, and seven species of reptiles were tallied in a total of 2,845 animal carcasses with an estimated weight of 418,279 pounds. Despite the fact that the bushmeat trade is illegal in protected areas, the numbers of animals killed annually make it clear the bushmeat trade is not sustainable. The town of Gamba had the highest ratio of bushmeat to population in the country.

Unsustainable hunting of bushmeat is often a more immediate and significant threat to the biological diversity in tropical forests than deforestation. To more fully understand the impact of the trade, the rate of exploitation of fifty-seven species of mammals from the Amazon Basin and thirty-one species of mammals from the Congo Basin were compared. Estimates were compiled from anthropological studies that reported animals killed and brought into settlements in the regions, along with the average number of animals consumed per person per year.

The rates at which species with specific body masses were exploited was significantly greater in the Congo than in the Amazon. Thus, mammals of the Congo Basin must annually produce approximately 93 percent of their body mass to balance current losses to hunting, whereas Amazonian mammals must produce only 4 percent of their body mass. It is estimated that over five million tons of meat from wild mammals feeds millions annually in both the Neotropical (0.15 million) and Afrotropical (4.9 million) forests. However, the estimates for the Congo Basin are four times higher than those previously calculated for the region, a finding that leads to the conclusion that the current rate at which bushmeat is extracted from the African rain forests is more destructive than previously thought. Here, one might inquire why people eat bushmeat.

Some people may eat it because they can afford it, but others choose it because it is a familiar and traditional source of meat that supplements a short supply from animal husbandry, confers prestige, tastes good, or simply adds variety. A 2002–2003 survey was conducted in 1,208 rural and urban households in Gabon in an attempt to understand the effect wealth and price had on the purchase of bushmeat.

Consumption of bushmeat, fish, chicken, and livestock was linked to increasing household wealth, but as the price of these commodities rose, consumption declined. Although the prices of substitutes did not significantly influence the selection of bushmeat, as the price of wildlife went up and its purchase fell, the acceptance of fish improved; fish and bushmeat thus appear to be dietary substitutes.

Now the question becomes, How does so much bushmeat get to the consumers in the first place? Little is known about the organizational dynamics of the trade or of the actors involved in it, and this lack of knowledge impedes the development of effective policies through which to protect the ecological services these wonderfully complex forests offer to all generations.

To rectify this lack, an investigation was launched into the structure and operation of a bushmeat-commodity chain that supplies a typical urban market, the city of Takoradi in southwestern Ghana, West Africa. Data collected from January through February 2000 uncovered 2,430 bushmeat transactions, which involved seventeen species from seventy actors who traded along the commodity chain: commercial hunters, farmer hunters, wholesalers, market traders, and chopbar (cafe) owners. Although bushmeat was freely traded among these people, the primary route for terrestrial mammals was from commercial hunters through wholesalers to chopbars. In contrast, only farmer hunters, market traders, and chopbars were involved in the trade of invertebrates.

Wholesalers captured the largest per capita share of the market because each wholesaler handled 4 percent of all sales, whereas the chopbars were the top vendors because, as a group, they made 85 percent of all bushmeat sales to the public. Variation in the price of bushmeat is explained largely by the cost of transport and the preferences of consumers. Transport costs were most significant for hunters and, as one would expect, were greatest on long journeys that involved large loads. Nevertheless, the hunters obtained the greatest income per pound of bushmeat sold.

This type of simple commodity exchange does not, however, explain the whole bushmeat trade. To explore the complicated interlinkages of the informal bushmeat economy in times both of political stability and of armed conflict, a study of the trade was undertaken in Garamba National Park, Democratic Republic of Congo. The investigation focused on the sale of protected and unprotected species in urban and rural markets, with special emphasis on the commodity chains that supplied the markets.

During peacetime, protected species (predominantly elephants and buffalos) were poached in the park. Automatic weapons were required to hunt these large mammals, weapons supplied to hunters by the military officers who controlled the urban trade. During wartime, however, the military officers fled, leaving behind an open-access system, which encouraged a massive exploitation of protected species. Consequently, the sales of protected species increased fivefold in the urban markets.

In contrast, traditional chiefs, who administered the village markets, discouraged the use of such weapons. Therefore, rural markets remained relatively stable because of the continued authority of the village chiefs, which indicates that sociopolitical factors can be important in controlling the processes that drive species extraction. In addition, traditional authorities can be potentially valuable partners in protecting the ecological sustainability of African forests.[25]

As grim as the bushmeat trade may sound, we humans are continually changing the landscape in which we live—in that we have no choice. The ways in which we change the landscape, however, depend on the level of our consciousness of the effects we cause when our thoughts are translated into decisions, which are in turn converted into actions. There is a caveat, however: whereas nature is governed by impartial, biophysical principles, people are stirred by their desires, which are always biased in some respect. Nevertheless, both nature and humans cause far-reaching effects in time and space through their actions.

4

The Never-Ending Stories of Cause, Effect, and Change

The one who tells the stories rules the world.–Hopi proverb

Everything—living creatures, plants, air, water, rocks, time, and space, *everything*—exists in relationship to everything else. Each action you take is like dropping a pebble into a quiet pool of water. A pebble's impact on the water's surface creates concentric rings flowing outward from the center, touching everything in their path. The farther the rings travel from the epicenter, the wider and more diffuse they become. Sharp eyes might catch their visual disappearance, but no witness will observe their ultimate dissipation because the rings continue to exist in everything they have touched.

As the rings of cause flow outward from the collision of pebble and water, they become the rings of effect. In turn, each ring of effect becomes a cause when it intersects with a ring of effect from a pebble dropped in another part of the pool. In this way, each cause becomes an effect that becomes a different, often a synergistic cause that initiates still another effect, and so on ad infinitum. This insight echoes an observation by Leonardo da Vinci: "In rivers, the water that you touch is the last of what has passed and the first of that which comes; so with present time."

Accordingly, all things in nature are neutral when it comes to any kind of human valuation. Nature has only intrinsic value in that each component of a forest or an ocean—be it a microscopic bacterium, a towering eight-hundred-year-old tree, or a huge whale—is allowed to develop its evolved structure, carry out its biophysical function, and interact with other components of the ecosystem through their interdependent processes and feedback loops. No component is more or less valuable than another; each may differ in form, but all are complementary in function—whether on land or at sea.

Consider that an ancient tree rotting in a forest and a whale decomposing on the floor of an ocean are reinvesting the borrowed constituents of their bodies in the long-term productivity of their respective ecosystems. Here, they are entering the atomic interchange, where the shared biological capital vested in timeless atoms is borrowed and given up, only to be on loan again. Each atom represents the cosmic currency endlessly used in the long corridors of time. Today, an atom may be part of the dead whale on the deep-ocean floor. In the far memory of the universe, it may have constituted a bit of enamel on the tooth of a carnivorous dinosaur. In a dream of the future, it may grace the iridescence of a South American butterfly or the brilliant feather of a Nepalese sunbird.

In each case, the atoms, which for an instant in cosmic time composed a living being, continue their journey as they participate in, and thus create, the perpetually widening ripples of eternal stories about the continuum of relationships within the one story, which began with the birth of the universe—the story that contains all the other stories.

The Never-Ending Story in a Forest

It is critical to understand that vegetation in a forest both invests and reinvests biological capital into the ecosystem when it dies. Vegetation not only uses elements from the soil as nutrients but also converts energy from the sun via photosynthesis in order to grow. The elements (old energy) are reinvested into the soil when a plant dies, whereas the energy captured from the sun (newly accumulated energy) is invested in the soil for the first time.

A tree may die standing, only to crumble and fall piecemeal over decades to the forest floor, or it may fall directly to the ground as a whole tree. Regardless of how it dies, the standing dead tree and fallen tree are only altered states of the live tree, which means that the large, live old tree must exist before there can be a large standing dead tree or a large fallen tree.

How a tree dies is important because its manner of death determines the structural dynamics of the habitat its body provides. Structural dynamics, in turn, determine the biochemical diversity hidden within the tree's decomposing body as ecological processes incorporate the old tree into the soil from which young trees will grow.

The falling of the ancient tree, such as a Douglas-fir, forever alters the forest in which it lived and starts an irreversible chain of events that can last for centuries. This time lapse is due in part to the character of the available wood, which varies greatly in different parts of the tree. Proteins are concentrated in the living tissues of the cambium, or inner bark, which is the most easily digested part of dead tree. Carbohydrates, however, are predominant in the sapwood, which is moister and more digestible than the drier heartwood, composed, as it is, of lignin. Each portion of a fallen tree therefore supports a characteristic group of insects adapted to a specific microhabitat. The numbers of any one species are regulated by the quantity and quality of their food supply. The cambium furnishes the most nutritious food. The area of next greatest importance is the sapwood, then the heartwood, and finally the outer bark.

Bark beetles, such as the Douglas-fir beetle, are the first to take up residence within a newly fallen tree, where wood-boring beetles later join them. When wood-boring beetles, such as the golden buprestid, penetrate a fallen tree and begin to thrive within it, nature's system of checks and balances is also activated.

At first, this system is composed primarily of predaceous beetles. In Douglas-fir, for example, adult red-bellied checkered beetles prey on the adult Douglas-fir beetles, and larval checkered beetles prey on larval Douglas-fir beetles, a process that helps keep the number of bark beetles in check.

During the tree's first two years on the ground, a seemingly continual variety of animals use it for shelter, food, and foraging, and as perches. And throughout each year the surrounding vegetation continues to grow and change, gradually adding another dimension of ever-increasing diversity to the fallen tree.

When a large tree falls, it creates a notable hole in the forest canopy. Without further disturbance, the increased light striking the ground "releases" the shade-tolerant understory trees to grow and, in time, fill the hole. (Shade-tolerant means that a plant can survive in the shade of another plant; when the shade is removed, however, the plant responds to the available light with increased growth.) The extent of an opening can be considerably enlarged if a falling tree starts a domino effect—the successive uprooting and breaking of neighboring trees. A large, intact tree is more likely to knock over the trees it strikes as it falls than is a decayed snag (a standing dead tree), even a large one. A snag will most likely break or shatter when hitting a large tree, but it may knock down a small tree if it strikes the tree directly.

The response of ground vegetation to the falling of a tree varies in many ways depending on how the tree falls and on the tree's size, species, and health, as well as on the characteristics of the surrounding forest and its topography. Understory vegetation, both existing and potential, is likely to be released when a large tree falls because the extensive opening created in the canopy admits light to the floor of the forest. There are also resources that plants can use to grow and places where they can become established: first, on the mineral soil of the newly exposed root-wad; second, on the fallen tree itself; and third, as the trunk decays, in the wood under the bark.

A newly fallen tree interacts only passively with the surrounding forest because its interior is not accessible to plants and most animals. But once fungi and bacteria, which are smaller than the wood fibers, gain entrance, they slowly dissolve and invade the cells, whereas wood-boring beetles, carpenter ants, and Pacific damp-wood termites chew their way through the wood fibers. Meanwhile, other organisms, such as plant roots, mites, springtails, amphibians, and small mammals, must await the creation of internal spaces before they can enter.

With time, open areas develop within a fallen tree through biophysical processes. As the tree's decomposition continues, these interior spaces increasingly enable the flow of plant and animal populations and communities, air, water, and nutrients between the fallen tree and its surroundings. For example, the water-holding capacity of a large fallen tree varies by day, season, year, decade, and century, thereby adding yet another dimension of diversity to the forest.

A tree cracks and splits when it falls and then dries. Microbial decomposition breaks down the cell walls and further weakens the wood. Wood-boring beetle larvae and termites tunnel through the bark and wood, inoculating the wood with microbes and opening it to colonization by other microbes and small invertebrates. Wood-rotting fungi produce zones of weakness, especially between the tree's annual growth rings, by causing the woody tissue laid down in spring to decay faster than that laid down in summer. Plant roots, which penetrate the decayed wood, split and compress the tree as they elongate and thicken in diameter.

Because of all this internal activity, the longer a fallen tree rests on the forest floor, the greater the development of its internal open areas. Most internal spaces result from biological activity, the cumulative effects of which increase not only through time but also from the synergistic feedback loops—insect activity promotes decomposition through microbial activity that encourages the establishment of rooting plants that use the stored nutrients in the old tree as though it was nature's grocery store.

Fallen trees thus offer myriad organisms multitudinous external and internal habitats, which persist across the decades, even as they change. Yet the casual observer might notice only a few mushrooms or bracket fungi. These structures, however, are merely the fruiting bodies produced by mold colonies, the filamentous, cellular threads of which run for miles within the tree. Many fungi fruit within the fallen tree, and thus they are seen only when the tree is torn apart. Even then, only a fraction of the fungi present might be noticed because the fruiting bodies of most appear for but a short time. The smaller organisms, not visible to the unaided eye, are also crucial components of the forest. We humans do not begin to grasp the notion that microbes and fungi change a forest just as surely as a raging fire, only inconspicuously and more slowly. Theirs is an unseen function that is just as critical and just as great as any in the forest. We are awed by a towering fir but not by a lowly bacterium and fungus, and yet it's these humble organisms that largely convert a fallen tree into soil and eventually into a new forest—a service essential to the health of the three spheres we call home.[1]

The Never-Ending Story in an Ocean

Nothing is wasted in our seas; every particle is used over and over again, first by one creature, then by another. In the spring, our ocean waters are deeply stirred and bring to the surface a rich supply of minerals ready for use by new life.–Biologist Rachel Carson, *The Sea around Us*

Not surprisingly, the communities that arise when a whale dies and sinks to the bottom of the ocean display underwater versions of the classical stages of succession and change seen in terrestrial ecosystems. But instead of grasses and forbs giving way to shrubs, which yield to trees that mature into a forest, dead whales nourish first scavengers such as the Pacific hagfish, then bone-eating zombie worms, and eventually clams, which use inorganic chemicals for sustenance.

In this first stage (which some researchers term the *mobile-scavenger stage*), the whale is largely intact but has hundreds of hagfish feeding on it. These eel-shaped fish, each about sixteen inches long, use their sharp, rasping teeth to scrape bits of meat off the carcass. They also grip the whale with their mouths, tie themselves in a knot, and use their bodies to loosen chunks of flesh.

In addition, Pacific sleeper sharks grab the whale and twist their whole bodies back and forth until they finally rip off a piece of flesh. In all, some thirty-eight species of scavengers have been observed in an open feast during this stage, and they do a good job when you consider that a whale's soft tissue accounts for approximately 90 percent of its weight. In fact, one whale, which weighed just over a ton (about 2,200 pounds), had the bulk of its flesh devoured in less than eighteen months.

The second, *enrichment opportunist stage* belongs to smaller organisms that scavenge the leftovers. These secondary scavengers include snails, amphipods that look like shrimp, and segmented worms. Around one whale carcass, which had been on the ocean bottom for almost two years, every 1.2 square yards of sediment hosted as many as forty-five thousand individuals, not counting the microbes.

At times, huge-celled bacteria form long filamentous lines that appear to the naked eye as a pale bacterial mat, looking almost as if it had snowed. There is also a segmented worm, affectionately called a snowboarding worm, that leaves a trail as it eats its way through the bacterial mat. In addition, many other segmented worms, called polychaetes, show up during this second stage. Although related to earthworms, those species that congregate around whale carcasses are much more diverse than their terrestrial cousins.

Finally, there are bone-eating zombie worms, which get their nutrition by sending a tangle of green, rootlike coils into the whale's bones. Inside of this green tangle reside rod-shaped bacteria that break down the complex, organic compounds of which the whale's skeleton is composed.

When the hordes of wee creatures have reduced the whale to nothing but a pile of bones, the third, *chemoautotrophic stage* begins, wherein organisms like bacteria begin to digest portions of the skeleton. Many of the larger organisms that appear during this stage carry their own sulfide-metabolizing bacteria, such as the vesicomyid clams. These clams don't eat in the usual sense, but rather get their nutriments from sulfide-metabolizing bacteria that live in their gills. Also, a species of mussel can amass a population of more than ten thousand individuals on the skeleton of a single whale, in addition to which a species of polychaete worm forms such dense colonies around whale skeletons they resemble lawns of orange grass.

The fourth and final stage is called the *reef stage* because, with the nutritional component exhausted, the community shifts to being composed of undersea animals that require craggy structures as habitat. At this point, a whale's skeleton acts much like anchorage.

The carcass of a whale settling on the ocean bottom offers as much food as would normally be delivered by the regular rain of detritus in two thousand years.

Moreover, some whale carcasses in the third stage are still bristling with chemoautotrophs after seventy to eighty years of resting on the deep-ocean floor.[2]

In sum, what goes on inside and around the decomposing body of a dead tree or whale is a manifestation of hidden biological and functional diversity. That trees become injured, diseased, die, and fall to the forest floor is therefore critical to the long-term structural and functional health of the soil, and so of the forest. The forest, in turn, is an interactive, organic whole defined not by its respective parts but rather by the interdependent functional relationships of those parts in creating the whole—the intrinsic value of each piece and its complimentary function. The counterpart of a fallen tree is the sunken body of a dead whale slowly decomposing for nearly a century while it enriches the deep ocean. These processes are all part of nature's rollover accounting system—which simultaneously invests and reinvests biological capital in nature's health plan for the three spheres of life, mediated, as it were, through the atomic interchange.

The Never-Ending Story of a Decision

In contrast to nature's impartiality and intrinsic value, we humans are constantly introducing into the environment changes that represent our extrinsic money chase. Whatever we introduce, however, begins an unending story that's immediately out of our control. The production of ethanol as a biofuel from corn is an excellent example of this phenomenon. However, to balance the almost-euphoric government/industrial vision of this biofuel as the panacea for our demand for foreign oil (touted as a win-win decision for everyone), I will briefly focus on the unspoken environmental tradeoffs.

Proponents of mass-scale ethanol production from corn claim that it lessens our national dependence on foreign oil, makes automobile fuel less expensive (so we can drive our fuel-guzzling SUVs without having to change our lifestyles), increases profits for farmers, and harnesses the ability to grow corn as a sustainable crop (with no consequences for the health of the soil and its productivity). However, Alice Friedemann, a freelance journalist who specializes in energy, counters these glowing expectations: "Ethanol is an agribusiness get-rich-quick scheme that will bankrupt our topsoil." To back up her statement, Friedemann uses the example of Archer Daniels Midland (a billion-dollar giant in the agribusiness industry), which has relentlessly lobbied since the late 1970s for the use of ethanol in gasoline and now reaps record profits from corn ethanol, as well as government subsidies.[3]

Friedemann is hardly a lone voice in her critique of corn as biofuel folly, in which the exhaustion of finite resources, such as the fertility of available, arable soils, will be speeded up with no realistic prospect of restitution.[4] Wolfgang Haber of the Technische Universitäet in Munich, Germany, points out that although energy, food, and cultivatable land are interrelated, the absolutely decisive resource is farmable land, the increasing scarcity of which is totally underrated.

And the conversion of land for growing food to the production of biofuels will make good agricultural soils even scarcer—fast.[5]

Nevertheless, some people in the United States are apparently willing to starve already-hungry people, both at home and abroad, by converting food crops, such as corn, to the production of biofuel to perpetuate their love affair with the automobile. And this ignores the fact that many more people, living on the financial edge, will be pushed into food poverty. Despite those who are hungry, we, the affluent, stubbornly clinging to our egocentric, consumerist behavior. This economic callousness reminds me of an observation made by the Israeli statesman Abba Eban: "History teaches us that men and nations behave wisely once they have exhausted all other alternatives." Ultimately, however, wise behavior is a matter of consciousness, a social grace in critically short supply.

The bourgeoning human population coupled with the continual conversion of arable land to urban development is already straining food supplies. Moreover, population growth is compounded not only by increased infant survival but also by the longer life spans of adults, both of which put extra demands on available food. Here, the problematic linchpin is that already-limited staple foods, primarily grains, are being diverted from the table to the gas tank. When the economic principle of supply-and-demand-mediated pricing is added to the milieu, the obvious losers are the ones who can least afford it—ultimately, the children.

In addition, extremes in global weather patterns, such as prolonged droughts and severe flooding, are devastating vitally needed crops. And if climate models are correct, which I believe they are, the global situation will only get increasingly unpredictable as ecosystems adapt to changing conditions.

Then, because economic markets are driven by fear of insufficiency and loss, they trigger the tendency by those with monetary means or political power to garner and horde surplus food. Such a disaster mentality further enhances the specter of widespread famine and starvation—to say nothing of the biophysical devastation wrought by the ideological wars being waged around the globe.

Not surprisingly, there have been scattered riots over escalating food prices. These riots point to one inalienable fact: when frightened people are pushed to their psychological limit, they often become violent; such violence can only be destructive and thus worsens an already intolerable situation.

If large-scale commercial biofuels are to contribute to sustainable development, however, authentic, ecologically sound methods of farming and markets must provide concrete opportunities for a dignified livelihood and equitable terms of trade. Sustainable agriculture is a critical consideration because, as President Franklin D. Roosevelt admonished, "The nation that destroys its soil destroys itself."[6]

Of People and Soil

Some of the earliest written documents detailing the care and knowledge of soil were agricultural manuals. However, human behavior over the last couple of centuries has accelerated the rate at which soil is eroding, thereby rerouting the flow of nutrients.[7]

Today, the acres that constitute agricultural lands are finite, especially prime acres. Once lost to non-sustainable development, they are gone. Agricultural lands worldwide are being committed to housing developments, shopping malls, parking lots, and superhighways as if there were no tomorrow.

The loss of agricultural lands to municipal development is compounding an already growing problem, fewer and fewer acres to produce food for our burgeoning human population. This issue was addressed in 1798 by political economist Thomas Malthus, who was concerned about the yawning abyss between the rich and poor in the British Isles. Although Malthus wrote that the human population would at some point begin to overtake its available supply of food, he decried the use of contraceptives and was condemned as an alarmist. Nevertheless, he understood that if humanity did not control its population, nature would—and in ways most unpleasant. Consider that we humans added three times our number to the Earth's surface in the twentieth century and in doing so put more human pressure on each of the three spheres of its life-support systems than had existed in all prior recorded time.[8] Consequently, the ongoing degradation of soil is eroding the yields of crops and fostering malnourishment in many parts of the world.[9]

In 1900, for example, the population of the United States was 75 million. Today, depending on whether one has an accurate count of immigrants, legal and illegal, the population has swelled to somewhere between 281 million and 283 million people. It increased by 32.7 million during the decade of the 1990s because of a large wave of young immigrants and a birth rate that has overtaken the rate of death. This was the largest increase ever in the population of the United States in a single decade. Extrapolations indicate the population may increase to around 400 million by 2050 and to 571 million in 2100. Immigrants and their offspring who come here after the year 2000 will supply two thirds of that growth.[10]

Unfortunately, some people still think that the United States has endless open space available for development—and thus room for additional people. But how much of that land is arable and hospitable to farming or human habitation? How much of it has uncommitted, potable water readily available in a sustainable supply?

Many cultures have emphasized the trusteeship of the soil through religion and philosophy because it is the crucible in which the nonliving and living components of life are joined to form the great "placenta" of the Earth. The biblical Abraham, in his covenant with God, was instructed: "Defile not therefore the land which ye shall inhabit, wherein I dwell."[11] The Chinese philosopher Confucius saw in the Earth's thin mantle of soil the sustenance of all life and the minerals treasured by human society. A century later in Greece, Aristotle thought of soil as the central mixing pot of air, fire, and water, which formed all things.

In spite of the durability of such beliefs, most people cannot grasp their profundity because the ideas are intangible, on the one hand, and because the march toward technological specialization increasingly isolates us—the modern human—from nature and our place in it, on the other. The apparent invisibility of the soil stems from the fact that it's as common as air and, like air, is a birthright belonging to everyone and so to no one, and thus it is taken for granted. For example, the Federal Bureau of Soils stated in 1878 that "the soil is the one indestructible, immutable asset that the nation possesses. It is the one resource that cannot be exhausted."

"Soil is the most diverse and important ecosystem on the planet. Myriad biophysical and biochemical processes persist in parallel that are required to sustain all of the other trophic levels in the biosphere," according to soil scientists I. M. Young and J. W. Crawford of the University of Abertay in Scotland.[12] Although to many people soil seems "invisible" and thus indestructible, it's a seamless whole—the complexity of which seems all but unknown and beyond comprehension, even to soil scientists.

Whether people understand it or not, soil is important for at least seven reasons. First, soil is the basis for life, the stage on which the human drama and its many constructs are physically supported, enacted, and nourished. Second, soil plays a central role in the decomposition of dead organic matter, and in so doing adds to its store of potential nutrients. Third, soil stores elements that, in the proper proportions and availability, act as nutrients for the plants growing in it. Fourth, soil shelters seeds and provides physical support for their germination and roots; as the plants grow and mature, they produce seeds and so perpetuate the cycle. Fifth, soil is the nursery for spores of the microbes, as well as for decomposer and mycorrhizal fungi, the latter of which nourish the myriad plants. Sixth, soils of various kinds, acting in concert, are a critical factor in regulating the major elemental cycles of the Earth—those of carbon, nitrogen, sulfur, and so on. And seventh, soil both purifies and stores water.[13]

Human society is inextricably tied to the soil for reasons beyond measurable riches, for the wealth of the Earth is archived in soil, a wealth that nurtures culture even as it sustains life, as illustrated by the following quotes from 1938: "The social lesson of soil waste is that no man has the right to destroy soil even if he does own it in fee simple. The soil requires a duty of man, which we have been slow to recognize" (Henry Wallace).[14] "In the old Roman Empire, all roads led to Rome. In agriculture all roads lead back to the soil from which farmers make their livelihood" (G. Hambridge).[15]

Although these statements are as true now as the day they were uttered, to their detriment, people too soon forget. Consider that nearly 20 percent of the vegetated surface of the Earth was already degraded by human activities by the 1990s.[16] And, if not wisely used, modern machines and chemical agents will increasingly enable humanity to accelerate that degradation faster than nature can heal the old soil or create that which is new.

Unintentional fragility is imposed on ecosystems of the biosphere through direct and indirect pollution of soil and water. Soil, which is like an exchange membrane between the living components of the biosphere and the nonliving components of the litho-hydrosphere and atmosphere, is dynamic and ever-changing. Derived from the mechanical and chemical breakdown of rock, soil is built up by plants that live and die in it. It is also enriched by animals that feed on the plants, void their bodily wastes, and eventually die, decay, and return to the soil as organic matter. Soil, the properties of which vary from place to place within landscapes, is by far the most alive and diverse part of a biosphere. In addition, soil microorganisms are the regulators of most processes that translate into soil productivity. "Each soil," says author Hans Jenny, "is an individual body of nature, possessing its own character, life history, and powers to support plants and animals."[17]

The soil food web is a prime indicator of a healthy terrestrial ecosystem. But soil processes can be upset by changes like a decrease in the ratio of bacterial to fungal biomass; this decrease alters the fungal/bacterial activity in such a way that it reduces the number and diversity of protozoa and nematodes, thereby altering their community structure. Such disruptions can lead to a loss of vegetation or even the loss of human health.[18]

Biologists have for centuries studied patterns of plant and animal diversity at continental scales. Yet, until recently, similar studies were impossible to conduct for microorganisms, which arguably are the most diverse and abundant life forms on Earth. The variables that typically predict plant and animal diversity and community composition, such as moisture gradient, appear to be largely independent of geographic distance. Although the diversity and species richness of bacterial communities differ among ecosystem types, these differences could be explained largely by soil pH (the *potential of hydrogen*), not by geography.

Bacterial diversity was highest in neutral soils and lower in acidic soils, with soils from the Peruvian Amazon being the most acidic and least diverse. The biogeography of soil-dwelling microbial organisms is controlled primarily by edaphic variables (soil conditions) and thus differs fundamentally from the variables that account for the biogeographical distribution of macro organisms.[19]

Regardless of soil type, most of the terrestrial vegetation produced each year enters the decomposer system as dead organic matter, where the subsequent recycling of carbon and other nutrient-forming elements is a critical process for the functioning of ecosystems and the delivery of their goods and services. Within this paradigm, decomposition of litter from a particular species of plant changes greatly in the presence of diverse litter from coexisting species, even in the face of unaltered climatic conditions and litter chemistry. Most important, soil fauna determines the magnitude and directs the effects of decomposing litter from mixed species.

The species richness of the litter, coupled with the interactivity of soil macro-fauna (such as earthworms, beetles, and mice), determines the rate of decomposition in temperate forests. Put differently, the species composition of the litter affects the species composition of the decomposer organisms, which in turn affects the cycling of carbon and nutrients. Thus, ecosystems, which support a well-developed, soil-macrofaunal community, play a fundamental role in altering decomposition in response to the changing diversity of litter-producing species, and this alteration in turn has important implications for biogeochemical cycles and the long-term functioning of ecosystems, at least within the temperate zone.[20] That said, few events in history have been more closely tied to communities than their relations to the soil as the foundation of life.

Because of the ever-changing complexities of soil, we humans would be wise to develop the humility necessary to accept that we will never fully understand it; only then will we have the requisite patience to protect the organisms that perform the functions through which soil is kept healthy. Soil health cannot be maintained through applications of inorganic fertilizer, which not only disrupt the biophysical governance of the soil's infrastructure but also "addicts" soil to petrochemicals in order to grow the desired plants—for example, massive mono-cultures of corn in the same fields year after year after year. Furthermore, much of the fertilizer is lost as it leaches downward through the soil into the groundwater, which it then contaminates, because neither the soil nor the organisms in the soil's disrupted food web can retain all the added chemicals, such as nitrogen.[21]

Moreover, it is extremely difficult to stop the pollution of groundwater, espe-cially from inorganic fertilizers like those used to produce corn, which includes nitrogen.[22] "Once polluted," counsels ecologist Eugene Odum, "groundwater is dif-ficult, if not impossible, to clean up, since it contains few decomposing microbes and is not exposed to sunlight, strong water flow, or any of the other natural purification processes that cleanse surface water."[23]

In some cases, adding fertilizer even acts like a biocide, killing the organisms in the soil's food web, thereby further degrading the soil. In addition, the so-called inert components of chemical compounds are not tested for toxicity and can recombine with other chemicals in the soil or the groundwater or both to become toxic. Here it is critical to understand that "inert" is an industrial euphemism whereby the illusion of chemical inactivity is suggested. A truly *inert* substance, however, is a biophysical impossibility in an interactive system. It's therefore much wiser to work in harmony with the soil and the organisms that govern its infrastructure because they are responsible for the processes that provide nutri-ents to the plants and thus the entire food web.

The development of soil depends on self-reinforcing feedback loops, in which soil microorganisms provide the nutrients for plants to grow, and plants in turn provide the carbon, in the form of organic material, that selects for and alters the communities of soil organisms. One influences the other, and both determine the soil's development and health.[24]

Even though protection of soil and its fertility can be justified economically, our human connection with the soil escapes most people. One problem is that tradi-tional, linear economics deals with short-term, tangible commodities, such as crops of corn, rather than with long-term intangible values, such as the future prosperity of our children. We will begin to see that the traditional, linear economic system is not tenable in the face of biological reality when we recognize that arable land is finite, and that every ecosystem has a limited, biophysical carrying capacity.

Those who analyze soil by means of linear economics determine the net worth of protecting the soil only on the basis of the expected short-term revenues from future harvests, thereby ignoring the fact that it is the health of the soil that produces the yields—not simply the soil's presence. In short, they see protection of the soil as a cost with no benefit because the standard method for computing soil-expectation values commonly assumes that productivity of the soil will either remain constant or will increase—but will never decline.

Given that reasoning, which is both short-sighted and flawed, it is not surprising that those who view the land simply as a means to an economic end seldom see protection of the soil's productivity as cost effective. But if we could predict the real effects of this economic reasoning on long-term yields, we might have a different view of the invisible costs associated with ignoring the health of the soil. For example, ecologist and historian Donald Worster points out that although "agriculture involves the rearranging of nature to bring it more in line with human desires, . . . it does not require exploiting, mining, or destroying the natural world."[25]

One of the first steps along the road to protecting the fertility of soil is to ask how the various ways humans treat an ecosystem affect its long-term productivity, particularly that of the soil itself. In turn, understanding the long-term effects of human activities requires us to be knowledgeable about the factors that contribute to ecosystem stability and productivity, such as habitat diversity and health. With such knowledge, we can turn our "often misplaced genius," as soil scientist David Perry rightly calls it, to the task of maintaining the resilience of the soil's fertility—thus buying an ecological insurance policy for our children.

After all, soil is a bank of elements and water that provides the matrix for the biological processes involved in the cycling of elements that become nutrients under the right conditions of concentration and availability to plants. In fact, of the sixteen chemical elements required for life, plants obtain all but three—carbon, hydrogen, and oxygen—from the soil. The soil stores these essential elements in undecomposed litter and in living tissues and recycles them from one reservoir to another at rates determined by a complex of biological processes and climatic factors.

As soil scientist W. C. Lowdermilk wrote, "If the soil is destroyed, then our liberty of choice and action is gone, condemning this and future generations to needless privations and dangers." To rectify society's careless actions, Lowdermilk composed what has been called the "Eleventh Commandment," which demands the full and unified attention of every gardener, farmer, rancher, forester, and urban developer if we are to fulfill our role as trustees of the soil for the benefit of all generations:

Thou shalt inherit the Holy Earth as a faithful steward, conserving its resources and productivity from generation to generation. Thou shalt

safeguard thy fields from soil erosion, thy living waters from drying up,
thy forests from desolation, and protect thy hills from overgrazing by thy
herds, that thy descendants may have abundance forever. If any shall fail
in this stewardship of the land, thy fruitful fields shall become sterile
stony ground and wasting gullies, and thy descendants shall decrease and
live in poverty or perish from off the face of the earth.[26]

Here, it is important to keep in mind that fertile soil—like good-quality
water—is one of the pillars of sustainable community. Destroy the fertility of your
community's soil, and nothing else you do will much matter because the soil is the
stage on which the quality of your community's life-play ultimately depends. And,
yet, soil scientists have been excluded from the discussion of corn-based ethanol
and its effect on soil.[27] I'm sure, however, that farmer and author Wendell Berry
speaks for them when he says, "The care of the earth is our most ancient and most
worthy and, after all, our most pleasing responsibility. To cherish what remains of
it, and to foster its renewal, is our only legitimate hope."

Corn as Biofuel

Author Richard Manning has the following to say about the importance of corn:

If you follow the energy, eventually you will end up in a field somewhere.
Humans engage in a dizzying array of artifice and industry. Nonetheless,
more than two thirds of humanity's cut of primary productivity [the
amount of green vegetation produced in a particular year] results from
agriculture, two thirds of which in turn consists of three plants: rice,
wheat, and corn. In the ten thousand years since humans domesticated
these grains, their status has remained undiminished, most likely
because they are able to store solar energy in uniquely dense, trans-
portable bundles of carbohydrates. They are to the plant world what a
barrel of refined oil is to the hydrocarbon world. Indeed, aside from
hydrocarbons they are the most concentrated form of true wealth—sun
energy—to be found on the planet.[28]

The advent of intensive maize (corn) agriculture among indigenous
American societies during late prehistory not only had profound effects on the
pre-Columbian landscape as a whole but also on the freshwater mussels, in particular *Epioblasma* sp. According to evidence from shell middens, the relative
abundance of these mussels has declined steadily during the last five thousand
years, a decline that could be interpreted as the result of either an increase in
direct human impacts on streams or of long-term, non-anthropogenic changes in
climate. Nevertheless, decline of these mussels increased significantly in the
southeastern United States about one thousand years before the present—a
decline attributable to the advent of large-scale, intensive maize agriculture. The
data suggest that such land use by early indigenous Americans wrought changes in

communities of freshwater mussels that were portents of the deleterious environmental effects intensive agriculture is causing today.[29] And what, you might ask, are some of those problems?

Today's large acreages of intensively farmed crops in the United States, which are addicted to toxic pesticides, host fewer species of birds than do smaller, organic farms. In Britain, where all species of insectivorous bats forage over agricultural habitats, their populations are declining—a phenomenon being seen throughout Europe, probably in response to growing agricultural intensification.

In addition to insecticides, the loss of habitat or its declining quality (or both) through fragmentation can also have strong, negative impacts on indigenous populations of insects, which in turn affect insect-eating birds and bats. For example, when prairie remnants in Nebraska are converted to agriculture, there is an overflow of generalist predatory insects, such as the ladybird beetles, that begin to consume the herbivorous insects indigenous to the prairie ecosystem. In other words, populations of native insects decline when confronted with an increasing loss of habitat to agriculture, and there is then a corresponding upsurge in populations of predatory species having a generalist proclivity. This phenomenon is not confined to the United States, however, and may be partially responsible for the decline of British bats.

A study shows that insect abundance and species richness, including that of moths, was significantly higher on British organic farms than on today's intensive, inorganic farms. Insect abundance was also considerably higher in pastures and around water on organic farms than in the same habitats on conventional, corporate-style farms. Moreover, the activity of the bats that mainly ate moths was directly correlated with the abundance of moths. Data suggest that agricultural intensification has a profound impact on nocturnal insect communities and thus on bats, whose distribution is limited by their food resource. Not surprisingly, therefore, a reduction in the availability of their prey through escalating agricultural intensification will adversely affect bat populations. Conversely, less intensive farming benefits populations of British bats by maintaining structurally diverse habitats, which in turn support a wide selection of insect prey, including those species that are important prey for a number of rare bats.[30]

And it's not just in Europe and the United States that intensive farming is having a negative impact. In Colombia, South America, for example, the amount of land planted in illicit crops, such as coca and opium poppies, grew an average of 21 percent per year over a five-year period and may account for half the total area deforested in 1998.[31]

Given current trends in the expansion of illicit crops and the narrow endemicity of some species of birds, the conversion of forests for such crops may result in the extirpation of several species. Although a number of areas in Colombia are threatened by illicit crops, the largest forested areas imperiled by these crops occur in Amazonia and the Amazonian foothills of the East Andes.[32] Elsewhere in South America, the risk of nest predation for open-cup-nesting birds is relatively high in the present agricultural landscape of Chile, a finding that indicates that much of the available wooded areas (forest edges, narrow corridors) offer poor nesting habitat, although it may be suitable for foraging and traveling.[33]

Advocates of corn-based biofuel tout the possibility of genetically engineering corn to boost its ethanol production—as though that would reduce the amount of acreage committed to corn monocultures. Most supporters of genetic modification downplay the difference between the genetic engineering of organisms and time-honored selective breeding. They claim, for example, that the only difference is that engineering crops is more precise, faster, and cheaper. Although this claim would seem to be good news, it's ecologically misleading and genetically irresponsible.

"Experiments have shown," writes Ricardo Steinbrecher, a genetic scientist and member of the British Society for Allergy, Environmental, and Nutritional Medicine, "that a gene is not an independent entity as was originally thought." Genetic engineers increasingly want to transform plants and animals from organisms with inherent novelty into predictable, "designed commodities," while ignoring the many unknown hazards.

Steinbrecher cites the example of a 1990 experiment in Germany in which the gene for red coloration in corn was transferred, together with a gene for antibiotic resistance, into the flowers of white petunias. The researchers expected a localized field of twenty thousand red-flowering petunias. The genetically engineered petunias turned red all right, but also had more leaves and shoots, a higher resistance to fungi, and lower fertility. These unexpected results were completely unrelated to the genes for color and antibiotic resistance. Such results (both unrelated and unexpected) have been termed *effects*, which, by their very nature, are totally unpredictable.[34]

In this case, the pleiotropic effects (many effects produced by a single gene) were clearly visible and easily identified without molecular analysis. But what happens if pleiotropic effects are not so obvious, if they, in a clandestine manner, affect the composition of proteins, the expression of hormones, or the concentration of nutrients, toxins, or allergens? Who is going to monitor all the possible pleiotropic effects before a genetically engineered plant is introduced into the environment or placed on our dinner plates? There are neither regulations nor voluntary guidelines and practices with which to check for pleiotropic effects; this lack makes cross-species cloning—including between humans and cattle, rabbits, and other species—particularly arrogant.[35] And even if there were regulations governing pleiotropic effects, how would one know what to look for?

In addition to pleiotropic effects, there are simple biological and ecological limitations to what is and is not possible. Consider, for instance, that a 1997 report by the World Bank indicated that a 1.5 to 1.7 percent per year increase in the yields of grain could be expected. With such a rosy outlook, the World Bank, an arm of the International Development Association, projected a surplus in the capacity of agriculture throughout the world as a whole, accompanied by declining prices of food.

However, the Worldwatch Institute, an independent, globally focused environmental-research organization in Washington, D.C., came up with a different outcome from the World Bank's analysis. The difference is that the economists of the World Bank based their predictions on simple extrapolation, arguing that because yields grew along a linear path from 1960 to 1990, they would continue to do so. Two aspects of their logic stand out as particularly problematic: the economists' view of cyclical ecological systems was linear, and they construed the increase in the yield of grain as an independent variable, which cannot exist in an interdependent living system.

Thus, while extrapolating trends in the yield of grain worked well in previous decades, it will not work as it once did. The ecological limiting factors to the ever-increasing production of grain, which always loomed invisibly on the horizon, are becoming undeniably manifest. To illustrate, the robust 2.1 percent per year increase in the yields between 1960 and 1990 was replaced with a 1.0 percent increase between 1990 and 1995.

Although one can argue that a five-year increment is too short to establish a clear trend, it may, nevertheless, be a strong portent of the future. Clearly, when a country's farming practices fall short of their potential, the yield can be rapidly improved—that is, until the ecological limits are reached. Beyond that, no amount of money, ingenuity, water, or fertilizer can force more out of the soil and thus the crop. By the same token, after the biological limits of genetic manipulation have been reached, no amount of money or ingenuity can force more out of the crop.[36]

As with everything else, agricultural intensification has consequences. But people keep trying to push nature into ever-higher production. Corn, it turns out, is one of the most energy-intensive crops when it comes to the amount of fertilizer it requires, and farmers are applying seven times the amount of synthetic nitrogen as they did in the late 1960s. As with everything else, agricultural intensification has consequences. Although the production of grain has doubled since then, largely because of the widespread use of synthetic fertilizers, pesticides, and intensive irrigation, the current rate of increased agricultural output is unsustainable, as evidenced since the late 1980s by diminishing returns in crop yields, despite the increased application of fertilizers. A common strategy for reducing dependence on fertilizers is the rotation of nonleguminous crops (those not in the pea family) with leguminous crops (those in the pea family, such as clovers, peas, beans, and peanuts), which capture atmospheric nitrogen via symbiosis with nitrogen-fixing rhizobia bacteria.

Here, one must understand that the root system of a plant is as complicated in its reactions with the matrix of substances and myriad organisms that surround it as any aboveground part of the plant. The fine roots constitute the most active part of the system in acquiring water and nutrients, while the multitude of root tips are sites of intense chemical activity that strongly modifies the soil they contact by mobilizing reluctant ions, immobilizing toxic ions, as well as coating the soil particles with mucilage and selecting the microflora. Moreover, the entry of nitrogen-fixing bacteria into a root hair requires molecular recognition by the plant. The entry of nitrogen-fixing bacteria into the root hairs of these plants is critical because nitrogen is often the element that determines what plants grow where and how much they produce. Even though air is composed of 78 percent nitrogen, neither plants nor animals can use it in gaseous form, even when it passes through a plant's leaf or an animal's lung.

Atmospheric nitrogen is composed of a pair of tightly bound atoms (N_2), which no living thing composed of cells with a nucleus can break by itself. Yet a simple life form, the humble microbe, has the ability to sever the chemical bond and render the elusive atmospheric nitrogen into user-friendly ammonia. This type of root endosymbiosis makes a vital contribution to plant nutrition and fitness worldwide.

The most well-understood plant-microbe symbiosis is between bacteria and legumes. The nitrogen-fixing bacteria enter the plant through its tiny root hairs, which ultimately become pinkish, bulging, nodular nitrogen factories. The pinkish color represents botanical "hemoglobin," which is akin to the oxygen-carrying molecules in our blood.

The formation of a nodule begins with a plant's release of flavonoid compounds into the soil, where the bacteria secrete molecules known as "nod factors." Even faint traces of these exudates cause a dramatic movement of calcium within the root hairs. Then, often within seconds, calcium floods into the root-hair cells, where it continually spikes into substantial concentrations for an hour or so. If everything functions as it should, the tiny root hairs begin to form hooks, which curl around the bacteria. The curled root cells of many legumes open an internal tunnel that guides incoming bacteria to the area of tissue that will eventually bulge into a nodular nitrogen-fixing factor.[37]

There is, however, a subset of organochlorine pesticides, agrichemicals, and environmental contaminants that inhibit or delay the recruitment of rhizobial bacteria to the roots of host plants, which then produce fewer root nodules; rates of nitrogen fixation are thus lowered, and the overall yield at harvest time is eventually reduced. The environmental consequence of a farmer's and the soil's addiction to synthetic chemicals is compromising bacterial nitrogen fixation, thereby increasing dependence on synthetic nitrogenous fertilizer, while simultaneously reducing soil fertility and increasing the long-term non-sustainability of crop yields and the demand for fossil fuels.[38]

In addition, the species diversity of local, native plants generally declines in response to soil enrichment with artificial, nitrogen-rich fertilizers. In a study of more than 900 species across nine terrestrial ecosystems in North America, the risk of losing species because of fertilization ranged from 60 percent for the rarest species to 10 percent for the most abundant. Perennials, species with nitrogen-fixing symbionts, and those of native origin also experienced increased risk of local extinction after fertilization, regardless of their initial abundance. Whereas abundance was consistently important across all systems, functional mechanisms, such as nitrogen-fixing symbionts, were often system-dependent.[39]

Beyond the application of synthetic fertilizers, the natural gas used in their production accounts for 90 percent of the cost of the ammonia, which is the basis for the nitrogen fertilizer applied to corn. The pesticides and herbicides required to produce these vast monocultures are also gas-based petrochemicals. And then there is the substantial amount of diesel fuel needed to operate the farm machinery. And this says nothing of the enormous quantity of water this exceedingly thirsty crop requires—1,700 gallons for each gallon of ethanol produced.[40] Where might this water come from? Where does it go? Ask the Mississippi River.

As the water of the Mississippi River flows toward the Gulf of Mexico, collecting runoff from the Appalachian Mountains to the Rocky Mountains and everywhere in between, it passes through ten states, through massive agricultural fields and by numerous towns and cities; on its journey it gathers fertilizers and pesticides from the Corn Belt and leached sewage from the urban areas. By the time the Mississippi enters the Gulf, its current has been transformed into a conduit for chemical nutrients, and this enriched current stimulates massive blooms of algae every summer that strip the water of oxygen, thereby creating a huge dead zone. (A *dead zone* is an area that is virtually depleted of dissolved oxygen, from which aquatic life either flees or suffocates.) This dead zone, which is the size of Massachusetts (7,900 square miles), has existed since the 1970s and supports almost no life beyond algae and bacteria.

Today, after years of inaction, the problem is severe. What's more, much of the water entering the Mississippi comes from massive fields of corn, which is grown in soil with tile drains; consequently more nitrogen seeps into the river from corn fields than from fields of crops without drainage tiles. Therefore, making ethanol from corn not only will cause more cornfields to be planted but also will exacerbate the dead zone in the Gulf of Mexico—perhaps beyond repair.[41]

The dead zone in the Gulf of Mexico is not the only anthropogenic one however. The Chang Jiang River basin of China has the third largest discharge of water in the world, and it empties into the East China Sea from Shanghai, which is the fastest developing area of China. With the increasing nutrient load from the river, a severely hypoxic zone, on the order of 7,688 square miles, formed in the sea. Rather than coming mainly from the Chang Jiang River, the hypoxic zone developed because of decomposing organic detritus that was transported by the ocean current from the south.

Nevertheless, the dead zone is maintained by stratification between the large volume of fresh water from the Chang Jiang River and the salty water from the Taiwan Strait. This same phenomenon applies to other estuaries with large flows of fresh water and rapid-economic-growth drainage, such as the Pearl River basin. Furthermore, the hypoxic zone adjacent to the Chang Jiang estuary is much more sensitive than that outside the Mississippi River.[42] This is not the only hydrological challenge however.

In the Great Plains, where new ethanol plantations are being established, an unexpected environmental cost is exacted because groundwater is the only source for irrigation. As water soaks through the soil, it collects carbon dioxide from decomposing organic matter in the soil through which it percolates. According to Gwen L. Macpherson, a hydrogeologist at the University of Kansas in Lawrence, groundwater holds, on average, from ten to one hundred times as much carbon dioxide as water in lakes and rivers.[43]

Thus, when groundwater is pumped to the surface, the carbon dioxide escapes into the air, where it adds to the growing supply of greenhouse gases. Nonetheless, people have been pumping about 178 cubic miles of water from below ground annually and thereby have been releasing approximately 331 million tons of carbon dioxide into the atmosphere every year. Although the volume of carbon dioxide released from the groundwater is a small percentage of that produced from the combustion of fossil fuels, it is about three times the amount spewed from the throats of volcanoes, which are a natural source of the greenhouse gas.[44]

However, a long period of drought has affected the aquifers in such areas as Nebraska to the point that water tables have dropped about fourteen feet since the late 1990s. In fact, shortages of water now occur in most of the midwestern and south-central states and increasingly in the southwestern states, such as Nevada.[45] It is precisely this kind of circumstance that caused Kofi Annan, secretary-general of the United Nations, to warn that "fierce competition for fresh water may well become a source of conflict and wars in the future."[46]

The projections of the World Bank are thus irresponsible because they permit governments to become complacent about the value of soil, the availability of water, the sustainability of the food supply, and the treatment of agricultural land as a commodity that can be subdivided for houses, paved over for shopping malls, or otherwise frittered away with self-centered impunity. The results of this kind of thinking are visible in the Central Valley of California, where housing projects march unimpeded up the valley, and consume some of the world's finest farming land. In China, the government is paving over millions of acres of agricultural lands so automobiles can replace bicycles. And the fertile rice lands of Indonesia are being converted to golf courses.

Losing farmland to other uses is not the only irreversible problem; losing water is another. In Texas, where farming has historically relied on irrigation, 14 percent of the irrigated area has been lost since 1980 as a result of depleting the aquifer. Water for irrigation is also being lost in California, Kansas, and Oklahoma. Water is being diverted from irrigation to cities in the Hebei Province of China to satisfy the soaring urban and industrial demands for the precious liquid. And, in the agricultural areas surrounding Beijing, farmers have been prohibited since 1994 from using water stored in reservoirs because all the region's water is now preempted to quench the capital city's growing thirst.[47]

In addition to the misguided projections of the World Bank and problems with the use of inorganic fertilizers is the fact that manufacturing these fertilizers concomitantly produces carbon dioxide from the burning of fossil fuels. In turn, the world's oceans absorb carbon dioxide from the atmosphere in a direct air-to-ocean exchange, which theoretically and ideally reaches a sort of equilibrium.

Oceanic Effects of Carbon Dioxide

The global oceans are the largest natural reservoir of carbon dioxide. They absorb about one-third of the carbon dioxide we humans spew into the atmosphere every year. Although this process is extremely slow, taking hundreds to thousands of years, once dissolved in the water a carbon atom can remain there for decades or centuries depending on the depth in the ocean in which it is located. However, anthropogenic carbon dioxide now penetrates the whole water column of the North Atlantic Ocean.

Moreover, there's a strong possibility that dissolved carbon dioxide in the ocean's surface waters will double over its pre-industrial levels by mid-century and will be accompanied by greater acidity as well as by a decrease in the carbonate ion. When carbon dioxide reacts with seawater, it produces carbonic acid, which can be thought of as the soda-water effect. This change in seawater chemistry will have profoundly negative effects on those calcium-secreting organisms in the world's oceans that depend on calcium carbonate for the production of their shells (mollusks, including planktonic mollusks, and marine algae) and skeletons (corals).

On a global scale, the alterations in surface-water chemistry from the anthropogenic deposition of nitrogen, sulfur, and dissolved inorganic carbon are relatively slight compared with the acidification caused by the oceanic uptake of anthropogenic carbon dioxide. The impacts are more substantial in coastal waters, however, than in the deep ocean. In coastal areas the ecosystem responses to acidification could have severe implications for people, especially those who rely on the seas of the world for food.

Over time, these changes will send ripples throughout the marine food web, from the microscopic plankton to the plankton-feeding whales and all life in between. As the ocean gets warmer and more acidic, the amount of dissolved oxygen will diminish accordingly, as will the building blocks for coral and other calcium-secreting organisms.

In fact, these species already have a reduced ability to produce their protective shells and supportive skeletons. In addition, the increase in carbonic acid is even now beginning to dissolve the shells and skeletons once they are produced and is also making them increasingly susceptible to wear and erosion. Decreased calcification will no doubt compromise survival of these organisms and could shift marine flora and fauna toward noncalcifying species. For example, the common periwinkle (a small marine snail) normally grows extra-thick shells when living among crabs, but if the water is too acidic, the snail's ability to produce a thicker-than-normal protective shell is disrupted.

"You don't have to believe in climate change to believe that this is happening," says Joanie Kleypas, an oceanographer with the University Corporation for

Atmospheric Research in Boulder, Colorado. "It's pretty much simple thermodynamics." According to Kleypas, "Acidification is more frightening than a lot of the climate change issues" because it's much harder to turn around. "It's a slow-moving ship, and we're all trying to row with toothpicks," she observes.[48]

Direct Human Influence on Oceans

Although the oceans of the world seem immutable, no area is unaffected by human influence. In fact, 41 percent of the oceans have been seriously degraded by multiple human factors; to name a few: fishing commercially a mile below the surface of the water with high-tech gear; pollution; commercial shipping; military sonar; offshore oil exploration, extraction, and the inevitable spills. Less than 4 percent of the oceans can be classified as areas of very low anthropogenic impact, and they are mainly near the poles.[49] Even so, the Adèlie penguins shifted their diet of primarily fish to a diet of predominantly Antarctic krill within the past two hundred years because of human activities.

During the nineteenth century, commercial sealers hunted Antarctic fur seals to near extinction. With the precipitous loss of these prodigious consumers of krill, Adèlie penguins shift their diet to more easily obtained prey. What's more that slaughter was followed in the twentieth century by the widespread killing of krill-eating baleen whales, which enabled the tiny crustaceans to proliferate essentially unchecked, until there is today a surplus of krill in the Southern Ocean.[50] Clearly, the overexploitation of the Antarctic waters has left its mark, even if that human signature is faint by today's standards.

As we humans attack the biosphere through such activities as overfishing the oceans, we alter how the biosphere relates to both the atmosphere and the litho-hydrosphere.[51] In doing so, we are changing the interactions among the three spheres—a change mediated through the myriad self-reinforcing feedback loops of the global climate.

Consider, for example, the overexploitation of the large predatory marine fishes, such as sharks and tuna, which allows the populations of smaller, plankton-feeding fishes to proliferate. At some point, their numbers become large enough to dramatically reduce the amount of phytoplankton and thus the ocean's ability to absorb atmospheric carbon dioxide; in turn, these changes affect global warming.[52] That said, curbs on fishing such species as big-eye tuna and yellowfin tuna until their populations are larger than those required to maintain a sustainable yield could, within biological limits, lead to maximum profits from fisheries.[53] But then, warming oceans affect the major wind patterns, which affect the direction of ocean currents, which is shifting dead zones in the oceans and causing them to grow.[54]

Shifting Winds

In turn, the stronger, more persistent winds that are expected to accompany a warming climate will shift the dynamics of the biosphere through such phenomena as increased areas of drought and their prolonged duration. Could these winds bring a repeat of the 1930s, when arguably the most severe drought of the past century gripped almost two-thirds of the United States, as well as parts of Mexico and Canada, and when numerous dust storms occurred in the southern Great Plains?

The Great Plains receives most of its precipitation as rain during the spring and summer. The moisture is carried by the westerly trade winds, whose currents of air flow around the equator, then move westward from the Gulf of Mexico over the North American continent and out over the Pacific Ocean, picking up additional moisture as they go. They then circle back, carrying their precious life-giving water in the upper atmosphere toward the United States. But two types of events can disrupt this cycle of the trade winds.

First, if the surface of the Pacific Ocean is cooler than normal, it causes the returning air to be dryer, to flow at a lower altitude, and thus to be more likely to bring less rain to the Great Plains. Second, when the Atlantic gets warmer than normal, it heats the air just over its surface, which makes it expand and rise. Thus, instead of the trade wind's usual moisture-laden onshore flow from the Gulf of Mexico, dry, high-pressure air creates an offshore flow that blows eastward from the continent out over the Atlantic.

Therefore, as the soil of the Great Plains dried out during the Dust Bowl years, evaporation diminished accordingly, which foreshadowed even less rain. Although this self-reinforcing feedback loop is likely to have intensified the drought, decades of poor agricultural practices were also to blame.

The grasslands had been deeply plowed and planted with wheat, which offered up a bountiful crop when there was adequate rain. While the farmers kept plowing and planting as the droughts of the early 1930s deepened, nothing would grow. With the ground cover gone, and nothing to hold the soil in place, winds whipped across the fields raising billowing clouds of dust into the skies, which could darken for days. Even the most well-sealed homes could have a thick layer of dust inside. In some places, the dust would drift like snow, covering farmsteads.

As a result, more than two million farmers were forced to abandon their land, as somewhere in the neighborhood of a billion tons of topsoil blew eastward and southward, mostly in large, black clouds. On April 14, 1935, known as Black Sunday, twenty of the worst Black Blizzards occurred throughout the Dust Bowl. At times, the clouds blackened the sky all the way to Chicago, although much of the soil blew out to sea and was deposited in the Atlantic.[55]

Drought, in turn, slows an area's vegetative growth and thus diminishes the area's ability to absorb atmospheric carbon dioxide; the rise in global temperatures is thereby augmented, and the rising temperatures warm the litho-hydro-sphere, melting glaciers and increasing ocean temperatures, both of which raise the level of the world's oceans, one by adding water and the other by expanding the existing water.

Moreover, although the vegetation in North America annually absorbs millions of tons of atmospheric carbon dioxide, it does not keep up with the prodigious emissions of the planet-warming gas produced by automobiles, power plants, the service industry, cement manufacturing, and other activities.[56] The fact that plants absorb excess carbon dioxide is not always beneficial however.

Enter the Insects

Elevated levels of atmospheric carbon dioxide can profoundly affect environmental feedback loops between crop plants and the insects that eat them, and may promote yet another form of global change: the rapid establishment of invasive species. For example, elevated carbon dioxide increases the susceptibility of soybean plants grown under field conditions to the invasive Japanese beetle. In a study, leaf tissue was analyzed for longevity-enhancing antioxidants because increases in dietary antioxidants can increase lifespan, as in fact if does for the beetle. Moreover, under experimental conditions, females consuming foliage with higher than normal levels of carbon dioxide laid approximately twice as many eggs as those eating foliage under ambient circumstances.

Like many plants, soybeans that get munched by insects produce a surge of defensive chemicals, and therein lies the problem. The beetle bonanza comes about because the increased carbon dioxide in the foliage impairs the soybean's normal defense against ravenous insects by depressing the expression of genes related to the plant's chemical ability to jam the beetle's digestive enzymes. Thus, by altering the components of leaf chemistry, other than sugar content, elevated carbon dioxide may increase populations of Japanese beetles, as well as other insect "pests," and so increase their negative impact on crop productivity.[57]

A similar situation may be developing in the cotton trade, where a cotton aphid seems poised to become a serious problem because of enhanced survivorship under a global regime of increased carbon dioxide that simultaneously delays the development time of its main predator, the lady beetle.[58]

What's more, if the predicted increases in low-altitude ozone in this century are accurate, they will stifle the growth of vegetation in many regions and thus cause carbon dioxide to accumulate more rapidly than expected in the Earth's atmosphere.[59] Because we humans do little to curb our materialistic appetites and our burgeoning population, our behavior further increases the atmospheric temperature. These increases affect the litho-hydrosphere and so the biosphere and thus the atmosphere in a never-ending, self-reinforcing feedback loop that is detrimental to a good quality of life as we know it. And this review covers but a pittance of the dynamic effects that accompany currents in the ocean of air as they circumnavigate the Earth.

Currents in the Ocean of Air

Air—everyone's birthright—can be likened to the key in a Chinese proverb: to every man is given the key to the gates of heaven, and the same key opens the gates of hell. Air is the key to both life and death as it circumnavigates the globe.

Air as the key to life carries the spores of fungi and the pollen of various trees and grasses to the reproductive benefit of the species. It also transports dust and microscopic organisms over great distances. In fact, if it were not for these air currents circling the Earth, the Amazonian jungle would starve to death.

The wind-scoured, nearly barren southern Sahara Desert of North Africa feeds the Amazonian jungle of South America with mineral-coated dust from the Bodélé Depression, which is the largest source of dust in the world. During the Northern-hemisphere winter, winds routinely blow across this part of North Africa, where they pick up 700,000 tons of dust on an average day and sweep much of it across the Atlantic. Approximately twenty million tons of this mineral-rich dust fall on the Amazon rainforest and enrich its otherwise nutrient-poor soils. The Bodélé Depression accounts for only 0.2 percent of the entire Saharan Desert and is only 0.05 percent of the size of the Amazon itself.[60]

Although air currents carry life-giving oxygen, water, and life-sustaining dust to the Amazon, they also transport the "key to death"—a human legacy made visible. In addition to the carbon dioxide affecting the world's oceans, toxins from such areas as the notoriously polluted air of Mexico City hitchhike on the wind across the Gulf of Mexico toward the United States, where the forest edges in fragmented landscapes function as significant traps for airborne nutrients and pollutants from both near and afar.[61] Nearby areas include agricultural lands—such as monocultures of corn for biofuel—and urban settings, whereas faraway places could be Mexico City or Beijing. What is more, forest edges effectively concentrate these chemical fluctuations below the canopy, where they can have cascading effects on soil-nutrient cycling, microbial activity, seedling dominance, and other ecological processes.[62] And even this is somewhat mild compared with what is happening in our national parks.

A report of a $6 million study by the U.S. National Park Service, titled *Western Airborne Containments Assessment Project*, documents the fact that pesticides, heavy metals, and other airborne contaminants—seventy in all—are literally raining down on twenty national parks and monuments from Denali in Alaska and Glacier in Montana to Big Bend in Texas and Yosemite in California. Over time, the pollutants enter the atmosphere, are flushed out with rain and snow, only to reenter it again and be flushed out again—and again and again—but each time at a higher elevation. The toxins range from mercury produced by power plants to such industrial chemicals as PCBs, dieldrin, and DDT (both dieldrin and DDT are banned insecticides).

In addition, contaminants in fish from the eight parks studied all exceed the safe threshold for human consumption. The parks most affected are Sequoia and Kings Canyon (California), Mount Rainier and Olympic (Washington), Glacier (Montana), Rocky Mountain (Colorado), Gates of the Arctic and Denali (Alaska), as well as Alaska's Noatak National Preserve. As well, mercury exceeds the safe limits for fish-eating wildlife at all eight parks, and DDT is in dangerous amounts for fish-eating wildlife in Glacier, Sequoia, and Kings Canyon national parks.[63]

Findings like these cause some people to tout the blend of 85 percent ethanol and 15 percent gasoline as good for air quality. In contrast to this euphemistic outlook, however, a study by Mark Z. Jacobson, an atmospheric scientist at Stanford University in California, uncovered little difference between the ethanol blend and straight gasoline with respect to emitting pollutants. According to Jana B. Milford, an environmental engineer at the University of Colorado, Boulder, Jacobson's study "should remind policy makers and others to be really skeptical about claims that E85 [the ethanol/gasoline blend] will improve air quality."[64]

This skepticism is well founded considering the aforementioned key to death found in Plastic Lake, Ontario, Canada, which, regardless of its name, is remote from any point source of pollution, such as a plastics factory or chemical plant. Nevertheless, as the pH in the lake decreased from 5.8 (acidic) to 5.6 (more acidic) over a period of six years, the resident population of northern crayfish became extinct, despite having exhibited a population size typical of those in the nearby Canadian Shield lakes. This is one of several documented examples of biotic impoverishment caused by the long-range aerial transport of strong acids.[65] Crayfish are not the only victims of pollutants, however.

As the level of pollution increases, air quality decreases, which causes chronic, adverse effects on lung development in city children from the age of ten to eighteen years. Moreover, diesel exhaust from buses, trucks, and farm equipment is a major component of air pollution throughout the world and is linked to lung cancer.[66] Such conditions are particularly bad in many regions of Asia, where brown clouds of smoke and soot from slash-and-burn agriculture and the combustion of fossil fuels blanket large areas. In addition to outright pollution of the air people breathe, these clouds enhanced lower-atmospheric solar heating by about 50 percent.

Taking into account the vertically extended atmospheric clouds of pollution over the Indian Ocean and Asia, circulation models suggest that brown clouds themselves contribute as much to the regional warming of the lower atmosphere as do increases in anthropogenic greenhouse gases. The air temperature between 1,650 feet and approximately two miles in altitude is 33 degrees Fahrenheit warmer than it would be without the pollution. Moreover, roughly 90 percent of the heating is attributable to soot.[67]

However, rain and snow scrub many pollutants from the air and deposit them in the soil and open waters, where they begin the journey to the oceans of the world. A case in point is the breezes that carry agricultural chemicals from the Central Valley of California high into the Sierra Nevada Mountains. There, endosulfan, a much-used insecticide, is scrubbed from the air and accumulates in the lakes and streams in sufficient concentrations to threaten a number of species of frogs and toads.[68]

Clearly, we humans directly affect the atmosphere and both directly and indirectly affect the soil and water—the litho-hydrosphere. However, although the spread of point-source pollution is scientifically predictable, its path of dissemination is not necessarily intuitive. If, for example, we choose to clean the world's air, we will automatically cleanse the soil and water to some extent because airborne pollutants will no longer exist to be extracted by rain and snow. If we then choose to treat the soil in a way that allows us to grow what we desire without the use of artificial chemicals (and if we stop using the soil as a dumping ground for toxic wastes and avoid overly intensive agriculture), the soil can once again purify water by filtering it. If we then discontinue dumping toxic effluents into the ditches, streams, rivers, estuaries, and oceans, they too can begin to cleanse themselves and regain some of their former health. That said, it's unlikely the oceans will ever fully regain their previous condition.

With clean and healthy air, soil, and water, we can also have clear, safe sunlight with which to power the Earth. Clean air is the absolute bottom line for social-environmental sustainability and, therefore, long-term human survival. With the eventual repair of the ozone shield, we can enjoy a more benign—and perhaps predictable—climate than we now have. In addition, effective population control can tailor human society to fit within the world's biophysical carrying capacity.

A population in balance with its habitat will reduce demands on the Earth's resources. Reduced competition for money—learning the true meaning of "enough"—might foster the cooperation necessary to heal our landscapes and thereby allow them to provide the maximum possible biodiversity. Protecting biodiversity translates into the gift of choice, based on sustainable ecological services, which in turn offers hope and dignity for all generations.

Genetic Engineering

Is genetic engineering a legitimate choice for creating designer corn—or anything else, for that matter—to meet our desired production of biofuel, despite ecological limiting factors? Perhaps we can, but it's a gigantic environmental gamble—the outcome of which may be ecologically disastrous and uncontrollable. To examine this gamble, let's begin with a new buzzword, *transgenic technology*, which is taking one part of an organism's genes and placing it into the genes of another organism. Transgenic technology is far from precise because genes are not machines that can simply be snipped from their host and placed somewhere else without opening the possibility of unpredictable, unwanted, and potentially uncontrollable results.

Consider that a trait not evident in one organism because it is genetically suppressed may come out of hiding, as it were, and become visibly expressed, or dominant, when working in concert with the full set of chromosomes present in a normal reproductive or germ cell of another organism. For example, DNA from genetically engineered corn has shown up in samples of indigenous corn in four fields in the Sierra Norte de Oaxaca in southern Mexico. This finding is "particularly striking," said University of California, Berkeley, researchers Ignacio Chapela and David Quist in 2001, because Mexico has had a moratorium on genetically engineered corn since 1998.

The fact is that corn genetically engineered to resist herbicides or to produce its own insecticides threatens to reduce the variety of plants in that region of Mexico because it may be able to out-compete the indigenous species. According to Quist and Chapela, the probability is high that diversity is going to be crowded out by these genetic bullies. This type of unwanted genetic transference is termed *genetic pollution*. In addition, the herbicide resistance could jump into weedy relatives and create super weeds that are beyond control. Furthermore, plants that have been genetically engineered to produce their own insecticide can have serious, deleterious effects on indigenous insects and microbes in the soil and thus have negative effects on indigenous plants.[69]

On top of that, some corn has been genetically engineered to produce crystal protein genes (from the bacterium *Bacillus thuringiensis*, referred to as Bt corn), which are encoding insecticidal endotoxins widely used for the development of insect-resistant crops. As is frequently the case with genetically engineered crops, Bt corn was symptomatically designed—but not systemically tested—before it was widely planted in the midwestern United States, often adjacent to headwater streams. Consequently, when pollen and detritus from the engineered corn entered the water, they unexpectedly reduced growth and increased mortality of aquatic insects that are important prey for aquatic and riparian predators, an outcome that has ecosystem-scale negative effects on critical self-reinforcing feedback loops.[70]

However, there is an even more potent genetically engineered corn, which is the result of fusing the endotoxin CryIAc (from the Bt bacterium) with the galactose-binding domain of the nontoxic ricin B-chain (RB). This fusion, designated BtRB, provides the toxin with additional binding properties that increase the potential number of interactions at the molecular level in target insects. Transgenic corn and rice engineered to express this fusion protein are significantly more toxic to insects than those containing only the Bt gene. They are also poisonous to a wider range of insects, including important pests that are not normally susceptible to Bt toxins. Killing vulnerable insects, however, can drive the evolutionary process of resistance to the Bt toxins.[71]

What, you might wonder, are the ecosystem-scale consequences of such poisonous plants, both of which are exceedingly thirsty and thus would transfer their engineered toxins into the water table should any part of them be recycled into the soil? This is an excellent question for which there is, as yet, no answer. Now, consider that economic trade appears to be a principal driver of policy with respect to the introduction of genetically modified organisms into non-industrialized countries.

When the different responses to unplanned imports of genetically modified organisms in Central America are compared with those in Africa, the contrast is strikingly clear. Where trade and environmental interests converge, as is the case in Africa, the implementation of a strong protective policy against the import of genetically modified organisms was decisive and swift. But in Central America economic trade and environmental interests do not overlap, which has resulted in a weak governmental response, as well as incremental policy shifts in favor of allowing the importation of genetically modified organisms.[72]

In the United States, however, a more mundane consequence of mass-producing corn as biofuel will be higher prices on the farm for animal feed and for many food items in the grocery store. The reason for the higher prices rests with corn as a basis for many foods, from beef to breakfast cereals. Corn feeds cattle, for example, and so affects the price of beef and all dairy products. Corn is also used in chicken feed and so affects the price of eggs and poultry. As well, corn is made into starch; is ground as meal for tortillas, hot mush, and cornbread; is distilled into liquor; and is packaged as cream corn, frozen corn, corn oil, corn syrup, and so on. Moreover, wheat and soy products will go up in price as farmers switch to growing corn. And these are only the visible prices we will all pay as a tradeoff for the manufacture of corn-based biofuel. There is an invisible price also—the far-reaching effects of the pollution associated with the production of corn.

Water as rain or snow washes and scrubs these pollutants from the air; it leaches them from the soil as it obeys gravity's call; and it carries them in trickle, stream, and river to be concentrated in the ultimate vessel, the combined oceans of the world—the depository of synthetic chemical compounds from human-generated sources.[73] Because oceans have no outlets whereby these pollutants can be flushed, they continually concentrate through the inflow of contaminated streams and rivers and through the evaporation and cycling of water from the ocean's surface, which is carried hither and yon by the currents of air. As the air-borne moisture condenses into drops of rain, it collects pollutants on its journey back to the ocean, where they can only become part of the on-going, self-reinforcing feedback loop of toxic chemical compounds and thereby affect such animals as sharks and dolphins, which store pollutants in their body fat, and polar bears, which suffer from pollutant-induced shrinking of their gonads.[74] Beyond these well-known animals and deeper in this vast, aquatic portion of the litho-hydrosphere live the coelacanths—ancient beings now in danger of extinction.

The Coelacanth

The coelacanth (pronounced SEAL-a-canth) is a rare fish that has survived deep in the Earth's seas almost unchanged for millions of years. The first captured coela-canth was caught in a deep-water gill net set for sharks about six hundred feet down off the mouth of the Chalumna River in southeastern Africa. In December 1938, Marjorie Courtney-Latimer, curator of a museum of natural history in East London, South Africa, went to the docks looking for interesting fish among the day's catch. There she found a 119-pound, lobe-finned fish that she described as "the most beautiful fish I had ever seen . . . a pale mauve blue with iridescent sil-ver markings."[75] Professor J.L.B. Smith described the fish as a new species in 1939 and named it *Latimeria chalumnae* in honor of Courtney-Latimer and for the Chalumna River.[76]

Upon examination by scientists, it was dubbed a living fossil because the remains of such creatures had been discovered only in rocks more than seventy-five million years old. At that time, the individual represented the only surviving species of coelacanths—a lineage of lobe-finned fishes that originated in the Devonian period, some 380 million years ago. It was thought to have become extinct, however, in the Upper Cretaceous period, around eighty million years ago, which is the date of the youngest fossil.[77]

How could this lineage of fishes have survived all that time without leaving a trace of its existence? A species can seem to disappear for three reasons: they are genuinely rare, they live in an uncommon habitat, or their remains do not fossilize well. In the case of coelacanths, all three reasons seem to apply, especially the latter two. They inhabit the "twilight zone" between five hundred to eight hundred feet deep in waters adjoining steep, rocky slopes of volcanic islands, where they cluster together in submarine lava deposits during the day. In this kind of habitat, sediment seldom settles fast enough to preserve a carcass.

Species that are typically low in numbers of individuals achieve their persistence through a variety of variation-reducing mechanisms. The one employed by the coelacanths is reliance on restricted "hot spots" of especially favorable habitat in which the local rate of growth is almost invariably strongly positive when the population is not crowded.[78] Indeed, these ancient fish are rigidly adapted to a couple of narrowly specific habitats, both of which are now threatened with drastic modification that may well cause the coelacanths to disappear into the great mystery from whence they came.

In the game of survival, the coelacanth has five ominous strikes against it: there are just two surviving species of a taxonomic group that was once considerably richer; it has not changed in millions of years; it is adapted to a specific habitat now threatened by human-caused pollution and human intrusion, such as severe pressure from local fishermen; it has a narrow resource base; and it has a poor ability to disperse.[79]

Since 1938, however, other coelacanths have been caught in deep water off the Comoros Islands, which lie between the coast of southeastern Africa and the northwestern tip of Madagascar. And on September 18, 1997, the wife of Mark Erdmann, an author of an article about coelacanths in Indonesia, saw one in Sulawesi (Celebes), Indonesia, being wheeled across a fish market on a cart. She barely had time to photograph the fish before it was sold.

Then, on July 30, 1998, Sulawesi fishermen dragged up a 4 1/2-foot-long, sixty-five-pound coelacanth that they had caught in a gill net set for sharks about four hundred feet down off the young volcanic island of Manado Tua in north Sulawesi. This specimen turned out to be a new species name *Latimeria manadoensis* (*manado* refers to the island and *ensis* means "belonging to"). Manado Tua is known to have submarine caves at about the same depth as those on the Comoros Islands, six thousand miles away.

All coelacanths are deemed to be endangered and are thus protected by the Convention on International Trade in Endangered Species of Wild Flora and Fauna. The reason for this status is the small population, an estimated five hundred individuals around the Comoros Islands, coupled with the low rate of reproduction (coelacanths bear live young). In the final analysis, however, we humans are the ones who are threatening the coelacanths' very existence through the chemical pollution of their deep-sea habitat.[80]

The continued survival of the coelacanth, after 380 million years in the deep sea, is suddenly threatened by major changes in its environment. These changes have been created by an upstart species (a global invader, as it were) that has been around for only five to eight million years—us.

What does it say about us, the human species, if we destroy the biophysical integrity of the coelacanth's habitat and its patterns of self-maintenance to the point of its extinction? It means that a whole, major line of evolution will suddenly disappear—forever. It means that all living individuals in the species, each one of which is the culmination of a 380-million-year chain of unbroken genetic experiments, will cease to be. How will the ocean ecosystem change with the loss of the coelacanths and their biophysical function as part of the system? Although such a pointless loss is, to me, unconscionable, it's just one more chapter in the never-ending story of irreversible cause and effect.

A team of scientists at the Virginia Institute of Marine Science in Gloucester Point, Virginia, found high levels of DDT and PCBs in the tissues of frozen specimens of coelacanths taken from the population off the Comoros Islands. "It's a very scary situation," John Musick, who headed the study, was quoted as saying. "It's even more alarming because if we lose the coelacanths, we're not losing a species, or a genus, or a family. We're losing a superorder—the last member of a species that dominated the world's ecology for millions of years." The loss of a superorder is, to scientists, the loss of a gigantic branch from the tree of life and thus an extant facet of the world in which we live.[81]

Some other ancient species, such as the North American opossum, are much less likely to become extinct because they meet nature's criteria for persistence. Persistence, in this case, means they live in environments that vary so much from day to day, month to month, and year to year that they're unlikely to meet anything in the future they have not already survived in the past. The living fossils are in much greater danger of extinction because they represent the only surviving species of a taxonomic group that was once considerably richer. Living fossils have an air of doom about them, as though they are living on borrowed time, holdovers from a different era. This category can also include plants, such as the remaining two species of the Pyrenean yam, a Tertiary relict that occurs in only one location in the world and is at risk of extinction.[82]

A comparison of the extinctions of birds, butterflies, and vascular plants in Britain shows that butterflies have experienced the greatest net losses. This decline in species is happening in all major ecosystems in Britain and is evenly distributed rather than occurring in just a few severely degraded regions.[83] Such a decline and ultimate disappearance of a population is a prelude to a species' extinction—which is forever.

Currently, 173 species of mammals are declining in numbers on six continents, where, collectively, they have lost over 50 percent of their historic ranges. This prologue to extinction is precipitated by the global loss of habitats caused by human activities. And the remaining habitats are increasingly fragmented into smaller and smaller "islands" with a severely reduced quality caused, in part, by anthropogenic pollution, such as the greenhouse gases, which are altering the climate. In fact, human alteration of the global environment is continually causing widespread changes in the distribution of organisms. These modifications in local biological diversity alter nature's biophysical processes and thus amend the resilience of ecosystems to environmental change. As with every species, regardless of size, its extinction (both local and total) represents a loss of its biological function, which has profound consequences for the ecological services we humans depend on for survival.[84]

There have been five major episodes of plant and animal extinctions over the last 440 million years, and each time it took upward of ten million years to recover species richness—each time with a different compositional arrangement of species and biophysical processes. Alteration of the global climate was a factor then, and it's a factor now in that climate change since the late 1970s has shifted the distribution and abundance of numerous species—and continues to do so.

The consensus among biologists is that we are now moving toward a potential sixth great extinction, ranging from the extinction of the smallest microorganisms to that of large mammals—some without our ever knowing they existed. This episode will be caused predominantly by the activities of a single species, however, us humans. Although scientists estimate that a minimum of ten million species inhabit today's world, they are disappearing between one and ten thousand times faster than they did over the past sixty million years.

Today, only a small fraction of the world's plants have been studied in detail, but as many as half of the species are threatened with extinction, primarily in the diverse tropical forests of Central America and South America, Central Africa and West Africa, and Southeast Asia. Moreover, nearly 5,500 species of animals are threatened with extinction. In addition, the International Union for Conservation of Nature's 2003 *Red List* survey of the world's flora and fauna indicated that almost one in every four mammalian species and one in eight avian species are threatened with extinction within the coming decades of this century.

Throughout most of geological history, new species seem to have evolved faster than existing ones became extinct, and so the planet's overall biological diversity has increased. But now evolution seems to be falling behind, in large measure because of our modern-day economic thinking.[85]

Three things plague us in Western culture: linear thinking, self-imposed constraints of time, and an unrelenting desire for instant gratification. We therefore spend most of our time looking for new areas of the world to exploit. In so doing, we gear our science and technology to efficiently wringing the wealth out of whatever dwindling resources we find. And the decisions we make today in our continual competition for control of the world's material goods will echo through the years and the lives of people for generations to come—consequences that will never be truly reversible.

5

Act Locally and Affect the Whole World

Only a people serving an apprenticeship to nature can be trusted with machines. Only such people will so contrive and control those machines that their products are an enhancement of biological needs, and not a denial of them.—Herbert Read, British philosopher

As a culture, we would do well to take an extended look in the rearview mirror at the degraded world we are leaving behind. Perhaps as a result of that closer look we might risk changing our minds about always seeking the unspoiled, which we then despoil; we might recognize a vast world waiting to be repaired—mended, as it were—in such a way as to once again yield up its wealth.

And if we would take the time to examine how and why we treat one another as we do, we might find that intense competition, which we take for granted as the only way to approach our natural resources, is merely a product of our thinking, one we would do well to change. Our thoughts about how things have to be are stuck in our minds with an adhesive called fear—fear of change, fear of loss, fear of being out of control—all of which makes us become self-centered in the money chase, our human concept of security in the form of wealth from material goods.

To dissolve the adhesive, we have to become other-centered, to get outside of ourselves and to work for the welfare of others, thereby increasing our own welfare but without focusing our attention on it. Repairing our damaged environment obviously carries us in this direction. But realizing its full value will not be easy—simple perhaps, but not easy.

Simplicity and ease are not synonymous. For example, changing our thinking is simple, like a snap of the fingers, but it might take twenty or more years to reach that state of mind because we each have many years invested in developing and defending the mechanisms whereby we cope with life. Taken together, these amount to our habitual systems of thought, belief, and behavior—basically, the routines of our life. In fact, life boils down to practicing relationships and the routines of doing so. A brief historical perspective might help us to understand why we, in the United States, think as we do.

When Europeans invaded the New World, they beheld a vast, rich continent, but did not see the land or its indigenous peoples—only the products of millennial processes, such as fertile soil, forage for livestock, timber, gold, slave labor, and abundant game animals. These products seemed both unlimited and free for the taking—as much as one could get hold of.

The invaders, beginning with the Spanish, coming from the pastoral scenes of Europe, saw what they wanted to see—a wild, untamed continent to be conquered, not a land to be nurtured. Why? Because they came from "civilized" countries and many felt superior to the "uncivilized" continent they confronted, with its dangerous "savages" and wild beasts. In line with a perfectly human tendency, their first inclination was to survive and then to seek psychological comfort by trying to re-create familiar surroundings from memory—all as a prelude to plundering the continent in the name of religion as exemplified by the Papal Bull issued in 1493 by Pope Alexander VI to justify the acquisition of personal wealth.[1]

The Europeans, including the French and British, brought their science and technology and relied on them, as they had in the past, to solve their social problems. What they did not understand, however, is that science and technology are human tools and, as such, are only as constructive or destructive, as conservative or exploitive as their users. Science and technology have no sensitivity, make no judgments, and have no conscience. It is neither scientific endeavors nor technological advances that affect the three spheres of life, but rather the thoughts and values of the people who use the tools.

Human social systems are governed by the same inviolate biophysical principles that quite literally "grew us" and thus control the survival and evolution of all living things. This statement is true even though a human society is composed of individually conscious and unique beings, each of which possesses a relative amount of free will. We have, in the short term, confounded this simple statement, however, by attempting to superimpose our human will onto nature's cycles within the biosphere.

That the biophysical principles govern human beings and their societies the same as they govern nature was not understood, or perhaps even considered, when the Europeans invaded the Western hemisphere. Little wonder they spoke grandly over the decades and centuries of "clearing the land" and "busting the sod," of "harnessing the rivers" and "taming the wilds." In keeping with this mentality, they begrudged the predators a right to life, and in the process became what they were against—the most voracious predators the Earth has ever hosted. And yet they only did the best they knew how to in their time and their place in history. How could they have done otherwise?

Today, however, we stand at a different time and a different place in history. We are present now, and we are making history now. Yet even today, at the dawning of the twenty-first century, we fail to understand or to accept that the biophysical principles by which the world is governed function perfectly, that only our perception of the way the world functions is imperfect. Our view is distorted because we focus solely on that portion of the world we intend to exploit—the products—and we ignore, even disdain, the biophysical processes that produce them. This warped sense of nature gave rise to the platform of deep ecology.

A group of Norwegian environmentalists, primarily the philosopher Arne Naess, introduced the term *deep ecology* in the early 1970s. The term is meant to characterize a way of thinking that approaches environmental problems at their roots, so that the problems can be seen as symptoms of the deepest ills of our present society.

The idea of deep ecology contrasts with *shallow ecology*, which I think of as material ecology or symptomatic ecology, because it merely addresses the symptoms through technological quick fixes, such as the requirement to install pollution-control devices and other regulations theoretically imposed on industry. It does nothing, however, to heal the problem, which lies in our thinking. Although new technologies and reforms in our current political system are much easier to implement than any fundamental changes in our thinking and our materialistic sense of values, these material solutions, and the people who propose them, are clearly avoiding the heart of the problem. This avoidance of the real issues faced by human society by those existing in spiritual bankruptcy, for which there is no Chapter Eleven protection, may ultimately cause the collapse of our social system.

In contrast to the symptomatic thinking of those who espouse material solutions to the systemic problems we humans are creating, nature is the embodiment of interactive parts that are unified by the novelty of the evolutionary process (particularly the spark of life), in which everything is always in the process of becoming something else. We industrialized humans, nonetheless, have chosen the reductionist metaphor of a machine not only for ourselves but also for our world. Although a machine has many parts, it has neither internal intelligence nor moral sense to guide it. In addition, the parts are unaware of their functions. And even though we can usually find or make one or more spare parts for a machine, we cannot do so with nature. Therefore, if polar bears become extinct, they are extinct forever. There is no way to reproduce one—no matter how noble the reason, diligent the attempt, or persistent the effort.

Thinking like machines is only one step away from living like machines. Such a synthetic lifestyle not only alienates us from ourselves and from one another but also alienates us from nature. In addition, a mechanistic lifestyle leads to economic problems through the separation of social classes and the accompanying philosophical dualities of either/or, right/wrong, us/them, and so on. Our synthetic, linear, mechanical thoughts and lifestyles also pit us against nature, and so our lives become increasingly complicated beyond the total complexity of nature.

And it is precisely because of our mechanistic thinking that we contend we can have more and more of everything simultaneously if only we can control nature—manage nature, as it were. In our drive to manage nature, we save the pieces for which we perceive a monetary value and discard those for which we do not. With this line of reasoning, we are simultaneously simplifying and disarticulating the biosphere by purposefully and accidentally losing pieces of it. This ill-conceived behavior results in redesigning our home planet even as we throw away nature's blueprint in the form of species and processes. In short, we focus so narrowly on the products—nature's ecological services—that we are systematically dismantling the processes that produce them.

We in Western society have become so linear and mechanical in our thinking and so irrational in the use of our knowledge we have forgotten that everything is defined by its relationship to everything else. In the end, we must accept that everything—*everything*—is a relationship that fits precisely into every other relationship and is constantly changing. Paradoxically, there is no such thing as a constant value, yet change is a constant process throughout the universe.

As human beings in Western society, we deal with and fit into this pattern of constantly changing relationships by thinking. We must therefore accept that any human influence in one of the three spheres—positive or negative—is a product of our thoughts, which, after all, precede and control our actions. We do nothing without first having the thought to do it. For example, the problem of pollution is not in the soil, the water, or the air, but rather in our mode of thinking (the cause); the problem only manifests itself (the effect of our thoughts) in the soil, the water, and the air.

We cannot, therefore, find a solution through science, technology, or politics without changing our thinking because all these things, which lie outside of ourselves, are the results of our thoughts, which lie within. Until we turn the searchlight inward to our own souls and consciously change our thinking, our motives, our attitudes, and thus our behavior, we will continue to compound the biophysical problems begun before a single human act was archived in history. One such problem is the "ownership" of water.

The First Ditch

In olden times, when people introduced something into the environment, they had not a glimmer of what they were doing to their environment or to the world at large. They were simply solving a problem, such as not having water close by. Nevertheless, everything that has ever been introduced into the environment has had effects that have reached beyond anyone's wildest imaginings. Perhaps one of the most influential introductions of all time was also the simplest—the first ditch and its never-ending story.

The first ditch was probably an idle scratch in the surface of the ground made by a child playing in a puddle of water after a rain or perhaps along a stream on some faraway afternoon in the dim past of humanity. The child had no grand scheme in mind while digging the little trench that allowed water to flow from where it was to where it would not otherwise have gone. It was a simple, innocent act with no outcome intended, but once the outcome became clear, the next little ditch had a purpose—to see if water would behave the same way a second time, and then a third, and then to see how far water would follow a ditch, and so on. With each experiment, the inquisitive, beginner's mind of the child enriched the child's knowledge of cause and effect and thereby gave the child a sense of control over water within the bounds of specific circumstances—which would be continually tested to find their limitations.

Somewhere in time a man or a woman had the budding idea and then the conscious thought of leading water from one place to another for a specific purpose—a purpose beyond playing or satisfying curiosity. That one thought, that one experiment in the control of water for a specific, practical end, forever changed the world and humanity's relationship to it. With the first purposeful ditch, water became a commodity that could be owned, as well as moved from place to place, stored, bought and sold, stolen, and fought over, and so the concept of water rights was born: Who had the first "right" to get the available water, how much, when, where, and for how long? With control of water, land became more and more valuable to individuals, family groups, communities, and ultimately to the nations of the world.

As the first ditch became many ditches, it allowed humanity and plants and animals to live in places that had previously been uninhabitable by those who needed water in close proximity. It helped give rise to agriculture and eventually led to such feats of engineering as the Suez and Panama canals, each of which physically connects one ocean with another. The first ditch irrevocably altered humanity's view of itself, its sense of society, and its ability to manipulate nature. But not all introductions in those faraway days were on purpose.

The Stowaways

We humans also have dramatic, unintentional, unforeseen impacts on terrestrial areas of our biosphere. To illustrate, let's consider Easter Island, which is a tiny, forty-three-square-mile piece of land in the South Pacific, 2,400 miles off the coast of South America. The island's oldest pollen dates back some thirty thousand years, long before the first people arrived. At that time, based on the pollen record, the island was forested with now-extinct giant Jubaea palms.

Polynesians settled on the island about twelve hundred to eight hundred years ago, when they began to gradually clear the land for agriculture and cut trees to build canoes. The island, while small, was relatively fertile, the sea teemed with fish, and the people flourished. The population rose to about three thousand or four thousand and probably remained relatively stable for several centuries. Eventually, trees were felled and cut into log lengths to transport and erect hundreds of stone statues, or *moai*, some of which are roughly thirty-two feet high and weigh as much as eighty-five tons.

Deforestation, which began shortly after the first people arrived, was almost complete five hundred years later, by the beginning of the eighteenth century. The pollen record shows that trees did not grow back to replace those cut. When the

Europeans discovered Easter Island in 1722, it was treeless and in a state of decline. Nevertheless, the Dutch explorer Jacob Roggeveen and the commanders of his three ships described the island as "exceedingly fruitful, producing bananas, potatoes, sugar-cane of remarkable thickness, and many other kinds of fruits of the earth." If the soil was rich enough for these plants, why then did the trees not grow back?

Not surprisingly, the Polynesians brought Pacific rat stowaways with them in their boats. As the human population expanded, and the people were busy cutting down trees, the rat population was keeping pace with the population of its human counterparts. In so doing, the rats ate more and more of the palm nuts, which prevented new trees from growing. The effects of drought, wind, and soil erosion could also have accelerated the island's deforestation. In addition, both people and rats exploited many of the island's other resources, such as its abundance of birds' eggs. The downward spiral had begun.

Deforestation meant there were no trees available to build canoes for fishing. Soil erosion led to reduced crop yields. And eggs of the sooty tern were probably exploited to such a point that the continual disturbance discouraged the birds from nesting on the island.

Fewer fish, eggs, and crops inevitably led to a shortage of food. Hunger, in turn, eventually brought the civilization to the brink of collapse. Today, all that remains of the original culture of Easter Island are the coastal statues, which once stood upright on specially built platforms but no longer do. Others lie abandoned between the volcanic quarries of their origin and their planned destinations, and still others remain unfinished in the quarries.[2]

To understand a rat-induced, landscape-level trophic cascade, such as might have taken place on Easter Island, a study was conducted on the Aleutian Islands to assess the impacts of introduced Norway rats on the densities of marine birds and the biotic structure of the rocky intertidal community. (A *trophic cascade* is the effect that a change in the size of one population in a food web has on the populations below it.) Data were gathered through surveys conducted on rat-free and rat-infested islands throughout the entire 1,180-mile archipelago.

Densities of birds that forage in the intertidal zone were higher on rat-free islands. But intertidal invertebrates were more abundant on rat-infested islands, where the cover of fleshy algae was reduced. Conversely, marine algae dominated the intertidal communities on rat-free islands.

Clearly, the invasive rats not only had reduced the population of marine birds directly through predation but also had indirectly affected the marine invertebrates by eating the algae in the rocky intertidal zone. Thus, a trophic cascade was created when the rats fed on the marine algae, which reduced the algae and thus shifted the structure of the intertidal communities on the rat-infested islands from one dominated by algae to one dominated by invertebrates.[3] This proved to be a situation similar in nature to that on Easter Island, where the rats ate so many palm nuts they influenced the island's deforestation and ultimate collapse.

A different kind of introduction, however, is altering the entire forests of the northern United States, namely the European night crawler. Although native worms were wiped out from the northern United States and Canada in the last glaciation, they persisted south of the ice sheet and permafrost. Consequently, the indigenous hardwood forest of the northern United States evolved earthworm-free over a period of ten thousand years. Even today, there is a line from Massachusetts to Iowa, north of which there are no native earthworms. Moreover, southern species of native worms have not advanced far into the once-wormless territory. But when Europeans colonized North America, night crawlers hitched a ride in soil, perhaps that used for ship ballast or in the root balls of plants.

European night crawlers are acknowledged to work wonders in the garden, where they aerate the soil and speed the release of nutrients as they consume organic material, such as fallen leaves. But as they continue to move northward, the worms are slowly changing the northern deciduous forests by eating their way through the leaf litter and duff required by the native plants in order to thrive. Duff is partially to fully decomposed organic material and is generally located between surface litter and mineral soil. Thus, a single invading exotic species—as humble as an earthworm—is indeed capable of altering an entire ecosystem.[4]

Just as a European earthworm changed an entire forest ecosystem, so domestic plants brought to Phoenix, Arizona, contaminated the once pollen-free air. A simple act by the people who moved to Phoenix to find relief from their allergies has placed Arizona among the top 10 percent of states in pollen count during the six-week season of allergies. Before urban sprawl began consuming the desert, the area around Phoenix was a haven for people who suffered from allergies. Doctors in the 1940s and 1950s sent patients there because the dry air was virtually pollen-free. But many of those people also brought with them their favorite plants, which subsequently matured and now fill the air with pollen during the spring of each year.

In addition, the dry climate causes pollen grains from nonindigenous plants to stay aloft, wafting on every zephyr. They are not washed from dry desert air as they are in nondesert areas, which get spring rains. Thus, the allergy sufferers themselves turned their own haven into their worst nightmare by not identifying and protecting the very environmental value that brought them to Phoenix in the first place—air virtually free of pollen.[5]

Getting rid of all the pollen-producing plants is no longer an option for the residents of Phoenix. Like so many other unintentional effects, the pollen is out of their hands and here to stay. But what, you might wonder, are some of today's larger human-caused effects?

The Invaders

Although the genus *Homo* (which includes humans) emerged only about 2.5 million years ago, and modern humans around 120,000 years ago, members of this genus have become remarkably adaptable and successful. Unlike most genera, which exist somewhere between five and ten million years before fading into extinction as other genera take over, we modern humans face no such immediate threat of extinction because we are extremely adaptable generalists.

On the one hand, a generalist can survive under a wide range of environmental circumstances, can use numerous kinds of energy, and can either adapt itself to a wide variety of conditions or can adapt a wide variety of conditions to itself. A specialist, on the other hand, is fitted to a highly specific set of conditions within its environment and can derive and use only limited kinds of energy to achieve certain ends.

As members of the human species, we are perhaps the most successfully adaptable generalists on Earth. People live in the frozen tundra and along the sea ice above the Arctic Circle, throughout the temperate forests and plains, in the hot deserts, and in the depths of steaming tropical jungles. We live on every continent and latitude between the two polar circles. Moreover, we have found cures for enough diseases to vastly increase our numbers and our longevity. In addition, we are generalists in the social sense because we have built and live in societies ranging from nomadic food-gathering tribes to sophisticated postindustrial civilizations and from raw-military dictatorships to grass-roots democracies.

There is a caveat to these comforting statements however. As the most invasive species on Earth, we are usurping an ever-increasing share of global resources, sometimes through our technological advances but most often by mere dint of our burgeoning numbers.[6] Therefore, we must count as a threat all the cumulative, unintended consequences of our technological developments, such as the increasing number of chemical-resistant diseases through which we are making our home planet unfit for our own existence. These cumulative impacts have ensured that our human shadow now covers the entire globe, thereby simultaneously foreclosing the ability of scientists to study pristine ecosystems in an attempt to understand nature.[7]

What might some of these unintended consequences be? Consider, for example, that doctors once predicted antibiotics would vanquish infectious diseases, but instead the "bugs" are rapidly confounding today's medicines. Overprescribing antibiotics is fast helping bacterial and viral disease organisms to mutate so medicines no longer work when they are really needed. Resistance to antibiotics is a worldwide situation that appears "to be on the verge of desperation," said Nobel laureate Joshua Lederberg, who chaired a panel of top bacterial experts convened to monitor the issue by the private Institute of Medicine at the request of the U.S. government.

Despite several years of repeated warnings, overuse of antibiotics is the main reason disease organisms are becoming immune to drugs—and the culprits are not just doctors and patients. Antibiotics are overused in animals raised for food. The new trend of putting germ-resistant coatings on toys, high chairs, and other items used by children may also be a problem, according to Gail Cassell, vice president of the drug company Eli Lily.[8]

In fact, James Hughes, director of the National Center for Infectious Diseases at the Atlanta Centers for Disease Control and Prevention, looking at one minute but important aspect of humanity's existence on Earth, says: "Today, we have only one drug to treat some infections. Once they become resistant to this drug, then we will basically be back in the pre-antibiotic era."[9]

The evolution of chemical-resistant diseases is but one infinitesimal cumulative effect of humanity's domination of the Earth. Few people realize that today's social-environmental problems are compounding more rapidly than and greatly transcend those of the past in both scale and complexity.[10] In addition, humanity's modifications of species and genetically distinct populations of organisms are substantial and are growing without bounds.

Albeit extinction and genetic modification are natural, biological processes, the current rate at which genetic variability in populations and species as a whole is being lost is far above the rate prior to the advent of modern human society. This loss is not only ongoing but also represents a totally irreversible, global change. At the same time, people are transporting species around the world, introducing them into new areas, where they disrupt existing systems. In the process, people are homogenizing the once-rich diversity of local indigenous species.

Although there is a fossil record of species invasions occurring in waves after geographic barriers had been lifted, these episodic events differed markedly from the human-assisted invasions of today, by which every region of the planet is simultaneously affected. Moreover, modern rates of invasion are several orders of magnitude greater than prehistoric rates; as a result, the potential for synergistic disruption of ecosystems by the current mass invasion, as well as its range of evolutionary consequences, is without precedent and should be regarded as a unique form of global change.[11]

As far as plants are concerned, ornamental horticulture has been recognized as the main pathway for plant invasions worldwide. In Britain, for example, the characteristics that increase the probability of a species escaping cultivation are being tall, being native to Europe, and being an annual. Climbing plants and species intolerant of low temperatures are less likely to escape. In contrast, the probability of establishment may be greater if a species belongs to a genus indigenous to Britain, and that likelihood increases as the number of continents in a plant's native geographical distribution increases.[12]

Today, however, the kinds of invasive species include such taxonomic groups as bivalve mollusks, earthworms, insects, fish, reptiles, birds, and mammals.[13] Considering that people have plied the world's oceans over the past five centuries, it's not surprising that human activities have led to the redistribution of a vast number of marine organisms, primarily through the movement of ocean-going vessels and the transport of products from the commercial fishing industry. Most biological surveys postdated these events, however, so the distribution of many of these now-cosmopolitan species is perceived to be the result of natural processes. This interpretation underestimates the role of humans in altering patterns of marine diversity through the distribution of ocean-dwelling organisms along the coastal margins of the world. Besides, available evidence suggests that introductions continue unabated on a large scale throughout the world.[14]

Regardless of local diversity, nonnative species are most abundant where human activity (economic prosperity, population density, and urbanization) is greatest: This relationship is particularly evident with freshwater fish because they are unable to move readily from one river basin to another on their own. The percentage of foreign fishes in 1,055 river basins worldwide attests to this strong link with adjacent human activity. Local economics is accountable for 70 percent of the distribution of these exotic fish. Nevertheless, the galloping pace of economic development throughout the industrialized countries of the world during the twentieth century and the likely increase of that pace in non-industrialized countries during the twenty-first century portend an increasing problem with invasive species.[15]

Many introductions of invasive, exotic species are effectively irreversible because once an introduced species becomes reproductively successful in a new area, eradicating it is both difficult and expensive. In addition, there are times when a newly arrived exotic actually triggers a population explosion of an earlier alien by decimating native species. Such is the case of the European green crab and the eastern gem clam in Bodega Harbor near San Francisco, California.

The green crab was imported into the Bodega Bay area inadvertently around 1993 in boxes of fishing bait from New England. However, the eastern gem clam, which is barely half an inch across, hitchhiked from the East Coast to the West Coast in shipments of oysters in the late 1800s. The clams settled into San Francisco Bay and by 1960 had spread north to Bodega Bay, where they lived in small numbers among the native nutricola clams until the European green crab arrived and changed that.

The green crab, which is approximately three inches across, has long been recognized as a menace to marine systems other than its own. By 1994, the crabs were gobbling up the two native species of nutricola clams, which had lived in the harbor for millennia and had kept the inconsequential population of eastern gem clams in check. Consequently, the gem clams quickly occupied the decimated nutricola clams' vacant habitats and may eventually displace them altogether.[16] What's more, some introduced species have even more profound consequences than either the green crab or the gem clams, such as degrading the health of humans and indigenous species across continents.

After all, most infectious-disease-causing organisms are introduced as exotics over much of their geographical distributions, such as west Nile fever, which became established in the United States in 1999. Other introduced species cause economic losses amounting to billions of dollars; the zebra mussel is a well-publicized example. Some exotics disrupt biological processes and thus alter the structure and function of entire ecosystems. Finally, the introduction of exotic species, such as the fire ant, is a major driving force in the loss of indigenous populations of species, as well as whole species themselves.[17] And many invasive exotics find ready access into new areas along roads, commercial pipelines, and power-line rights of way, from which they spread and disrupt essential ecosystem services and processes, including autogenic succession, nutrient cycling, soil erosion, disturbance regimes, and the composition of plant and animal communities and their dynamics.[18]

Thus, from an ecological point of view, there is nothing environmentally friendly about a road, any road. Roads of all kinds have generally negative effects, in part because they largely determine the patterns of land use, predominantly by fragmenting habitat.

In addition, the process of construction kills sessile and slow-moving organisms, injures organisms adjacent to a road, and alters physical conditions beneath a road. After a road is in use mortality from collisions with vehicles affects the demography of many species, both vertebrates and invertebrates. For instance, a disproportionate road mortality of female freshwater turtles on nesting migrations can skew population sex ratios.

Beyond mortality from collisions, roads affect animal behavior, thus altering home ranges by acting as barriers to movement, creating noise, and providing artificial light by night. They can affect reproductive success, an individual's ability to escape, and its physiological state. As well, roads can also influence the health of wildlife by spreading disease agents and hosts or by generating environmental conditions that sustain these agent and host populations, especially in ponds that are in close proximity to roads.

Roads change soil density, temperature, soil-water content, light levels, dust, surface waters, patterns of runoff, and sedimentation, as well as adding heavy metals (especially lead), salts, organic molecules, ozone, and nutrients to roadside environments, such as ditches; from there these materials begin their journey to the nearest stream, river, estuary, and ocean. Besides being a source of chemical pollution, roads are a major cause of landslides and account for high levels of sediment in streams and rivers, which is associated with reduction in fish habitat, including that of salmon.

Roads promote the dispersal of exotic species by altering habitats in the form of migration corridors, thereby stressing native species. In a sense, they also form corridors that promote increased hunting, fishing, passive harassment of animals, and landscape modifications.[19]

In the Republic of Congo, for example, road density is closely linked to market accessibility, economic growth, exploitation of natural resources, habitat fragmentation, deforestation, and the disappearance of wildlands and their associated wildlife. Roads established and maintained by logging concessions intensify the hunting of bushmeat by providing access to relatively unexploited populations of forest wildlife and by lowering the cost of transporting the bushmeat to market.[20]

Moreover, the ecological effects of roads are underestimated, in part because available sources of data do not include the full road network and also because of the vast network of "ghost roads" created by recreational use of off-road vehicles.

In addition, the road-effect zone (the area where negative effects are found) extends from about 330 feet to 3,280 feet into the surrounding habitat, with an average of about 600 feet.[21]

Roads serve one purpose—and one purpose only—to allow human access to an area in a vehicle of some kind. There are now more than 385,000 miles of major roads in our national forests alone. Deny vehicular access to land, especially public lands in the United States, and the outcry is almost instantaneous. Yet scientific evidence is continually reinforcing the fact that the ecological impacts of roads are more extensive than previously thought.

"Imagine you are looking at a beautiful and richly detailed painting when suddenly all of the colors begin running together into blotchy grays," says ecologist Daniel Simberloff from the University of Tennessee in Knoxville. "Something similar is happening to the world's plants and animals. The flow of exotic organism[s] is increasing rapidly as the world becomes more interconnected. Without action, a growing army of invasive species will continue to overrun the United States, causing immense economic and ecological damage."[22]

Simberloff is speaking about the movement of exotic species into new areas, where they are causing damage to such ecosystems as grasslands, wetlands, forests, and even oceans. "The world's habitats are rapidly being homogenized," admonishes Simberloff.[23]

Once something, such as an exotic species, is introduced, the results are out of our hands and usually out of our control. Whereas the introduction of an exotic species adds diversity in a sense, the results may cause a drastic loss of indigenous diversity, especially if the species is invasive by nature and overruns the habitat by displacing indigenous species.

The cost to the U.S. public is billions of dollars annually because around one-quarter of the value of agriculture goes to controlling foreign plants. In addition, public lands and waterways are being greatly impaired for recreational use by aggressively invasive aquatic plants, not to mention the non-indigenous species that pose major risks to human health, such as the Asian tiger mosquito from Japan, which is spreading in the United States and is carrying with it encephalitis, yellow fever, and dengue fever.

Although plants and animals have been piggybacking on human travelers for millennia, the rapid growth of current travel and trade, as well as the wholesale displacement of entire human populations because of wars, has created an entirely new problem.[24] In the past, when an organism stowed away in the cargo hold of a ship, it not only had to survive for months but also had to be transported from the ship into an entirely new environment, where it had to survive. Today, however, millions of tons of cargo and hundreds of millions of people travel annually on commercial airliners, and this kind of extensive travel greatly increases the chances that a foreign organism will survive.

The brown tree snake, for example, has already devastated populations of forest birds on the island of Guam and has traveled to Honolulu in the wheel wells and cargo bays of airplanes. A giant African snail that has ravaged agricultural crops on many Pacific islands was carried by a boy from Hawaii to Florida, as a gift to his grandmother. The Asian chestnut-blight fungus arrived in New York City in the late nineteenth century and killed almost all American chestnut trees along the eastern coast of the United States; the tree was the most common species in many forests prior to the introduction of the fungus. And the list goes on and on.

Eradication of these exotic species is not only impossible most of the time but also expensive because chemical, mechanical, and biological controls, which can sometimes minimize economic and ecological damage, are frequently harmful to beneficial species as well as to us humans.

Unfortunately, the laws that restrict entry into the United States or individual states are generally designed to use what Simberloff refers to as *blacklisting*, which means that an organism must first be proven detrimental before anything is done about prohibiting its spread. More effective would be what Simberloff calls *whitelisting*, which means that all potential introductions of foreign species would be subjected to scrutiny; such oversight in turn would stimulate the scientific research necessary to determine ahead of time if and when a given species has the potential to be destructive if introduced into the United States.

Another obvious requirement to help control the spread of unwanted exotic species is an early warning system with the capability of responding rapidly before a species reaches a stage beyond control. Even if we had such a system, which we do not, the greatest barrier to an effective response concerning invasive exotics, according to Simberloff, is public nonchalance.

In addition, a number of special-interest groups, such as importers of exotic plants and animals and owners of pet shops, work to minimize barriers to the importation of exotic species, despite the harm escapees could cause. Such self-centered myopic behavior is, in my opinion, totally irresponsible because an escapee often causes a severe, long-term, ecological problem as well as necessitating an ongoing, astronomical cost to the public in trying to control it. But without an educated public and educated legislatures, says Simberloff, the special-interest groups undermine the ability of government agencies to blacklist or whitelist harmful species.[25]

There is, however, another aspect to this issue—the supply side. Lawmakers in Hawaii have taken steps toward protecting their colorful, indigenous marine fish by limiting the number that can be captured and sold to supply pet shops and home aquariums throughout the United States. Although a committee in the House of Representatives rejected an outright ban on capturing the fishes along the west coast of the island of Hawaii, it adopted a plan to set aside about a third of that area as a sanctuary for the protection of tropical fish. In October 1999, part of the same area became a 650-foot-deep no-capture sanctuary for all fish.

The issue pitted some people in the state's $50 million tourism industry, which needs the fish to entertain snorkelers and scuba divers, against those in the $10 million industry that captures the tropical fish to supply pet shops and collectors throughout the United States. But in the long term, said Jack Randall, a retired ichthyologist who worked at the Bishop Museum in Hawaii, the sanctuary would benefit both industries, even if it causes economic pain in the short term. Randall went on to explain that the protected area will give fish more of an opportunity to grow larger than they now have time to do; such protection is important because the larger the fish are, the more young they produce.[26]

In fact, some people, such as James Bohnsack, a research fisheries biologist with the National Marine Fisheries Service in Miami, Florida, contend that no-capture marine sanctuaries should be viewed as "the controls, and everything else is the experiment." Bohnsack goes on to say that, by allowing fishing throughout the ocean, "we've been conducting a giant, uncontrolled experiment over the entire ocean for years." The experiment, according to scientists and policymakers, is the introduction of non-indigenous animals and plants into marine ecosystems, where they are a major threat to biodiversity. Although government agencies have struggled to control alien species on land and in freshwater for decades, the control of exotic species of marine organisms is in its infancy.[27]

To the extent special-interest groups undermine the protection of either indigenous species from being overexploited or ecosystems from being damaged by the invasion by foreign species, they are stealing diversity and thus a potentially better quality of life from future generations. I say this because the continual escape of invasive exotics, as well as the depletion of indigenous populations to supply the pet trade, will increasingly impoverish the ecosystems our children, their children, and their children's children will inherit.

The upshot is that the rates, scales, kinds, and combinations of change being wrought by humanity are fundamentally different from those at any other time in history. We are changing the composition, structure, and function of the Earth more rapidly than we are perhaps able to understand—or grasp the consequences of.[28] Here, a critical lesson in prudence might be usefully gleaned from the notion that we, like a parasite, cannot afford to kill our host planet—Earth. The sooner we begin to accept and face these problems, the more options we will have not only for ourselves but also to pass to our children. After all, these choices are decisions we make about the "commons."

Trespassing on the Commons

We humans have jointly inherited the commons, which is more basic to our lives and well-being than either the market or the state.[29] We are "temporary possessors and life renters," wrote British economist and philosopher Edmund Burke, and we "should not think it amongst [our] rights to cut off the entail, or commit waste on the inheritance."[30] (An entail is the restriction of the future ownership of real estate to a particular descendant, through instructions written into a will.)

Despite the wisdom of Burke's admonishment, the commons is today almost everywhere assaulted, abused, and degraded in the name of economic development as corporations are increasingly hijacking (euphemistically termed "privatizing") both nature's services and every creature's birthright to those services. For example, the greatest threats to the world's oceans are probably increasing temperature, destructive fishing in deep waters, and point-source pollution, some of which is simply human sewage.[31] Pollution also despoils the air, defiles the soil, and poisons the water. Noise has routed silence from its most protected sanctuaries. City lights hide the stars by night. Urban sprawl, the disintegration of community, and attempts to control, engineer, and patent the very substance of life itself are all part of the economic raid on the commons for private monetary gain.

"Corporations," says author David Korten, "are pushing hard to establish property rights over ever more of the commons for their own exclusive ends, often claiming the right to pollute or destroy the regenerative systems of the Earth for quick gain, shrinking the resource base available for ordinary people to use in their pursuit of livelihoods, and limiting the prospects of future generations."[32]

This is not to say that all corporations are bad. But it is to say that both corporations and the market must have boundaries to keep them within the realm of sustainable biophysical principles, human competence, and moral limits. "The market economy is not everything," asserted conservative economist Wilhelm Ropke in the 1950s. "The supporters of the market economy do it the worst service by not observing its limits," says author Jonathan Rowe.[33] And it is by ignoring the moral limits of the market economy that we, the adults of the world, create poverty and increasingly mortgage the birthright of all future generations—beginning with our own children and grandchildren.

For the children's sake we must understand, acknowledge, and remember that we, in the biosphere, live sandwiched between the atmosphere (air) and the litho-hydrosphere (rock and water). As a reminder, our lives depend on two great oceans: one of air and one of water, both of which have currents that circumnavigate the globe. Because each sphere is inexorably integrated with the others, if we degrade one, we degrade all three. And we are currently busy polluting all of them as if there were no tomorrow.

Pollution not only is destroying the global commons worldwide but also is "trespassing" onto the local commons and private property. I say trespassing advisedly because loud noises, the unwanted glare of lights at night, someone else's property littered with human junk that is visible from one's home, the stench from a nearby factory, and industrial chemicals fouling water in a private well are all examples of pollution caused by someone else, somewhere else—pollution that crosses the boundary into such commons as the seven seas, national parks, city parks, as well as private property, all without the owner's permission.

Consider as illustrative the uninvited contribution to the commons of the drug manufacturers in Patancheru, near Hyderabad, in southern India—a major production site of generic drugs for the world market. In Hyderabad, the industrial plant that processes effluent from the ninety large pharmaceutical manufacturers in Patancheru discharges the highly contaminated water into a stream that eventually joins the Godavari River, the second largest in India to empty into the Indian Ocean, which it then contaminates. The released water contains astronomical amounts of antibiotics, along with large concentrations of analgesics, drugs for hypertension, and antidepressants. Furthermore, in keeping with a common practice, the treatment plant mixes raw human sewage with contaminated effluent, which contains enormous quantities of antibiotics that will encourage the evolution of bacteria to resist these same antibiotics.[34]

Ultimately, therefore, pharmaceuticals, ranging from painkillers to synthetic estrogens, are entering the waterways of the world, and thus the global commons, through human excreta, hospital and household wastes, and agricultural runoff, as well as from water-treatment plants. Synthetic estrogens and their mimics are known to have negative impacts on the sustainability of populations of wild indigenous fish, as well as on the developmental processes of amphibians, in streams that receive polluted water from municipal wastewater-treatment plants.[35] Granted, this sort of trespass is silent, but its sibling trespasser is noise.

Silence Please

Once upon a time, silence could be found throughout much of the world, especially in the high-mountain snows of winter and in the great, still expanses of the world's deserts. Today, however, silence is a rare and elusive part of the commons—from the highest mountain to the depths of the deepest ocean. In fact, the world has gotten so noisy, even beneath the ocean, that the ability of many sea creatures to seek food, find mates, protect their young, and escape their predators is severely compromised. The effects of underwater noise can be likened to being trapped in the center of an acoustic traffic jam, where the din comes simultaneously from all sides. In deep water, where marine animals rely on their sense of hearing, the noise is especially harmful.

Noise from supertankers and military sonar equipment, as well as from the explosions of seismic exploration for offshore oil, scrambles the communication signals used by dolphins and whales, which causes them to abandon traditional feeding areas and breeding grounds, change direction during migration, and alter their calls. They also blunder into fishing nets. In fact, the global unintentional catch—"bycatch," in today's vernacular—of marine mammals is hundreds of thousands and is likely to have significant demographic effects on many populations. In addition, dolphins and whales can no longer avoid colliding with ships on the open seas, where international shipping produces the most underwater noise pollution, with few regulations to control it. And the military acts as though controls are beneath it.[36]

Both within and beyond the sea, certain levels of noise—unwanted sound—negatively affect our own health, as well as that of our pets, such as zebra finches, which forego fidelity to a mate as sound blares.[37] Some of the other problems associated with noise pollution are loss of hearing; chronic stress; sleep deprivation; high blood pressure; mental distractions, with the resultant loss of enjoyment and productivity—all of which are part of a declining quality of life. Moreover, noise may be unwanted because it prevents us from hearing a bird's song, the sigh of an autumn breeze in our garden, or the song of a waterfall.

When I first built my garden pond, one of the primary joys was the music of the water falling over the little rock wall I had made for it. The waterfall sang freely in those early days, and my wife, Zane, and I could hear its song from every corner of the garden in the back of our house. In fact, we would open our bedroom window at night and listen to the water sing. There are no words to describe the inner feeling of peace and well-being conveyed by the love song of the waterfall.

We humans live in the "invisible present," wherein things change so slowly that we don't notice the tiny, cumulative effects of their continual transformation. Although I knew this phenomenon and had written about it in other books, I did not realize that we were once again to experience the creeping invisibility of daily change.

The change of which I am speaking began with the background noise of increasing traffic as the town grew. With the prosperity of the 1990s, home-improvement projects seemed to spring up everywhere, adding to the din. Then came the insidious leaf blowers, lawnmowers, and finally a fleet of helicopters, which were headquartered at our local airport and flew incessantly in the vicinity of our house. As noise was added to noise, it became harder and harder to hear the waterfall.

We first noticed that the waterfall's song could not be heard when we opened the bedroom window and, listening, failed to detect the splashing water. Next, the outer corners of the garden became devoid of its song. Over time, we had to get closer and closer to the pond in order to hear the music. Then came the time we could barely hear it when we sat on our bench, a scant eight to ten feet away. Finally, the urban noises penetrating our garden became so intrusive that even when we stood next to it, we could not hear the waterfall.

When the voice of the waterfall was drowned out by the ever-increasing noise pollution, some of the spiritual essence disappeared from the pond. In the end, therefore, I removed the waterfall rather than have its inaudible presence be a constant reminder of the waning quality of life that daily besets us.

Noise pollution is one of society's growing concerns because it increasingly affects the quality of everyday living—especially if one lives in the flight path of an airport; within a few miles of a railroad crossing; next to an increasingly busy street; near an athletic field, a university fraternity, or ongoing construction. And there seem to be few places one can escape from it.

As urban sprawl claims more and more of a community's landscape, the collective noise of human activities increasing invades the once-quiet sanctuary of private homes—often to the discomfort and frustration of its inhabitants, both human and nonhuman. In some places, city revitalization is also part of the problem. Trespassing noise can interrupt sleep, meditation, and conversations, distract students in class, entertainment, and the dignity of funerals, as well as diminish recreational experiences. And if the trespass of noise were not enough, its effects are exacerbated at night with the growing pollution of light.

Dim the Lights

There was a time in the 1940s and 1950s when the night sky of the Willamette Valley in western Oregon winked with the light of a million stars. In those days, the Milky Way was visible from almost everywhere in the south end of the valley, but no more. Now, only the brightest stars can be seen, even on the darkest of nights, because of light pollution. Thus, we, in the United States, are losing the only portal to the wonder of nature that is open to virtually everyone, everywhere as part of the global commons.

This doorway to the heavens can feed our sense of wonder, which in turn opens our mind to possibilities rather than keeping it confined to what we are certain we know. The ability to view the stars gives people another way of connecting with nature, so critical if we are to arrive at sustainable solutions. And lights not only hide the stars but also blind us.

Glare from lights can simply be uncomfortable or annoying, such as the glare from excessively bright security systems in residential areas. But glare can prevent a motorist from seeing a pedestrian in dark clothing because the driver is blinded by the glare of bright lights from an oncoming vehicle. In addition, the ever-increasing output of lights from commercial and residential settings (such as commercial parking lots and private security systems), as well as from roads, affects our quality of life and our safety.

The most pervasive form of light pollution, however, is *urban sky glow*, which is caused by artificial light passing upward, where it reflects off of submicroscopic particles of dust and water in the atmosphere. First noted as a visual problem by astronomers, it is no longer just their issue. In fact, urban sky glow, which can be seen more than a hundred miles away from large cities, is beginning to seriously destroy our ability to experience the nighttime sky in some of our national parks.

In addition to the diminished wonder and enjoyment engendered by gazing at the twinkling stars, light pollution is a rapidly expanding form of human encroachment on other species, particularly in coastal systems, where it alters the behavior of sea nesting turtles. It also affects the foraging behavior of Santa Rosa beach mice, which tend to avoid artificially lit areas. In fact, this artificial phenomenon appears to be driving some strictly nocturnal species, such as the California glossy snake, toward extinction. According to zoologist Robert Fisher of the U.S. Geological Survey in San Diego, California, "It might be that you can protect the land, but unless you can control the light levels that are invading the land, you're not going to be able to protect some of the species."[38] Moreover, the glare of lights at night exposes the unpleasant view of human garbage that is increasingly scattered through many urban areas, as well as rural areas, as urban dwellers deposit their junk rather than pay to have it properly disposed of.

Visual Pollution

Historically, pollution has referred to the human introduction of noxious substances into the environment that impaired a given ecosystem's ability to function by disrupting its biophysical processes. From a visual point of view, the items people introduce into the environment that are generally considered to be eyesores are termed a *blight*, which is any cause of spoilage, damage, or ruin, especially in urban areas. I use the term *visual pollution*, however, because it results from the same disregard for the beauty and biophysical integrity of planet Earth as do the other kinds of pollution.

Visual pollution includes such things as discarded garbage along roads; unkempt, junk-ridden properties; tangles of aboveground power, phone, and cable lines; cell-phone towers; and the ubiquitous signage—all of which progressively degrade the aesthetic quality of our life. The foregoing discussion has dealt with a potpourri of issues; let's now consider a very solid one—the Aswan High Dam in Egypt.

The Damnable Effect of Dams

Global biodiversity in river and riparian ecosystems is created and maintained by the geographic variation in stream processes and fluvial disturbance regimes, which, in turn, largely reflect regional differences in geology and climate. The extensive network of dams constructed by humans has greatly diminished the seasonal and year-to-year variability in the streamflow of rivers, thereby altering ecologically important biophysical dynamics in the continental- to global-scale drainage basins. The cumulative effects of modifying regional-scale environmental templates are largely unexplored.

To understand the environmental effect of this dam network, 186 long-term streamflow records on intermediate-sized rivers across the continental United States were examined. This study showed that dams homogenized the flow regimes on third- through seventh-order rivers in sixteen historically distinctive hydrologic regions over the course of the twentieth century. Such homogenization occurs chiefly through controlling the magnitude and timing of the seasonal high and low flows. However, no evidence for homogenization was found on the 317 free-flowing, reference rivers, despite documented changes in regional precipitation throughout the period.

The estimated average density of dams in third- through seventh-order rivers in the United States is one every thirty miles; such a high concentration of dams indubitably has the continental-scale effect of homogenizing regionally distinct environmental templates. What is more, the so-called reclamation of our national rivers has created the illusion of human control over their flow and thus, to many with economic interests, favors the spread of urban development. Both development and the accompanying introduction of non-indigenous species will, as always, exact an irreversible toll on the locally adapted, native biota.[39]

Although dams can provide considerable economic and social benefits, their placement and construction must be grounded in sufficient knowledge of the river and its catchment basin to account for long-term ecological consequences. Dams are highly individualistic, and similar physical circumstances may elicit dramatically different responses. The effects of a dam in time and space can be considerable and may become apparent only after a long time. For purposes of illustration, I shall discuss one dam with which I have some personal experience, the Aswan High Dam in Egypt, with additional examples from other places.[40]

While I was working as a vertebrate zoologist with a scientific expedition in Egypt in 1963 and 1964, a representative of the Egyptian Ministry of Agriculture spent time with us as we worked just north of the Sudanese border along the Nile. One day, three of us from the expedition tried to help this man understand that building the Aswan High Dam across the Nile River was an ecological mistake. He could not, however, see beyond the generation of electricity and irrigation, which was the official argument of the government for constructing the dam, behind which would be Lake Nasser—named after Gamal Abdel Nasser, then president of Egypt.

We explained that building the dam would increase the geographical distribution of the snails that carry the tiny blood fluke that causes the debilitating disease schistosomiasis from below the existing Aswan Dam (built by the British in the early 1930s at the town of Aswan) south to at least Khartoum in the Sudan, several hundred miles above the new, yet to be completed Aswan High Dam. At that time, it was still safe to swim above the existing Aswan Dam, where the water was too swift and too cold for the snails to live, but it was not safe to swim, or even catch frogs, in the water below the dam, where the snails already lived.

We told him that the Nile above the high dam would fill with silt, which would starve the Nile Delta of its annual supply of nutrient-rich sediment and affect farming in a deleterious way. We even conveyed to him that the dam could easily become a military target for the Israelis, as German dams were targets for the British during World War II. However, all our arguments were to no avail.

The engineers building the Aswan High Dam had intended only to store additional water and to produce electricity, which they did. Nevertheless, deprived of the nutrient-rich silt of the Nile's annual floodwaters, the population of sardines off the coast of the Nile Delta in the Mediterranean diminished by 97 percent within two years.[41] In addition, the rich delta, which had been growing in size for thousands of years, is now being rapidly eroded by the Mediterranean because the Nile is no longer depositing silt at its mouth.[42]

A similar loss of fish and fisheries is currently beginning to happen in the East China Sea, as a result of the Three Gorges Dam across the Yangtze River in China. Beyond the effect the Three Gorges Dam is having in the East China Sea, however, it's wreaking a devastating influence on the Yangtze River, the most species-rich river in the Palearctic region. (The Palearctic is the biogeographic region of the Arctic, including the immediately adjacent, temperate regions of Europe, Asia, and Africa.) The Yangtze River hosts 162 species of fish in the main channel of the upper river, of which 44 are endemic and thus under serious threat because of the dam. Of these species, 24 may survive in tributaries, 14 have an uncertain future, and 6 have a high probability of becoming extinct once the dam is filled.[43]

Dams in the United States also affect instream species, such as the freshwater mussels in the Little River of southeastern Oklahoma, where dams deplete their abundance. There is a gradient of mussel extinction downstream from impoundments in this river such that a gradual, linear increase in mussel-species richness and abundance is associated with increasing distance from the main reservoir. The only sites that contain relatively rare species are those farthest from the dam. These same trends are apparent, although much weaker, below the confluence with the inflow from a second reservoir. Nevertheless, the overall abundance of mussels is greatly reduced in relation to successive dams, a finding that suggests that considerable lengths of stream are necessary to overcome the negative effects of impoundments on populations of some freshwater mussels.[44]

The story is somewhat different with the Colorado River clam in the Colorado River of the United States and Mexico. Although a large river, the Colorado is so extensively dammed and diverted that only a fraction of its previous flow still reaches its estuary. The Colorado River clam was once the most abundant species to inhabit the delta of the Colorado River, but now just a small population survives near the river's mouth. In addition, the relative abundance of its empty shells decreases with increasing distance from the mouth of the river, a finding that indicates that this clam is dependent on the river's fresh water. The clam's decline in abundance is probably due to the post-1930 decrease in the flow of the river into its estuary.[45]

To return to the Aswan High Dam, until it was built, the annual sediment-laden waters of the Nile added a little less than a sixteenth of an inch of nutrient-rich silt to the farms along the river each year. Now that the new dam has stopped the floods, the silt not only is collecting upriver behind the dam, thus diminishing its water-holding capacity, but also is no longer being deposited on the riverside farms, thus decreasing their fertility. In addition, because irrigation without flooding causes the soil to become saline, the Nile Valley, which has been farmed continuously for five thousand years, may have to be abandoned within a few centuries. Also, schistosomiasis has indeed spread southward to the Sudan. What right, I wonder, does one nation have to knowingly cause the spread of a highly infectious disease into another nation in the name of economic self-interest, or for any other reason, without the receiving nation's permission? What's more, who gave the Egyptian government the right to wipe out a culture?

The Nubians, whom I got to know, were a community of beautiful people living many miles south of Aswan on small farms sandwiched between the east bank of the Nile and the Eastern Desert of restless sand and outcroppings of ironstone. Their village was neat and clean, and dinner-sized plates, decorated with designs around their borders, were embedded in the outer mud coating of the doorways. The people had a wonderful sense of humor, were quick to laugh, and seemed genuinely pleased that I delighted in playing with their children, and vice versa. For their part, the children had a good sense of self and of each day as an adventure to be lived to the fullest.

But the Aswan High Dam changed all that. The Nubians were moved inland from the bank of the Nile, whose quiet flowing waters and silent guardian desert had been a part of their lives for centuries. In place of their freely spaced, cool, airy, self-designed, and self-constructed mud-brick homes, they were put into government-built, look-alike, minimum-quality housing. Gone was the peaceful silence of the desert, along with its clean air. Gone were the songs of the birds in the shrubs along the Nile's banks. Vanished was their experience of the still, black nights, ablaze with crisply visible stars, including the magnificent Southern Cross. Their freedom of choice stolen, many Nubians could not adjust to the loss of their culture, their gentle way life, and simply died.

How, I wonder, does one justify telling a whole people, a whole culture, that they have no value? How does one tell children that all their tomorrows are sealed in the concrete of a dam?

There is another consequence of the Aswan High Dam, one I would never have thought of, even though I had studied the mammals along the Nile. The Nile annually flooded the many nooks, crannies, and caves along its edge, killing the rats whose fleas carry bubonic plague. Because the floods no longer occur, the rat population has soared, and bubonic plague once again poses a potential threat.

I learned about this unexpected consequence from Wulf Killmann of the Deutsche Gesellschaft für Technische Zusammerarbeit, whom I met in Malaysia. As we talked about the effects of dams on rivers and oceans, I told him about my experience in Egypt. Killmann then told me that he had been part of a project to figure out how to control the ever-growing population of rats, which had become a serious health problem.

But that is not all; the saga of the Aswan High Dam continues, according to R. G. Johnson of the Department of Geology and Geophysics at the University of Minnesota. "If the Mediterranean Sea continues to increase in salinity," says Johnson, "shifting climatic patterns throughout the world may cause high-latitude areas in Canada to glaciate within the next century."

When the Aswan High Dam cut off most of the annual flow of the Nile to use it for irrigation, it greatly reduced the amount of fresh water that entered the Mediterranean. In addition, evaporation from the surface of the Mediterranean is increasing because of global warming. Consequently, a larger amount of fresh water is being lost to human activities and evaporation than is being replaced by rainfall and the inflowing of freshwater rivers. Hence, the Mediterranean is becoming increasingly saline, and that salinity is being modified at the Strait of Gibraltar, where the waters of the Atlantic and the Mediterranean mix. Barring a significant change in the regional circulation of the atmosphere, Johnson contends that the two human-caused losses of fresh water from the Mediterranean (the Aswan High Dam and global warming) will cause the salinity to increase for some time.

The higher salinity in the Mediterranean will lead to more of the Mediterranean flowing into the Atlantic through the Strait of Gibraltar, which will modify the high-latitude oceanic-atmospheric circulation and, in effect, initiate new glaciation. This hypothesis, says Johnson, arises from his 1997 study of the climatic conditions and inferred changes in the oceanic-atmospheric circulation that probably triggered the last glaciation.[46]

The hypothesis, which is presented here in a simplistic form, works something like this: Leaving Gibraltar, the more saline, and thus heavier, water of the Mediterranean sinks and mixes with the very cold, deep water of the Atlantic, moving northward until it enters the northern gyre, a great circular vortex. As the fast-flowing water of the Mediterranean approaches the shallow banks north and west of Ireland, it comes to the surface by upwelling. The upwelling apparently acts like a fluidic switch that deflects the relatively warmer surface water of the Atlantic past Greenland into the colder Labrador Sea off the eastern coast of Canada, which in turn becomes warmer; the warmer water, in connection with cloudy, cooler summers, causes increased precipitation around Baffin Island and other regions in northern Canada, which in turn causes sheets of ice to grow while cooling the Nordic seas and northern Europe.

Johnson says that "today's climate may be close to the threshold for new glaciation" because the large plateau areas of Baffin Island are already covered with semi-permanent snowfields that expanded during the historic Little Ice Age, which lasted from about the mid-sixteenth century to the mid-nineteenth century; during this time cool summers and extremely severe winters were frequent in northern Europe. Initiation of new growth in the ice sheet is of grave concern, says Johnson, because of the strong positive feedback from the enhanced electromagnetic radiation reflected by the white surface of a growing ice sheet, termed the *albedo* effect. In addition, increasingly dense cloud cover could ostensibly lock in the beginnings of an ice age despite global warming. The ultimate consequence, warns Johnson, might be a combination of two extremes in which strong global warming in the lower latitudes would nourish the rapid expansion of ice sheets in Canada and Eurasia.[47]

Although there are many unknown variables, Johnson says that if his conceptual model is "approximately correct, a new ice age can be avoided if a partial dam is constructed on the sill across the strait 25 miles west of Gibraltar." The idea, which again is presented in simplistic form, is to limit the outflow of the Mediterranean to something like 20 percent of today's rate of flow, which would remove the faster flowing water of the Mediterranean from traveling northward to the shallow banks off Ireland and thus diminish the upwellings. With the upwelling diminished, the warm surface water, now diverted into the Labrador Sea, would once again enter the Nordic seas. Canada would remain dry, and Europe's climate would remain mild and relatively stable.[48]

I find in this scenario an interesting problem, one I see arising again and again. We humans introduce something, such as the Aswan High Dam, into the environment, where it provides some benefits to humanity (electricity and additional irrigation) while simultaneously causing untold, unknown, even unimaginable problems. But when the problems begin to manifest themselves, rather than removing the cause—the dam—we propose to remedy the problems by introducing more of the same—another dam—into the environment. I have never seen a second dam fix the ecological problems created by the first dam.

Yet what would happen if the Aswan High Dam were removed—negating the perceived need for a new dam across the Strait of Gibraltar? Although the floods would once again begin fulfilling their many ecological roles, there would be an immediate problem of how to deal with all the silt trapped behind the dam. Then there is the question of what to do with all the steel, concrete, and other materials of which the dam is built. What would happen to all the economic investments and technological developments that have over the years sprung into existence because of the dam? How would the Egyptian people replace the social benefits engendered by the dam? What physical developments, constructed since the dam was built, would renewed, seasonal flooding destroy?

Even if all conceivable questions could be answered and most of the effects could be to some extent reversed, there is at least one effect of the dam that is final. The Nubian culture, in which I found such beauty and joy, would still be extinct. Therefore the question is, How reversible in reality are the effects of the Aswan High Dam? Here, an observation by the Russian-born, Nobel-Prize-winning chemist and physicist Ilya Prigogine is opportune. Prigogine points out that all large processes in the real world, particularly all chemical processes, are irreversible. Reversibility of processes, Prigogine contends, always corresponds to idealization. And he is correct. Because change is a constant process, *nothing* is truly reversible.

Our Changing Climate

Although I have no proof that today's climate is being affected by the Aswan High Dam, I do know that irrefutable evidence points to a change in global climate. Glaciers are melting worldwide, and birds are migrating northward. In North America, both insectivorous and granivorous birds are expanding their geographical distributions northward, which means the northern limit of birds with a southern distribution shows a significant shift of 1.5 miles per year.[49] These data are corroborated by a study of changes in the abundance of Central European birds.

Although it is known that changes in the use of land and in climate have an impact on ecological communities, it is unclear which of these factors is currently the most important. To answer this question, changes in the abundance of birds with different breeding habitats, latitudinal distributions, and migratory behaviors were examined by using data from the semi-quantitative *Breeding Bird Atlas of Lake Constance*; Lake Constance borders Germany, Switzerland, and Austria.

Changes in the regional abundance of the 159 coexisting species of birds from 1980–1981 to 2000–2002 were influenced by all three variables. The numbers of farmland birds, species with northerly ranges, and long-distance migrants declined, whereas birds that inhabit wetlands and species with southerly ranges increased in abundance. A separate analysis of the two decades—1980/1981–1990/1992 and 1990/1992–2000/2002—definitely indicated that climate change had an increasingly significant effect over time.

Although latitudinal distribution was not significant in the first decade, it became the most significant indicator of changes in abundance during the second. Moreover, this is the first study to suggest that our changing climate is now having a greater influence in determining population trends of birds in Central Europe than is the continuing alteration of the modern landscape.[50]

6

Repairing Ecosystems

Only when the last tree has died, and the last river has been poisoned, and the last fish been caught will we realize we cannot eat money. –Cree proverb

Ecological restoration is the thought and the attempt to put something into a prior position, place, or condition. That much is clear enough. But why should we humans bother trying to put something back the way we perceive it to have been? Why try to go backward in time when society's push is forward, always forward? The answer draws on two paradoxes: backward is sometimes forward, and slower is sometimes faster.

In our drive to maximize the harvest of nature's bounty, we—especially in the United States—typically strive for an ever-increasing yield of products, and we intensively alter more and more acres worldwide to that end. What we need, however, is a sustainable yield, which we cannot have until we first have a sustainable ecosystem, such as a forest or an ocean, to produce the yield.

In practice, we tend to think it a tragic economic waste if nature's products, such as wood fiber or forage for livestock, are not somehow used by humans but are allowed instead to recycle in the ecosystem, compost as it were. And because of our paranoia over lost profits (defined as economic waste), we extract far more from every ecosystem than we replace.

We will, for example, invest capital in a crop, but not in maintaining the health of the ecosystem that produces the crop. This type of investing is part of our Western, industrialized tradition and thus is engrained in our culture. As a result, much of the litho-hydrosphere, the biosphere, and the atmosphere are being degraded through overexploitation because people insist the stock market should be like a bullet train going full speed on an endless, straight track of limitless, natural resources that are free for the taking—a biophysical impossibility.

Resource Overexploitation

According to a song popular some years ago, freedom is equated with having lost everything and thus having nothing left to lose. In a peculiar way this sentiment speaks of an apparent human truth. When we are unconscious of a material value, we are free of its psychological grip. However, the instant we perceive a material value and anticipate possible material gain, we also perceive the psychological pain of potential loss. This sense of potential loss affects governments at all levels, not just individuals, as can be seen in Brazil.

Economic projects in Brazilian Amazonia share many common characteristics that lead to severe impacts on the region's natural ecosystems. Top-down political decisions ensure that desired economic projects are inevitable before any environmental studies can be made—or even despite negative tradeoffs that are already known. In addition, government officials frequently renege on commitments to protect natural habitats and tribal areas. Thus, environmental measures are often merely symbolic and serve to tranquilize public concern during a project's period of vulnerability prior to its becoming a politically irreversible fait accompli.[1]

The larger and more immediate the prospects for material gain, the greater the political power used to ensure and expedite exploitation because not to exploit is perceived as losing an opportunity to someone else. And it's this notion of loss that people fight so hard to avoid. In this sense, it is more appropriate to think of resources managing people rather than of people managing resources. Here, logging in tropical forests is illustrative.

Despite abundant evidence that both the damage to forests and the financial costs of logging can be reduced substantially by training workers in sustainable techniques, such as preplanning skid trails and directional felling, destructive logging practices are still common in the tropics. The principal reason poor logging practices persist in the face of potential cost savings is that lower-impact logging restricts access to steep slopes and prohibits ground-based yarding of timber on wet ground. Both types of restrictions are probably seen as synonymous with reduced income because of the decreased number of trees that can be cut. Under such conditions, loggers likely opt for their old logging methods out of economic self-interest.[2]

Moreover, subsidies to large farmers in Latin America tend to be associated with low productivity of the land coupled with excessive deforestation. Government officials and those seeking such positions are prone to accept contributions or bribes and offer subsidies to farmers in exchange for their political support. However, the tradeoff is that farmers then adopt inefficient modes of production as a mechanism for capturing proffered subsidies. Land use in Latin America indicates that subsidy schemes have been counterproductive—they distort and constrain development and trigger excessive depletion of natural resources in at least nine countries.[3]

Historically, then, any newly identified resource is inevitably overexploited, often to the point of collapse or extinction. Such was the case in the sixteenth century with the pearl-oyster beds off the coast of Cubagua, Venezuela, where the pearl oysters were replaced by the turkey-wing mussel. The oyster's depletion was the result not only of overexploitation in a short period of time but also of the ecological stress the exploitation generated. Consequently, the turkey-wing mussel out-competed the pearl oyster and thus prevented its recovery.[4]

A similar situation has occurred in near-shore benthic assemblages of mollusks in southeastern Tasmania over the past 120 years. Based on data collected from sediment cores at thirteen sites in water ranging between twenty-six and sixty feet deep, the average number of species and individuals declined with every two-inch slice of a core from 150 individuals of twenty-one species in 1890 to just 30 individuals of seven species in 1990. The decline corresponded with the rise of dredge fishing for scallops, which ultimately forced the fishery to close.

Here, the major concern is that although the loss of both species and numbers had not been previously recognized, they extended throughout the entire sixty-two-mile area of coast in the study. Given that various types of shellfish harvesting and other anthropogenic impacts are virtually ubiquitous for the coastal zone but are not monitored for their effects, major losses in the current biodiversity of mollusks may be globally widespread—yet unnoticed and thus ongoing.[5]

Such overexploitation is based on the perceived entitlement of the exploiters to get their share before others do and to protect their economic investment. Moreover, the concept of a healthy capitalistic system is of one that is ever-growing, ever-expanding, but such a system, and the capitalistic ventures that fuel it, is no more sustainable biologically than is the turtle trade in China. Indeed, the exploitation of turtles and tortoises for today's market in Asia contributes to a crisis in extinction of global proportions. Thus, it serves as a contemporary example of an unsustainable capitalistic venture based on a biologically renewable resource.

Although mainland Southeast Asia has long been regarded as a mecca of diversity for turtles and tortoises, little is known about them in Laos, Cambodia, and Vietnam (formerly known as French Indochina) because biological investigations were limited prior to World War II. Since then, decades of civil unrest, political instability, and military conflict have largely prevented fieldwork. Nevertheless, turtles and tortoises face continuing exploitation for food and medicinal markets in Laos, Cambodia, and Vietnam, where hunters in rural villages capture them for local consumption or to sell to traders who periodically visit villages to purchase wildlife. Although turtles are eaten locally and traded in Laos, Cambodia, and Vietnam, most are exported through Vietnam to markets in southern China to appease the people's insatiable demand for the turtles' meat in soup and their shells for use in traditional Chinese medicine.

In China, there are over one thousand large, commercial turtle farms, collectively worth over a billion U.S. dollars.[6] The scale of these lucrative operations, especially those pertaining to endangered species, pose a major threat to the survival of China's diverse turtle fauna.[7] This threat stems primarily from the fact that turtle farmers are the primary purchasers of wild-caught turtles; they buy them to increase their overall stock of adult animals and to secure wild breeders. Wild breeders are important because successive generations of farm-raised turtles exhibit a marked decrease in reproductive capability. The reliance on individuals captured in the wild demonstrates that turtle farming is not a sustainable practice.

As populations of wild turtles decline, it will become increasingly difficult to supplement farm stock from the wild.

Even with the inevitable crash in the farming of native turtles, the depleted wild populations will still face overexploitation because there is an entrenched cultural demand for wild-caught meat. The nutritional properties of wild animals are promulgated by the practitioners of traditional Chinese medicine and are thus deeply ingrained in the national psyche. Consequently, wild-caught turtles fetch significantly higher prices than those raised on farms, and no amount of captive breeding will decrease the insistence on obtaining wild turtles for consumption.

China is developing rapidly, and the escalation of turtle farming has followed the path of other capitalist ventures since the economic reforms of the 1980s. The fusion of China's growth with the utilitarian attitude of the Chinese toward nature clearly emphasizes the aforementioned fear of losing short-term profits, even at the cost of rendering long-term biodiversity unsustainable—as the history of economics has so often demonstrated. In the case of Chinese turtles, the farmers are grabbing the last vestiges of wild populations to process for the soup pot.

Even if the Chinese government could alter the unsustainable practices of turtle farming, it's unlikely that black-market turtle farms could be effectively regulated because established, well-heeled farmers are continuing to purchase turtles secured in the wild whenever possible. In effect, they are opting to earn profits as long as they can, regardless of the ecological outcome. In the long term, therefore, turtle farms serve a single function—to generate profit for a few entrepreneurs, despite the social-environmental consequences.[8]

As with any renewable natural resource, the non-sustainable exploitation of turtles has a built-in ratchet effect, which works in this way. During periods of relative economic stability, the rate of harvest of a given renewable resource (wild turtles) tends to stabilize at a level that economic theory predicts can be sustained through some scale of time. Such levels, however, are almost always excessive because economists take existing unknown and unpredictable ecological variables and convert them, in theory at least, into known and predictable economic constant values in order to calculate the expected return on a given investment from a sustained harvest.

During good years in the market or in the availability of the resource or both, additional capital investments are encouraged in harvesting and processing because competitive economic growth is the root of capitalism and the enhancement of personal profits. But when conditions return to normal or even below normal, the individual or industry, having over-invested, typically appeals to the government for help because substantial economic capital is at stake—including potential earnings. If the government responds positively, it encourages continual overexploitation.[9]

The ratchet effect is thus caused by unrestrained economic investment to increase short-term yields in good times and strong opposition to losing those yields in bad times. This opposition to losing yields means there is great resistance

to using a resource in a biologically sustainable manner because there is no predictability in yields and no guarantee of yield increases in the foreseeable future.
In addition, our linear economic models of ever-increasing yield are built on the
assumption that we can in fact have an economically sustained yield. This contrived
concept fails, however, in the face of the biological sustainability of the yield, a concept that, on a global scale, was missed in the Kyoto Protocol of December 1997.

The Kyoto Protocol caps neither the emissions of greenhouse gas at a level
that will achieve climate stability nor economic growth based on thresholds of
biophysical sustainability.[10] David Ehrenfeld, a professor of biology at Rutgers
University, puts it this way:

> Criticisms of globalization have been largely based on its socioeconomic
> effects, but the environmental impacts of globalization are equally
> important. These include acceleration of climate change; drawdown of
> global stocks of cheap energy; substantial increases in air, water, and soil
> pollution; decreases in biodiversity, including a massive loss of crop and
> livestock varieties; depletion of ocean fisheries; and a significant increase
> in invasions of exotic species, including plant, animal, and human
> pathogens. Because of negative feedback from these changes, the future
> of globalization itself is bleak. The environmental and social problems
> inherent in globalization are completely interrelated—any attempt to
> treat them as separate entities is unlikely to succeed in easing the transi-
> tion to a post-globalized world.[11]

Then, because there is no mechanism in our linear economic models of ever-
increasing yields that allows for the uncertainties of ecological cycles and variability or for the inevitable decreases in yields during bad times, the long-term
outcome is a heavily subsidized industry. Such an industry continually overhar-
vests the resource on an artificially created, sustained-yield basis that is not
biologically sustainable. When the notion of sustainability does arise, the overexploiting parties marshal all scientific data favorable to their respective sides as
"good" science and discount all unfavorable data as "bad" science, thereby politi-
cizing the science and largely obfuscating its service to society.

Because the availability of choices dictates the amount of control we feel we
have over our sense of security, a potential loss of money is the breeding ground
for environmental injustice. This is the kind of environmental injustice in which
the present decision-making generation steals from future generations by overexploiting a resource rather than facing the uncertainty of giving up some potential
income. Still, there are six important lessons to be learned from the historical
overexploitation of natural resources: (1) emphasize quality rather than quantity,
(2) recognize that loss of sustainability occurs over time, (3) recognize that
resource issues are complex and process driven, (4) accept the uncertainty of
change, (5) stop perceiving loss as a threat to survival, and (6) favor biophysical
effectiveness over economic efficiency.

Lesson One: Emphasize Quality Rather Than Quantity

Maximizing the quality of whatever we do with the Earth's finite resources will always conserve them, thereby spreading nature's wealth among more people and generations. Conversely, maximizing the quantity of any material withdrawn from the Earth's finite supply to feed the insatiable appetite of today's consumer economy can only squander nature's limited wealth. This said, we must choose because we cannot maximize both quality and quantity simultaneously, as exemplified by bottom-fishing in the ocean.

Bottom-trawling and bottom-dredging are two of the most disruptive and widespread human-induced physical disturbances to seabed communities worldwide. They are especially problematic in areas where the interval between events of dredging or trawling is shorter than the time it takes for the ecosystem to recover. Extensive areas can be trawled from 100 percent to 700 percent per year or more, and such a large amount of trawling affects the cycling of nutrients.

The frequency and extent to which nitrogen and silica in the bottom sediment are resuspended in the water column by trawling and dredging has important implications for regional nutrient budgets. Trawling may also produce changes in the successional organization of soft-sediment infaunal communities. (An *infaunal community* is composed of aquatic animals that live in the substrate of a body of water, especially in the soft bottom of an ocean.) This type of bottom-fishing can decrease habitat complexity and biodiversity, as well as enhance the abundance of opportunistic species and certain prey important in the diet of some commercially important fishes.

Bottom-trawling and the use of other mobile fishing gear on the seabed are, in a manner of speaking, similar to clear-cutting a forest, which is recognized as a major threat to biological diversity and economic sustainability. Structures in benthic communities, while generally much smaller than those in forests, are just as critical to structural complexity and thus to biodiversity. Nevertheless, mobile fishing gear can have large and long-lasting effects on benthic communities, including the young stages of commercially important fishes, although some species benefit when structural complexity is reduced.

Use of mobile fishing gear crushes, buries, and exposes marine animals and structures on and in the substratum, thereby sharply reducing structural diversity. Its severity is roughly comparable to other disturbances that alter biogeochemical cycles. Recovery is often slow because recruitment is patchy and maturation can take years, decades, or even longer for some structure-forming species, such as corals.

Recent technological advances (such as rockhopper gear, global positioning systems, and fish finders) have all but eliminated natural havens safe from trawling. The frequency of yearly trawling on the continental shelf is orders of magnitude higher than the frequency of other severe seabed disturbances. In fact, trawling covers an area equivalent to perhaps half the world's continental shelf each year, or 150 times the land area that is clear-cut on an annual basis.

In addition, fishing gear, which is used over large regions of continental shelves worldwide, can reduce habitat complexity by smoothing the microtopography of the bottom, removing pebble-cobble substrate with emergent epifauna, and eliminating species that produce structures, such as burrows. (*Epifauna* are animals that live on the surface of sediments or soils.) The effects of mobile-fishing gear on biodiversity are most severe in areas least affected by natural disturbance, particularly on the outer continental shelf and slope, where damage from storm waves is negligible and biological processes, including growth, tend to be slow.[12]

Lesson Two: Recognize That Loss of Sustainability Occurs over Time

A biologically sustainable use of any resource has never been achieved without first overexploiting it, despite the lengthy catalog of disastrous historical examples (warnings, if you will) and the vast amount of contemporary data. If history is correct, resource problems are not environmental problems but rather human ones that we have created many times, in many places, under a wide variety of social, political, and economic systems, as exemplified by the whaling industry.

The whaling industry is illustrative because it exists in many nations, some of which persist in killing these huge creatures despite their precarious hold on existence, as initially acknowledged in 1946 by the fifteen nations that signed the "International Convention for the Regulation of [Commercial] Whaling." Nevertheless, the politics of contemporary whaling are increasingly contentious and the effects far-reaching. A case in point: the Galápagos population of sperm whales illustrates the substantial negative impacts severe exploitation can have on an animal population well outside the range of its pursuit and for at least a decade after hunting has ended.

Although it was generally expected that whale populations would rebuild following the end of whaling, such is not the case with the sperm whales that visit the waters off the Galápagos Islands. In fact, the population is dwindling for two reasons: the whales' migration into productive but depopulated waters off the Central and South American mainland and the virtual elimination of large breeding males (in their late twenties and older) from the region because of their being hunted for years. Although other factors may be involved, both the high rate of emigration and the low rate of recruitment are probably related to heavy whaling in Peruvian waters, which ended in 1981.[13]

Lesson Three: Recognize That Resource Issues Are Complex and Process-Driven

The fundamental issues involving resources, the environment, and people are complex and process driven. The integrated knowledge of multiple disciplines is required to understand them. These underlying complexities of the biophysical systems preclude a simplistic approach to ecosystem manipulation. In addition, the wide, natural variability and the compounding, cumulative influence of continual human activity mask the results of overexploitation until they are severe and largely irreparable within a human lifetime or ever.

For example, the overall species richness of birds in Bogor Botanical Gardens, West Java, Indonesia, isolated since 1936, declined by 59 percent (from ninety-seven species to forty) by 2004,; of these the forest-dependent birds had declined by 60 percent (from thirty species to twelve). Large-bodied birds were particularly prone to extinction prior to 1987, whereas the seven species of forest-dependent birds that attempted to become established after 1987 perished. These dynamics are a result of this woodlot's reduction in area and subsequent isolation, as well as the continuity of intense human use and perverse management: removal of the understory layer within the woodlot simplified its basic structure and thus reduced its ability to function as a quality habitat.[14]

Our management of the world's resources is always to maximize the output of material products—to put conversion potential into operation. In so doing, we not only deplete the resource base and degrade habitat but also produce unmanageable by-products, often in the form of hazardous wastes. In unforeseen ways, these by-products (which are really unintended products) are altering the way our biosphere functions, usually in a negative way. Such by-products include hazardous chemicals in our drinking water and several veterinary drugs with which farmers inoculate their livestock but which could kill scavengers—those species that clean our environment.[15] Although habitat contamination by humans may seem distant from the ongoing dynamics in the Bogor Botanical Gardens, the economic incentives that drive exploitation destroy the quality of habitat wherever it is.

Lesson Four: Accept the Uncertainty of Change

As long as the uncertainty of continual change is considered a condition to be avoided, nothing will be resolved. However, once the uncertainty of change is accepted as an inevitable, open-ended, creative process, most decision-making is simply common sense. Consider that common sense dictates that one would favor actions having the greatest potential for reversibility, as opposed to those with little or none. Such reversibility can be ascertained by monitoring results and can be instituted by modifying actions and policies accordingly.

Lesson Five: Stop Perceiving Loss as a Threat to Survival

We interpret the perceived loss of choice over our personal destinies as a threat to our survival. This sense of material loss usually translates into a life-long fear of loss, which fans the flames of overexploitation through unbridled competition in the money chase and top-down, command-and-control management of natural resources.

As the human population grows, with a corresponding decline in the availability of natural resources, the pressure grows to increase top-down, command-and-control management of those resources. The fallacy of attempting to control ecosystems through management is that we humans are not in control to begin with—and never will be. We are, therefore, destined to fail whenever we attempt to enclose nature in a designer straightjacket: witness tornados, hurricanes, and fires.

Nevertheless, our socioeconomic institutions are inclined to respond to nature's erratic or surprising behavior by exerting more direct control. Command and control, however, usually results in unforeseen consequences, both for ecosystems and for human welfare, in the form of collapsing resources, social and economic strife, and the continuing loss of biological diversity—along with the ecosystem services such diversity provides. Moreover, if the potential variability of an ecosystem's behavior is reduced through command-and-control management, the system becomes less resilient than it was to perturbations, and the outcome is an unwanted biophysical disaster.

The ultimate pathology of fear and arrogance emerges when resource-management agencies lose sight of their original purposes and become obsessed with their initial command-and-control success. To protect their accomplishment from unwanted scrutiny, they eliminate research and monitoring and focus on the efficiency of control rather than the effectiveness with which they discharge their original mission. And in the process they become increasingly isolated and inflexible. Simultaneously, through overcapitalization, society becomes dependent on the command-and-control ideology of management, while demanding ever-greater certainty in supply, even as it ignores the visible warning signs of developing ecological change or collapse.[16]

Thus, people with the command-and-control ideology tend to think that any resource not converted into some sort of immediate profit is an economic waste, and therefore they view salvage logging and preemptive thinning as the only viable alternative to a biophysical disturbance, such as a hurricane or a beetle infestation. Nevertheless, forest managers initiate substantial changes in ecosystem structure and function when they initiate salvage logging in areas affected by windstorms, fires, or other ecological impacts. Similarly, harvesting potential host trees in advance of insect infestations or disease or preemptively thinning or cutting forests in an attempt to improve their resilience to potential stress and future disturbances may fill coffers in the short term, but they come at the long-term ecological expense of lost biological capital in the soil bank.

Despite dramatic physical changes in forest structure that result from hurricanes and insect infestation, little disruption of biogeochemical processes or other ecosystem functions typically follows such disturbances. The natural restructuring of a forest after such disturbances provides a long-term, biophysically rich context of habitat diversity and landscape heterogeneity, which is often lacking because of centuries of cultural land use.

Clearly, there are valid, short-term reasons for salvage or preemptive logging, such as a financial desire to capture an immediate opportunity for unexpected profits before the possibility deteriorates, to enhance human safety, or to shape the characteristics of a forest's composition and structure to fulfill the product mode of an economic tree farm. However, if one were to catalog the long-term ecological benefits derived over space and time from leaving a forest alone when it is affected or threatened by nature's disturbances, it would quickly become apparent that allowing nature to take its course by doing nothing, which is a viable alternative, results in overall benefits that greatly outweigh those that accrue from any active management strategy.[17] Implementing such strategies allows us to think we are somehow in control of events, which clearly we are not.

Put simply, interactive systems perpetually organize themselves, with infinite novelty, to a critical state in which a minor event can start a chain reaction that leads to destabilization and collapse, such as that of a forest following a fire. Following the disruption, the system will begin reorganizing toward the next critical state (e.g., forest succession), and so on, and so on indefinitely.

Another way of understanding this phenomenon is to ask, If change is a universal constant and nothing is either static or reversible, what is a natural state? In considering this question, one soon begins to realize that the conceptual balance of nature, in the classical sense, is untenable—disturb nature and nature will return to its former state when the disturbance is removed.

Although one may perceive the pattern of vegetation on the Earth's surface to be relatively stable, particularly over the short interval of one's lifetime, in reality the landscape and its vegetation are in a perpetual state of dynamic imbalance with the forces that sculpted them. When these forces create novel events that are sufficiently rapid and large in scale, they are perceived as disturbances.

Perhaps the most outstanding evidence that an ecosystem is subject to constant change and ultimate disruption, rather than existing in static balance, comes from studies of naturally occurring external factors that dislocate ecosystems. For a long time, ecologists failed to consider influences outside ecosystems. Their emphasis was on processes internal to an ecosystem even though what was occurring inside was driven by what was happening outside.

Climate appears to be foremost among these factors. By studying the record laid down in the sediments of oceans and lakes, scientists know that climate has fluctuated wildly over the last two million years, and the shape of ecosystems with it—witness what is going on today around the world. The fluctuations take place not only from eon to eon but also from year to year and season to season and at every scale in between; thus, the configuration of ecosystems is continually creating different landscapes in a particular area through geological time.

Lesson Six: Favor Biophysical Effectiveness over Economic Efficiency

As an economy grows, natural capital, such as air, soil, water, timber, and marine fisheries, is reallocated to human use via the marketplace, where economic efficiency rules. The conflict between economic growth and the conservation and maintenance of natural-resource systems is a clash between the economic ideals of efficiency and the realities of biophysical effectiveness. This economically driven divergence creates a conundrum because traditional forms of active conservation require money, which, in the United States, is highly correlated with income and wealth. That notwithstanding, the conservation and maintenance of biodiversity in all its forms will ultimately require the cessation of economic growth.

Thereafter, the ultimate challenge will be to maintain biodiversity, especially in the wake of globalization, because the number of threatened species is related to per capita Gross National Product (GNP) in five taxonomic groups in over one hundred countries. Birds are the only taxonomic group in which numbers of threatened species decreased throughout industrialized countries as prosperity increased. Plants, invertebrates, amphibians, and reptiles showed increasing numbers of threatened species with increasing prosperity. If these relationships hold, increasing numbers of species from several taxonomic groups are likely to be threatened with extinction as countries increase in prosperity.[18]

Wealthy, industrialized nations, such as the United States, left such a large ecological footprint in the last four decades of the twentieth century that the damage adds up to more than the poor, non-industrialized nations owe in debt. In fact, the rich nations have caused upward of $2.5 trillion in environmental damage to poor countries. The well-off disproportionately affect the poor through such things as exacerbation of climate change (a negative outcome to which China is now also contributing), depletion of stratospheric ozone, agricultural intensification and expansion, deforestation, overfishing, and destruction of mangroves.[19]

For biologist Ehrenfeld:

> Criticisms of globalization have been largely based on its socioeconomic effects, but the environmental impacts of globalization are equally important. . . . Because of negative feedback . . . , the future of globalization itself is bleak. The environmental and social problems inherent in globalization are completely interrelated—any attempt to treat them as separate entities is unlikely to succeed in easing the transition to a post-globalized world.[20]

At length, therefore, the abuse of nature requires repair, hence the notion of restoration.

Restoration, as We Currently Think of It

Basically, restoration—as it is generally thought of—helps us to understand how a given ecosystem functions. As we strive to put it back together by reconstructing the knowledge of times past, we learn how to sustain the system's ecological processes and its ability to produce the products we valued it for in the first place and might value it for again sometime in the future.

Similarly, restoration helps us understand the limitations of a given ecosystem or a portion thereof. As we slow down and take time to reconstruct what was, we learn how fast we can push the system to produce products on a sustainable basis without impairing its ability to function.

Thus, the very process of restoring the land to health is the process through which we become attuned to nature and, through nature, to ourselves. Restoration, in this sense, is both the means and the end, for as we learn how to restore the land, we heal the ecosystem, and as we heal the ecosystem, we heal the deep geography of ourselves. Simultaneously, we also restore both our options for products and amenities from the land and those of future generations. This act is crucial because our moral obligation as human beings is to maintain the welfare of our children and those beyond. To this end, maintaining healthy, viable ecosystems is an expression of the heart and the spirit of taking care of the Earth as a biological living trust. I use the word *spirit* on purpose because it is derived from the Greek word for breath, which denotes life.

We, as citizens of the Earth, must learn to understand and accept that the sustainability of a forest, a prairie, or any ecological system for that matter, is an ever-elusive prize, which, like a horizon, continually retreats as we advance. The dance of approach and retreat causes me to think of sustainability as the duty of each generation to pass forward to the next as many positive opportunities for safe-keeping as is humanly possible. This notion requires clarity of mind because it means that we, the adults, must finally come to grips with the fact that each generation is obligated to pay its own way—beginning with us, here, now. The cost of our presence on Earth must be accounted for in how we treat the ecosystems that we, like all generations, are obliged to rely on for our survival. By this, I mean all debts incurred by the generation in charge must be paid by that generation—not passed forward as an ecological mortgage to encumber the social-environmental welfare of those who are young or unborn.

To achieve the level of consciousness and the balance of energy necessary to maintain the sustainability of ecosystems, we must focus our questions—social and scientific—toward understanding the biophysical principles inherent in the governance of those systems and our place within that governance. Then, with humility, we must develop the moral courage and political will to direct our personal and collective energy toward living within the constraints defined by those principles—not by our economic/political ambitions. To this end, William Greider, a veteran reporter and columnist for the *Nation* and the *Washington Post*, is of the opinion "that there is nothing inherent to the functional principles of capitalism that requires it to be . . . [exploitive]; that's a value choice made by people who have power within the system."[21]

Rethinking the Concept of Restoration

The biophysical systems we are redesigning by our existence in and our interaction with our surroundings are continually changing the environment—all of it, if in no other way than through the generalized pollution of atmosphere. Consequently, conditions prior to the arrival of Europeans in North America are irrelevant because the compounding environmental influences of today's burgeoning human population and its so-called permanent developments have, in many ways, limited the possibilities of ecosystem restoration. Added to our current environmental dilemma is the fact that indigenous populations were much smaller and often more nomadic than our contemporary mega-populations. Moreover, the ecological systems with which we daily interact are becoming ever further removed from the types of biophysical balances that characterized pre-European conditions.

Our challenge today is to mature sufficiently in personal and social consciousness to recognize a functionally healthy and sustainable ecosystem when we see it and then to maintain it as such. Beyond that, we need to repair functionally degraded ecosystems to the greatest extent we're capable of. Achieving sustainability is a process, a journey toward the ever-increasing consciousness that we humans must acquire in order to learn how to treat our environment for the benefit of all generations. Sustainability is not an absolute—not a materialistic endpoint, but rather a lifetime journey of growing consciousness.

If you wonder why some people appear unwilling to begin this journey, I think British author George Monbiot has put it well: "There are several reasons why we do not act. In most cases, the personal risk involved in the early stages of struggle outweighs the potential material benefit. Those who catalyse revolution are seldom the people who profit from it."[22] Another reason people stick to the status quo is the perceived opportunity to make substantial amounts of money. According to the Intergovernmental Panel on Climate Change, the hottest debate has been in the realm of agriculture, where predictions suggest that crop yields will rise in some areas, at least under certain conditions. "The avowed possibility for substantial monetary gains has caused some political factions and business interests to dismiss all relevant data on the environmental hazards of global warming and tout the benefits of climate change [to agriculture]."[23] However, Jonathan Foley, an environmental scientist at the University of Wisconsin–Madison, warns that "if the whole world begins to look like Iowa cornfields, we'll have to take an even larger share of [the] global biological production into human hand, and that leaves a lot less for other things. And those other things won't be just pretty butterflies and tigers and charismatic animals, they'll be things that matter to us, like the things that clean our water, preserve our soils, clean our atmosphere, and pollinate our crops."[24]

Although sustainability is not a condition in which a biophysical compromise can be struck, the social decisions leading toward sustainability often necessitate conciliation. Seeking sustainability to a degree, an apparently innocuous concession, defeats sustainability altogether. Leave one process out of the equation or in some other way alter a necessary feedback loop, and the system as a whole will be deflected toward an outcome other than the one originally intended, which calls into question what some people think of as the balance of nature.

The Balance of Nature

This so-called balance of nature is a figment of the imagination, something conjured to fit our snapshot image of the world in which we live. In reality, nature exists in a continual state of ever-shifting disequilibrium—irrespective of human influence. Disequilibrium means that an assembly of plant-animal communities is a dynamic process in which the composition of species changes as some move into and colonize a new area, others increase in abundance, while yet other species decrease or go locally extinct. Although the coalescing of species in a community is driven partly by intrinsic interactions, such as competition and predation, extrinsic forces, such as physical disturbances, epidemics of disease, and the colonization of new species, play a part. Extreme climatic events are examples of severe, but infrequent, physical disruptions that can differentially affect certain species, thereby altering the community's composition, interactive feedback loops, and thus the use of available resources.

Here, a sheet flood at Portal, Arizona, is illustrative. Sheet flooding is caused by comparatively shallow water flowing over a wide, relatively flat area, which typically does not have the appearance of a well-defined watercourse. It is especially dangerous because even when standing in an area subject to sheet flooding, one often does not find it obvious that the area could become inundated.

A massive downpour and subsequent sheet flooding dramatically reduced the population of six of the eight species of seed-eating rodents that were present at Portal, Arizona, before the flood. These species included the banner-tailed kangaroo rat, Merriam's kangaroo rat, and Ord's kangaroo rat, all of which suffered dramatic mortalities despite having been the historically dominant members of the community. Conversely, the flood caused no detectable mortality in either Bailey's pocket mouse or the desert pocket mouse. This shift in species composition resulted in the immediate, dramatic, and long-lasting reorganization of the rodent community, a permanent shift of unprecedented magnitude to a new interspecific structure dominated by pocket mice. So both biological and physical perturbations can reset the structure and dynamics of a community on a new, relatively stable trajectory. One kind of biotic influence is the invasion of a new species (either native or exotic), which can be facilitated by a catastrophic physical event, such as the aforementioned flood.

Although the reassembly after the flood did not change the identity of the four most abundant species, it did change their interactions. The long-term increase in the number of Merriam's kangaroo rats in response to the increasing shrubby vegetation (an increase caused by climate change) and the decline of its larger competitor, the banner-tailed kangaroo rat, were reversed after the flood. But the ensuing decline of Merriam's kangaroo rat ultimately allowed the pocket mice to dominate the community.

These dynamics indicate that the flood-caused differential mortality altered the preexisting hierarchy and allowed formerly subordinate species surviving the event to dominate the new hierarchy. The best explanation for these changes is that resident individuals within a species had the advantage of incumbency. In this instance, the loss of incumbency not only altered interspecific competition but also had profound, long-lasting effects on community structure because these desert rodents have established home ranges, stores of seeds, and burrow systems. In addition, they rely on acquired knowledge of their territory for finding and securing food, as well as avoiding predators. Incumbency also facilitates a resident's defense of its territory and helps to ensure that its offspring, or at least an individual of the same species, will inherit the home range when the resident dies.

Although the incumbency advantage acts to stabilize the dynamics of both the population and the community, the sheet flooding had a destabilizing effect in that it caused the wholesale mortality of the dominant rodents. In fact, the flood, which mostly eliminated the incumbents, largely equalized the competitive interactions, thus allowing individuals of previously subordinate species to colonize the area and establish territories. Moreover, they had available the resources the original residents had stored and defended. The ultimate effect of the flood was to facilitate the immigration of previously rare, native species in a manner similar to that of other disturbances, such as road construction, in aiding the colonization of exotic, invasive species. So disrupting the incumbency advantage is one way an extrinsic physical circumstance can interface with intrinsic competitive processes and thereby alter the rules of assembly and engagement in the wholesale reorganization of a community. Similar responses occur in plant communities in response to fire and grazing.[25] Other types of responses to catastrophic events further demonstrate the disequilibrium of natural systems, such as those of cottonwood trees, a caddisfly, and a giant waterbug.

Cottonwood trees, which once grew in profusion along the banks of western streams and rivers in the United States, where they provided shade, woody debris, and nutrients to the aquatic-terrestrial interface, have all but disappeared, to the detriment of the ecosystems they served. Cottonwoods require the bare, scoured banks that result from floods in order for their seeds to germinate and grow, despite the fact that some of the trees die as a consequence of the flooding. Today, because of flood-controlling dams, cottonwood trees are dying out in many areas.

There is a caddisfly that inhabits a stream system in the mountains of Arizona, where it is subjected to the extremely violent force of flash floods that occasionally scour out the stream channels. The caddisfly, in turn, has evolved through the generations to metamorphose from the immature, aquatic state into the winged, adult phase during a period that is almost perfectly timed to miss the most common season of flooding. This behavior keeps enough of the population out of harms way to perpetuate the species.

Finally, a giant waterbug that lives in some desert streams has adapted over the last 150 million years to "read" the weather and make a mass exodus from a stream that is about to experience a flash flood. During the exodus, the waterbugs literally climb the canyon walls to escape the dangerous waters but return to the stream within a day.[26]

If the balance of nature can now be dismissed as a viable hypothesis, where does that leave us humans in relation to the concept of a natural ecosystem?

We Humans Are a Natural Part of Nature

There has been an increasing emphasis in recent years on "natural" ecosystems, as though only those devoid of visible human influence qualify. This idea has been perpetuated by writers who created the romantic myth that indigenous Americans somehow had the wisdom and self-control to live in perfect harmony with nature, taking only the bare minimum of what they needed to survive and, by inference, voluntarily keeping their own populations in check. It has also been assumed that predators and their prey were in a perfect balance, that nature's ecological disturbance regimes either did not exist or did not have any affect on the great American landscape until the Europeans invaded the continent—hence the idea of a *climax ecosystem*, one that is indefinitely stable.

With respect to our human influence on ecosystems, what sets us apart from our fellow creatures is not some higher sense of spirituality or some nobler sense of purpose but rather that we deem ourselves wise in our own eyes. Therein lies the fallacy. We are no better than or worse than other kinds of animals; we are simply a different kind of animal—one among the many. We are thus an inseparable part of nature, despite religious doctrine.

As a part of nature, what we do is natural even if it is often destructive. This is not to say our actions are wise, ethical, moral, desirable, or even socially acceptable and within the bounds of nature's biophysical laws. It is only to acknowledge that we will, of necessity, change what we call the natural world, and it is natural for us to do so because people are an integral part of the total system we call the universe. However, we may justifiably question the degree to which we change the world, the motives behind our actions, and the ways in which we make these changes. And it is our motive for redesigning our environment—spiritual humility or material arrogance—that is knocking at the door of our consciousness.

Today, many of us are trying somehow to reach back into human history to find our mythological roots and to recapture some primordial sense of spiritual harmony with nature. Our search is urgent because at some deep level we know we are destroying our only life-support system. We have an intuitive feeling that we humans as a whole have lost something we must find—our connection to planet Earth as a living entity. With this background, the question before us is, Why is restoration not possible?

Why Restoration Is Not Possible

This is an important question because, as we work to heal the land, we will advance our sense of consciousness and thereby rediscover our inseparable connection to nature. In the process, we will learn that cumulative circumstances have made it impossible to revert modern landscapes to those of old. This fact does not mean they cannot be healed, but rather that no ecosystem can be restored to some idealized prior condition. Moreover, the historical manipulation of an ecosystem toward a specific end has often resulted in a different long-term outcome. Therefore, wisdom and humility would dictate that the biophysical condition we choose to create is the repair of the functional integrity of an ecosystem, and the reason for doing so is to allow the system to once again produce its ecological services for the benefit of all generations.

Nevertheless, some people insist not only that ecological restoration is achievable but also that it should return New World ecosystems to pre-European times, a proposal that is neither feasible nor possible for four reasons: (1) change is a constant process, (2) Europeans fundamentally altered the New World, (3) no records exist of conditions prior to the arrival of the Europeans, and (4) we cannot go back in time.

Reason One: Change Is a Constant Process

Change is a constant, second-to-second, minute-to-minute process of ever-shifting relationships; such change honors the Buddhist notion of impermanence. This biophysical reality means there is no such thing as an independent variable, a constant value, or the possibility of anything being reversible—ever.

Reason Two: Europeans Fundamentally Altered the New World

We do not know what the conditions were prior to the European invasion, which began with Christopher Columbus and the Spanish in 1492. The first reason is self-evident; we weren't there. Moreover, we have no records.

By 1492, indigenous peoples had modified the extent and composition of the forests and grasslands through the use of fire. In addition, they rearranged microrelief through human-created earthworks. Agricultural fields were common in some areas, as were houses, villages, trails, and roads. Some of the environmental manipulations were so subtle Europeans mistook the altered landscapes for ones untouched by human hands.

Prior to the Spanish invasion of Florida in 1513, the indigenous population of North America, north of Mexico, was about 3.8 million people. The decline of indigenous peoples, once it began, was rapid and precipitous—probably the single greatest demographic disaster in history. With European disease as the primary killer (augmented by European atrocities), populations of indigenous peoples fell by 74 percent, to one million.

By the mid-1500s, the Spanish controlled land from the Carolina coast as far north as La Charrette (the highest settlement on the Missouri River—near today's Marthasville) and westward to at least San Francisco Bay in California, thereby exposing the indigenous peoples across the continent to European diseases. Decimation of the native population through conquest and the spread of European diseases affected the human-influenced landscape accordingly.

In the period between the decimation of the indigenous populations and the migration of significant numbers of Europeans westward, a significant environmental recovery took place, with a commensurate reduction of discernable indigenous cultural features. Some of these changes were already evident in the historical accounts of travelers as early as 1502–1503, when Columbus sailed along the north coast of Panama on his fourth voyage. During this voyage, his son, Ferdinand, described the land as well-peopled, full of houses, with many fields, open areas, and few trees. In contrast, Lionel Wafer, a pirate, found most of the Caribbean coast of Panama covered with forests and unpopulated in 1681. And so it was all over the Americas: forests grew back and filled in, soil erosion abated, agricultural fields became occupied by shrubs and trees, and indigenous earthworks were overgrown.

By 1650, indigenous populations had been reduced by about 90 percent in the hemisphere, whereas the numbers of Europeans were not yet substantial in 1750, when European settlement began to expand. As a result, the fields of indigenous peoples were abandoned, their settlements vanished, forests reestablished themselves, savannas retreated as forests expanded, and the subsequent landscape did indeed appear to be a sparsely populated wilderness.[27]

Prior to the invasion of Europeans, human impact on the environment was not simply a process of increasing change in response to the linear growth of the indigenous populations. Instead, the landscape was given time to rest and recover as people moved about; cultures collapsed; populations declined because of periodic starvation, disease, and war; and habitations were abandoned.

Human activities may be constructive, benign, or destructive, all of which are subjective concepts based on human values, but change is always continual, albeit at various rates and in various directions. All changes are, in addition, cumulative. Even mild, slow change can show dramatic effects over the long term. Although there was, of course, some localized European impact prior to 1750, thereafter (but especially after 1850) populations of European Americans expanded tremendously; they severely exploited the resources, greatly accelerating the modification of the environment in the process.

In particular, the settlers who introduced domestic livestock in the San Juan Mountains of Colorado left a dusty signature. Mineral aerosols in wind-blown dust from the soil's surface are an important influence on climate, as well as on terrestrial and marine biogeochemical cycles. The amount of dust can be affected by human activities, which alter surface sediments. The changes in the flow of regional- and global-scale wind-borne dust following the rapid expansion and settlement of human populations are not well understood, but a study of sediment cores from alpine lakes in the San Juan Mountains from the past five thousand years has yielded some regional-scale answers.

The chemical composition of the dust is not comparable to that of the surrounding bedrock, an indication that it came from hundreds of miles away. Such transport is not surprising because the winds of winter are known to blow dust from the deserts of California and Nevada to Colorado. Carbon and lead dating revealed that the average annual amount of dust deposited in the lakes after the 1800s was 500 percent more than the dust deposited prior to that time. This sudden increase appears around the time of a boom in ranching, when cattle and sheep spread across the landscape, overgrazing the native, erosion-controlling plants. There was a subsequent drop in the level of dust deposited, however, with the passage in 1934 of the Taylor Grazing Act.

Nevertheless, the dust has yet to settle. The past century has seen more than a five-fold increase in the deposition of agricultural chemicals from fertilizers: potassium, magnesium, calcium, nitrogen, and phosphorus. The infusion of these chemicals into the alpine ecosystem will have serious implications with respect to the alkalinity surface waters, aquatic productivity, and terrestrial nutrient cycling.[28]

Reason Three: No Records Exist of Conditions Prior to the European Immigration

Even with today's data, we have no way to discern the environmental conditions that existed prior to the introduction of livestock. There simply are no records. But even if data did exist, they would, at most, present a snapshot in time, a point along the continuum of change. Nevertheless, it was thought necessary to find a means through which to justify the exploitation of the land and the remaining indigenous peoples with moral impunity. Because such impunity required intellectual and political rationalization, the American myth was hatched.

The grand American myth in the United States is one of imagined pristine nature across an entire continent of wilderness filled with wild beasts and savages, which was probably not as difficult for Europeans to conquer as has been imaginatively conceived. The ignoble savage, nomadic and barely human, was invented to justify stealing the land from the few remaining indigenous North Americans and to prove that the indigenous peoples had no part in transforming an untamed wilderness into a civilized continent. When the Europeans walked into a forest, which they often described as parklands, they did not see the indigenous peoples creating the park-like forests through the use of fire, nor did they observe the prairie-like conditions of large, open valleys, such as Willamette Valley in Oregon or the savanna of Wisconsin, being maintained by the indigenous peoples, also with the use of fire.

In fact, the Indian use of fire may have been the most significant factor in designing the American landscape, but the British and French, who came after the Spanish, did not see the land as the Spanish had seen it.[29] When the British arrived on the scene they put the best spin on what they saw by assuming it was natural—which meant, and still means to many people, untouched by the defiling hands of humans.

Whatever the conditions were, they reverted toward the wild side between the time the Spanish landed on the North American continent in 1513 in what is now Florida and the time the British landed in the early 1600s. Some parts of the continent had reverted even more toward the wild side by the time Meriwether Lewis and William Clark made their historic trek of 1803–1806, and a few areas were wilder still by the 1840s, when the Oregon Trail was in full use.

Furthermore, returning something to a former condition, the old notion of restoration, is an oxymoron because whatever we create is new, although it may emulate—but only emulate—a prior condition, despite the amount of data in hand. Nevertheless, we can return to a given place and do our best to simulate what we understand it to have been like.

It is true that we can physically go back to a particular place, but we can never go back in time to who we were at a given moment in the past, and we can never go back to the particular circumstances that pertained at a specific time in the past. Therefore, whatever we create will be original and immediately entrained in the perpetual process of change and novelty.

Reason Four: We Cannot Go Back in Time

Even if we had an idea of what pre-European conditions were, we could not go back to them. To this end, Nigel Pitman, María Azáldegui, Karina Salas, and others wrote, "What biologists write about tropical forests today will be, in many cases, the only thing left of them in the future. The forest that twenty years ago held the world record for frog diversity is no longer standing, but young herpetologists can piece together what it might have looked like from scientific articles."[30]

For example, passenger pigeons were probably the most numerous birds in the New World prior to the European invasion. Population estimates from the 1800s ranged from one billion to almost four billion birds, but the population could have reached five billion; at the time, passenger pigeons constituted up to 40 percent of the total number of birds in North America. Their flocks, a mile wide and up to three hundred miles long, darkened the sky for hours and days as they passed overhead.

The pigeons inhabited the billion or so acres of forest that once covered the continent east of the Rocky Mountains, where they bred in colonies that could cover from 30 up to 850 square miles, with up to one hundred loosely constructed nests of small twigs in a single tree. Generally, one egg was laid and incubated by both parents. They tended their chick for two weeks, after which the entire flock would depart, leaving the flightless young dropping to the ground. After a few days, however, they would begin to fly and take care of themselves.

Passenger pigeons could have caused widespread, frequent disturbances in pre-European forests by breaking the twigs and limbs of the trees in which they roosted and nested, thereby creating the fine fuels that influenced the frequency and intensity of fires in the forests they inhabited. Moreover, their excrement would have nourished forest soil and plants.

Furthermore, the consumption of vast quantities of acorns by pigeons during their spring breeding season may partially account for dominance of white oak throughout much of the north-central hardwood region. The vast numbers of passenger pigeons undoubtedly did much to determine the species composition of eastern forests prior to the twentieth century.

Although their gregarious nature invited overhunting, no appreciable decline in pigeon numbers was noted until the late 1870s. Thereafter, it took only twenty-five years of relentless pursuit to complete their destruction. This dismal episode in American history was aided by the invention of the telegraph, through which the locations of flocks could be ascertained. Tens of thousands of individuals were killed daily from nesting colonies and shipped to eastern markets. The market hunting occurred simultaneously with the destruction of their forested habitat, which was cleared to make way for agriculture. As well, the expansion of northern red oak during the twentieth century may have been facilitated by the pigeon's extinction.

The passenger pigeon may be the only species for which the exact time of extinction is known. The last bird died in 1914 at the Cincinnati Zoological Garden—before any competent ornithologists could write an account of the species.[31]

Consequently, trying to restore an ecological condition for which we have no concrete data from a time we cannot recapture is an impossible task. Consider, for example, that there are far more people spread across the North American continent now than there were prior to the Spanish, French, and British invasions. As a result, the entire North American continent is polluted, a condition that dramatically affects what can be done in the name of ecological re-creation.

Researchers Scott Doney and David Schimel made this point clear: "Over a range of geological and historical timescales, warmer climate conditions are associated with higher atmospheric levels of CO_2, an important climate-modulating greenhouse gas. One emergent property is clear across timescales: atmospheric CO_2 can increase quickly, but the return to lower levels through natural processes is much slower. The consequences of human carbon-cycle perturbations will far outlive the emissions that caused them."[32] Hence, urgent action is required to curb the production of greenhouse gases in China and other fast-growing countries because China's carbon dioxide emissions are far outpacing previous estimates, which will make stabilizing atmospheric greenhouse gases much more difficult than it would have been before.[33]

Part of the pollution problem is caused by capitalism—a human invention that was foisted onto the North American continent by the European immigrants and is now used to fuel our insatiable appetite for material goods, even as the competition it spawns in the money chase destroys the ecosystems that sustain the economy. In addition, technological inventions designed to increase our ability to exploit nature have irreparably altered the entire landscape of the United States, even as they have alienated people from nature—a point aptly illustrated with the passenger pigeon and the advent of the telegraph.

Coupled with capitalism is the notion of the absolute rights attached to the concept of private property, both material and intellectual—a European precept in direct opposition to the indigenous practice of sharing rights to the use of communal land. This change by itself does much to preclude emulating the ecological connectivity necessary to re-create intact ecosystems, but all the changes mentioned have irreversibly altered the entire North American continent—to say nothing of the world at large.

With the foregoing in mind, William Schlesinger, a professor at Duke University in North Carolina, poses a critical question: "In the pre-industrial era, humans lived in concert with nature. No doubt it was a hard life, but it was sustained for centuries. The question we now face is whether we can live the way we aspire to today, without degrading the life support systems of the planet that would sustain us tomorrow."[34]

Basic Considerations in Repairing an Ecosystem

So what kind of re-creation will benefit us today and the children of tomorrow—and why? This is at once an intelligent, compound question and a wise one because it's both present- and future-oriented. Moreover, it raises another interesting question: If what we do is not ecological restoration, what is it?

A simple example of repairing something is mending a hole in a sock, a lesson my mother taught me over fifty years ago. To mend a sock, she had three items: a wooden darning egg (although an old-fashioned light bulb also works), a darning needle, and darning thread. With patience and dexterity, she wove the thread back and forth across the hole. Then, she turned the sock around far enough to weave the thread through the existing strands until the hole was repaired in a neat crosshatch. At this point, the mended portion of the sock was often stronger than the original fabric had been, which meant it took me longer to wear it out a second time. The sock was repaired, but not restored to its original condition. Its physical structure, however, was mended in a way that allowed the sock to continue functioning as a sock.

Another, more complicated illustration is a woman who suffers cardiac arrest and is "brought back to life." Clearly, she has been physically altered by the episode and psychologically changed by her nearness to death. Therefore, although a medical team can revive her, it cannot restore her to the condition she was in prior to her trauma. Even when a doctor performs a successful triple-bypass surgery, the functionality of the patient's system is repaired by a surgical creation, although the system itself—and thus the person's makeup—is different. Nevertheless, the system may function in a nearly normal condition for some years.

It is the same with ecosystems. We repair dislocated or otherwise broken parts by creating an "ecological bypass" in order to maintain the integrity of their processes. In so doing, we generate something other than what existed before. Let's take the North American prairie as a case in point. For nearly twenty million years, an unbroken swath of grassland, a thousand miles wide, stretched from northern Canada, through the midsection of the continent, to Mexico. Then, in less than a century after John Deere invented his steel-bladed plow in 1837, the American prairie all but disappeared. Today, only a fraction is left. The mixed and shortgrass prairie of the Plains states represents about 5 percent of the ecosystem's original extent, whereas less than 1 percent of the lush tallgrass prairie remains to the east. Most of the tallgrass prairie occurs as remnants in pioneer cemeteries and along old railroad rights-of-way.

Now the question becomes one of purpose: Why do you want to repair a patch of prairie? Schlesinger offers sound counsel:

> We must remember that we live in an integrated chemical system that spans only a thin "peel" about 20 km [12 miles] thick on the surface of planet Earth. How we manage that arena will determine the persistence and quality of life for every one of the species that now inhabit this planet with us. Many species will disappear; others will proliferate globally, bringing huge changes to the ecosystem functions that we have long regarded as "normal." Like it on not, *Homo sapiens* will be the supervisor of this arena. We can manage it well, manage it poorly, or through purposeful actions of terrorism and war, we can poison Eden. The chemistry of the arena of life—that is Earth's biogeochemistry—will be at the center of how well we do.[35]

A biologically sustainable ecosystem is a prerequisite for a biologically sustainable yield of the broad array of nature's interactive services and products on which our way of life depends. In turn, a biologically sustainable yield is a prerequisite for economically sustainable communities, which are a prerequisite for overall social-environmental sustainability. In addition, mended ecosystems would go a long way in counteracting global warming and the extreme weather it fosters.

For most people, however, the purpose of mending an ecosystem is to restore its vegetative beauty (the aesthetics of its biodiversity), which is at once the most notable and visual aspect whereby people connect with it. For them, the composition of species framed in a snapshot is important, particularly those in which they find pleasure, such as prairie flowers.

For people interested in native birds, however, the most important aspect of the mending process lies beyond mere plant-species composition. In this case, the composite structure of the ecosystem is critical because that structure both attracts the birds and serves them as habitat for feeding, reproduction, or both. The same can be said of people interested in butterflies, reptiles, small mammals, or any other group of organisms. In still other circumstances, someone may be concerned about the system's ability to capture and store water or in making a reinvestment of biological capital as a means of increasing soil fertility.

Each of these approaches is focused on retrieving a selected part of the prairie in an attempt to recapture an aesthetic snapshot, to repair a certain structural condition, or to maintain—and perhaps enhance—a process or function. In this sense, the outcome can be viewed as a commodity or amenity. Whatever the reason, each outcome will be different.

Nature's Dynamics

Ecosystems go through what is called *autogenic succession*, which means self-generating or self-induced succession. An example is the characteristic developmental stages a grassland or forest undergoes from bare soil to another grassland or forest—both above ground and below ground.

In considering how autogenic succession functions, a caveat needs to be introduced to counter the notion of discrete stages replacing one another in a pre-determined, orderly fashion. Rather than being distinct, the developmental stages form a complex continuum wherein each builds on the dynamics and biophysical nuances of the preceding one. Hence, no two areas ever develop in an identical fashion.

In addition to the relatively orderliness of autogenic succession, all living systems are visited by disturbance regimes, those events that stir the pot, as it were, and keep the systems healthy. A disturbance is any relatively discrete event that disrupts the structure of a population or community of plants and animals or that disrupts an ecosystem as a whole and thereby changes the availability of resources or that restructures the physical environment or that causes all three kinds of changes.[36]

A simple example is drifting trees in the ocean, which play an ecological role in structuring the biological communities of the outer shores, where rock-dwelling plants and animals compete for space on which to attach themselves. During the high-winter tides, drifting trees batter intertidal communities with sufficient force to smash the plants and animals against the rocks, thereby maintaining species richness through the creation of unoccupied spaces and the consequential flux in developmental stages.

The role driftwood plays in the processes of biotic communities on rocky shores is proportional to its abundance and size. How often an area is battered by large driftwood depends on its location. In a sheltered cove, for example, where such influences are absent, the dominant competitors gradually exclude the subordinate ones, thereby decreasing species diversity, which weakens the community as a whole.[37]

Ecological disturbances, which can range from a grazing cow or bison to a small grass fire or a major hurricane, are characterized by their distribution in space and the size of disturbance they make, as well as their frequency, duration, intensity, severity, synergism, and predictability. Physical features and patterns of vegetation often control terrestrial disturbances, such as those affecting prairies. The variability of each disturbance, along with the area's previous history and its particular soil, creates the existing vegetational mosaic.

Human-introduced disturbances, especially fragmentation of habitat, impose stresses with which an ecosystem is ill-adapted to cope. Not surprisingly, biogeographical studies show that connectivity of habitats within a landscape is of prime importance to the persistence of plants and animals in viable numbers in their respective habitats—again, a matter of biodiversity.[38] In this sense, the landscape must be considered a mosaic of interconnected patches of habitats, like vegetated fencerows, which act as corridors or routes of travel between and among patches of suitable habitat.

Whether populations of plants and animals survive in a particular landscape depends on the rate of local extinctions from an area of habitat and the rate with which an organism can move among existing patches. Species living in habitats isolated as a result of fragmentation are less likely to persist than are those living in connected habitats. Although this is generally true, there are some caveats for both plants and animals.

For example, many more species of plants than those with small populations may be vulnerable to genetic erosion—the loss of genetic diversity as a result of ongoing fragmentation processes. Because each individual organism has many unique genes, genetic erosion occurs whenever the gene pool of a species of plant or animal is diminished as individuals die without getting a chance to breed with others in their population. In this way, stored genetic variability is lost, which diminishes the remaining gene pool. This genetic loss is compounded when the plant in question is capable of self-fertilization or mating among related individuals, as opposed to real obligate crossbreeding. Therefore, many fragmented habitats are unable to support plant populations large enough to maintain genetic viability over time.[39] Clearly, modifying the existing connectivity among patches of habitat strongly influences the abundance of species and their patterns of movement.

The size, shape, and diversity of patches also influence the patterns of species abundance, and the shape of a patch may determine which species can use it as habitat. Conventional wisdom dictates that habitat fragmentation causes local extinctions of animal populations by decreasing the amount of viable, interior (core) habitat, while simultaneously increasing the effects of the habitat's edge. It is also widely accepted that interior-dwelling species are better off in bigger fragments because they have a larger amount of suitable habitat. Nevertheless, fragmented habitats in real landscapes have complex, irregular shapes. Irregularly shaped forest remnants in New Zealand, for example, consistently had populations of interior-dwelling species that were reduced by 10 to 100 percent as a function of the edge effects' influence on the species. In addition, species within a given habitat fragment tended to exist in small, disjunct populations.[40]

The interaction between the dispersal of a species and the pattern of a landscape determines the temporal dynamics of its population. Local populations of wide-ranging organisms may not be as strongly affected by the spatial arrangement of habitat as are more sedentary species. It is, after all, the relationship of pattern that confers stability on ecosystems—not the relationship of numbers. Stability flows from the relationships that have evolved among the various species. A stable, culturally oriented system, even a diverse one, that fails to support these ecologically co-evolved relationships has little chance of being sustainable. Here, it must be understood that plant-species composition is the determiner of structure and function, in that composition is the cause of structure and function rather than the effect—a concept that will be explained more fully later in this chapter.

Until the nineteenth century, when both bison and prairies were all but extirpated, bison ranged across the vast grasslands by the tens of millions. Prior to that time, fire and grazing by bison were the two factors that shaped this great midwestern ecosystem. Fires, started by lightening and indigenous Americans, attracted bison and other herbivores to patches of tender grasses, which they preferred as food, and so they avoided the tender broad-leaved forbs in the same area. In other words, bison grazed heavily on burned areas and avoided other acres altogether. These habitat patches were constantly moving, however, as fires came and went.[41]

This kind of selective grazing not only kept the grasses in check but also caused the chemical structures in plants to diversify. Because herbivores often fed on chemically similar plants, they imposed selective pressures on the plants to diverge or biased the assemblage of plants within a community toward chemical divergence. As specialization increased and the size of the area in which the plants grew decreased, plant communities tended to be more chemically dissimilar. At fairly local scales and where herbivores had a strong one-to-one interaction with plants, communities had a robust pattern of chemical disparity.

Coexisting herbivores feed selectively in yet another way. The mechanisms employed by coexisting, generalist herbivores can be thought of as complementary diversity. A mainstay of ecological theory is that coexisting species use different resources or, if they use the same resource, they use it differently. Such coexistence is usually attributed to a few species that operate within a fairly narrow latitude.[42] But what happens when multiple species of closely related herbivores share critical resources too generally? Species-specific nutritional niches provide a means whereby generalist herbivores might coexist despite broadly overlapping diets.

A study of seven closely related grasshoppers demonstrated that they eat proteins and carbohydrates in different absolute amounts and ratios even from the same species of plants. The grasshoppers' regulation of their protein-carbohydrate intake elucidates the active nature of dietary selection to achieve balanced diets that provide a buffer in the face of variable food quality.[43]

Taken all together, the selectivity of indigenous grazers kept the North American prairie habitat in a healthy mix of biodiversity among the various species of grasses, forbs, insects, amphibians, reptiles, birds, and mammals. Some birds, for example, focused on severely disturbed areas, others on largely undisturbed sites, whereas still others, such as prairie chickens, required a mix of habitats.

Ultimately, fire determined the initial landscape-scale pattern and then rearranged its configuration by controlling the grazing behavior of the bison. This dual set of disturbances created a habitat characterized by dynamic vegetational mosaics of various scales and duration over space and time. Allen Knapp, a plant ecologist at Colorado State University in Fort Collins, put it this way: "We have a romantic snapshot view of the prairie when Europeans settled it. But ecological systems are always dynamic, always changing."[44]

Therefore, it's critical that we become thoroughly aware of the silent, often hidden, values of a healthy environment and the huge cost of repairing any of its functions because such repair is all but impossible without losing still other hidden functions. Furthermore, it's imperative to understand that whenever we view an ecosystem, it's always in the present moment; this moment—the here and now—is therefore all we ever have. Patience with nature's timetable is thus a critical consideration because healing an ecosystem is a moment-by-moment, day-by-day endeavor in which the visible outcome of one's labor may be weeks, months, or even years away.

A Model for Repairing Ecosystems

I use mending a prairie remnant in the Midwest of the United States as a model for the discussion here to keep the concepts as simple as possible, albeit I use examples from other parts of the world to enhance our understanding of global commonalities.

To begin our discussion of repairing even one hundred acres of prairie, it is necessary to understand that a prairie remnant, although relatively simple compared with a tropical forest, is still complex, with as many as 150 to 180 species of plants. Moreover, reinvigorating a degraded area of prairie is both labor-intensive and costly. These considerations may cause some locations to be seeded with only a fraction of the species found in relic areas of a healthy prairie. In addition, certain species within the mix may be difficult to grow from seed, and thus the intended outcome is further compromised.

A common pattern emerging from studies of the relationship between plant diversity and ecosystem processes is that diversity increases productivity. The main mechanism increasing the input of nitrogen is the presence of nitrogen-fixing legumes. Even within a plant community lacking nitrogen-fixing legumes, however, a positive relationship exists between species richness and productivity, which is manifested a couple of years after plant establishment and strengthens over time. In addition, a complementary uptake of nutrients in space and time is important. Together, these mechanisms sustain consistently high productivity within high diversity.[45]

Nevertheless, attempts to mend native prairie for its intrinsic value as a functioning ecosystem have been unsuccessful. For example, areas of local prairie around the Fermi National Accelerator Laboratory in Batavia, Illinois, were selected for repair in 1975. Although the Fermi locations never achieved the biodiversity of the remnants, it wasn't until the 1990s that a comparative study revealed a decline in species richness in the manipulated areas but not in the relic patches of indigenous prairie.[46] Considering that the prairies of Illinois have virtually disappeared during the last 150 years because they were turned over to the production of soybeans and corn, is it possible that the soil is still too rich in nitrogen?

Evidence that this may be the case comes from a study of sixty-eight acid grasslands across Great Britain in which the long-term, chronic input of nitrogen from polluted air has significantly reduced the species richness of plants. Species richness declined as a linear function of the rate that inorganic nitrogen was deposited from the background levels of industrially polluted air. Species adapted to relatively infertile conditions are systematically reduced at high levels of nitrogen input. Similar studies of soil in central Europe show a 23 percent reduction in species compared with grasslands receiving the lowest levels of nitrogen from industrial air.[47]

Biophysical Dynamics to Consider When Repairing an Ecosystem

With the foregoing as background, it is time to consider the biophysical dynamics that need to be accounted for in healing an ecosystem. Whereas the commonalities of the following dynamics apply to every ecosystem, their visual scales differ widely. Even prairie remnants have vegetative structures of different sizes, depending on whether the parcel is shortgrass, tallgrass, or mix-grass prairie. Because the relatively small stature of prairie vegetation makes visualizing the dynamics that follow difficult, unless, of course, you are a ground squirrel or a mouse, I will use examples from ecosystems with bold structures, such as trees, or areas where familiarity with a species, such as elephants, might enhance understanding of a particular dynamic. These ecosystem commonalities are composition, structure, and function; cumulative effects, lag periods, and thresholds; habitat components and animal behavior; and habitat configuration, size, and quality.

Composition, Structure, and Function

The composition of an ecosystem consists of the number and kinds of organisms that grow in a particular area, as well as the length of time they live and then persist after death. The length of a particular organism's life, plus the length of time its body persists after death, is critical—particularly with long-lived and large-sized plants, such as trees. For example, a coast redwood tree may influence its habitat for more than two millennia, whereas a passing black bear may affect the habitat for only half an hour. The bear is a transient component while it's in the habitat, and even if it dies there, it is still a transient when compared with a redwood.

Structure, in turn, is an outcome of the composition of plants that grow in a particular locale because each individual and each kind of plant grows differently. The cumulative effect of how they grow creates the vegetative structure we see above ground as well as the structure below ground, which is unseen. The combined features of composition and structure allow certain functions to take place within a given area of a particular landscape, whether it be a desert, a prairie, or a tropical forest. As a simple illustration, let's examine the clearing of tropical forest and its effect on soil macrofauna in southeastern Amazonia.

As primary Amazonian forests are cleared, pastures and secondary forest occupy an increasing amount of space in the landscape; their presence has a variable effect on the soil macrofauna, particularly invertebrates, such as beetles, ants, termites, spiders, and earthworms. In one study, the richness of the soil macrofauna fell from seventy-six to thirty species per plot immediately after the forest was cleared, and the composition of the new community was different. Ants, termites, and spiders were most affected by the disturbance.

In plots situated where deforestation had taken place several years earlier, the effect was dependent on the type of land use—a pasture in which the grasses and forbs were continuously grazed, for example, or one where the land was allowed to lie fallow and thus could begin to recover toward a forest. The richness per plot in old clearings left fallow rose to sixty-six species, and the composition was closer to that in the primary forests than to that on land used in other ways. Although macrofaunal communities showed richness close to that in the primary forest in all fallow areas, the species richness of earthworms and beetles recovered only as areas next to the forest regained their forest cover. In contrast, species richness per plot remained low in pastures, just forty-seven species. The data show that clear-cutting the forest is a major disturbance to the soil macrofauna, whose recovery potential is much higher in fallow areas than in pastures, even six or seven years after logging. In Amazonia, therefore, areas that lie fallow after they are clear-cut may play a critical role in the conservation of soil macrofauna.[48]

Returning to North America, a simple management decision in southeastern Oregon provides an excellent example of what can happen to the composition, structure, and function of a remnant grassland when how to repair it is not understood. The only way to "rehabilitate" areas in southeastern Oregon following rangeland fires in the 1970s, while I was working in the area, was to plant crested wheatgrass (an exotic species from northern Asia) and call doing so "fire rehab." In fact, the habitual response of the federal-government range managers was to increase forage for cattle by planting rows of crested wheatgrass with a mechanical range drill, but this response was not based on any understanding of the ecosystem. Moreover, the seedings were touted as a way to control soil erosion, which was nowhere evident, as rill erosion was legion in the seedings.

The challenge began when the U.S. Department of the Interior, Bureau of Land Management, hired me to conduct a wildlife survey on 5.2 million acres. One of the first plants I encountered was hundreds of acres of crested wheatgrass. This tussock-forming grass (which grows in small, thick, coarse clumps) was planted in neat, well-spaced rows that were all but devoid of life other than the grass itself.

These uninhabitable areas had experienced the extirpation of virtually every indigenous species within their borders. In comparison, however, the areas surrounding the seedings were brimming with native life, and I wondered why. The answer occurred to me on a hot summer's afternoon, as I examined the seemingly endless seedings, and how simple—how very simple—the answer was. The grass had been planted in north-to-south rows, which allowed the blazing sun to perpetually scorch the bare soil between the rows as it traveled in a southerly arch from morning till evening.

Conversely, in a few areas, where the rows had inadvertently been planted east to west, the grass plants themselves created a physical barrier that shaded the ground from one row to the next. Here, plants and animals could survive, although less abundantly in both numbers and species than in the areas surrounding the seedings. Nevertheless, their survival radically decreased the amount of soil erosion in those areas because of the root systems of the indigenous plants, especially the native grasses.

In this case, a simple physical alteration in the orientation of the rows of exotic grass plants helped mend two functional problems—those of wildlife habitat and soil erosion, both of which shared the same self-reinforcing feedback loop of helping to stabilize the ecosystem of which they were a part. These are the most obvious problems inherent in the alien plantations, but the multitudinous acres of wheatgrass harbor subtle, long-term problems, some of which require decades to become apparent, by which time they're out of control.

Crested wheatgrass was imported during the drought-stricken Dust Bowl years of the Great Depression to augment livestock forage because local prairie grasses were failing to provide sufficient fodder for the cattle. The invasion of wheatgrass was sealed when farmers discovered that it withstood drought and overgrazing, had a long growing season, and made good hay. Today, through the unflagging efforts of ranchers and government agencies, such as the Bureau of Land Management, wheatgrass covers twenty-five million acres of prairie and shrubsteppe in North America north of Mexico. It is not, however, the problem-free panacea for the livestock industry that it is trumpeted as being.

An immediate, ecological difficulty lies in the propensity of wheatgrass to devote most of its energy above ground to the production of shoots, while maintaining only a meager root system. Indigenous grasses, in contrast, do not grow as tall, but they form prodigious networks of roots, which anchor soil in place and enrich it with nutrients and organic matter. As a result, soil in wheatgrass plantations contains significantly fewer nutrients and less organic matter than does soil in native grasslands.

The fact that native grasses put more nutrients into the soil and plant tissue than wheatgrass affects another component of the ecosystem, the sequestration of atmospheric carbon. When the prairie was converted to wheatgrass, a much more effective sink for carbon was lost—one that would tie up 480 million tons beneath rich native prairie, thereby removing it from the atmosphere. It seems, however, that the effects of introducing wheatgrass to increase short-term livestock feed extend well beyond the displacement of indigenous species and the reduction of diversity, to include the alteration of pools of energy and nutrients and how they flow within the prairie ecosystem.[49]

In these two scenarios (logging in Amazonian forests and crested wheatgrass seedings), the species composition of the vegetation was clearly the determining factor in how a given area functioned as habitat. Understanding these kinds of dynamics, however, requires patience because much of the ongoing change is initially unnoticed and thus unrecognized as cumulative effects.

Cumulative Effects, Lag Periods, and Thresholds

Nature, which has only intrinsic value, allows each component of a prairie to develop its prescribed structure, carry out its ecological function, and interact with other components through their evolved, interdependent processes and self-reinforcing feedback loops. No component is more or less important than another; each may differ from the other in form, but all are complementary in function.

Our intellectual challenge is recognizing that no given factor can be singled out as the sole cause of anything. All things operate synergistically as cumulative effects that exhibit a lag period before fully manifesting themselves. Cumulative effects, which encompass many little, inherent novelties, cannot be understood statistically because ecological relationships are far more complex and far less predictable than our statistical models lead us to believe. Our inability to recognize cumulative effects arises, in part, because we live in the invisible present.

The invisible present is our inability to stand at a given point in time and see the small, seemingly innocuous effects of our actions as they accumulate over weeks, months, and years. Consider that all of us can sense change—day becoming night, night turning into day, a hot summer changing into a cold winter, and so on. Some people who live for a long time in one place can see longer-term events and remember the year of the exceptionally deep snow when pronghorn antelope starved to death on the prairie or the year of the intense grass fires.

In spite of such a gift, only unusual people can sense, with any degree of precision, the changes that occur over the decades of their lives. At this scale of time we tend to think of the world as being in some sort of steady state (with the exception of technology), and we typically underestimate the degree to which change has occurred—like the amount of recovery exhibited by perennial grasses following the removal of domestic livestock. We are unable to directly sense slow changes, and we are even more limited in our abilities to interpret the relationships of cause and effect in these changes. Hence, the subtle processes that act quietly and unobtrusively over decades reside cloaked in the invisible present, such as gradual declines in habitat quality.

Changes in land-use patterns that alter habitat often have a delayed, negative effect on the species occupying the habitat; an effect may not be recognized for years, as demonstrated by a study of the cooperatively breeding acorn woodpecker. The study was conducted in Water Canyon, which is located in the Magdalena Mountains of south-central New Mexico.

Acorn woodpeckers rely on self-constructed sites in "granary trees" to store acorns for use during winter and spring. Most granaries, which consist of precisely made holes, are in dead trunks and limbs of narrowleaf cottonwood trees. Groups of woodpeckers with large storage facilities, which equate to high-quality territories, have greater reproductive success and better survival than do pairs or groups with poorly developed storage sites.

In the study, most territories, which had contained birds in 1985, were unoccupied by 1995. This drastic decline was associated with the gradual collapse of individual granary trees in the invisible present, a collapse that ultimately resulted in the loss of nearly all the large storage facilities. The lack of new, high-quality granaries from 1975 to 1995 was probably due in part to the distinctly bimodal age structure of the cottonwood trees—nearly all of which were either very young or old. The scarcity of middle-aged trees reflected a period of intensive grazing by cattle, during which production of young cottonwoods was suppressed because cattle tend to eat them.[50]

Hidden within this scale of time (1975 to 1995) are the interwoven threads of cumulative effects, lag periods, thresholds, and the various degrees of irreversibility. To understand the dynamics of cause and effect, we will visit the Serengeti-Mara Ecosystem, which is a component of the Serengeti Plain in Africa. Here, long-term ecosystem monitoring has highlighted the following complex of interactions.

The limitation of food has clear effects on regulating the populations of wildlife, particularly the migratory wildebeest and the non-migratory Cape buffalo. In turn, predation limits populations of smaller resident ungulates and small carnivores. Thus, systems can be self-regulating through the dynamics of food availability and predator-prey interactions. Interactions between African elephants and their food, for example, both allow and maintain the coexistence of savanna and grassland communities. However, with increased woodland vegetation, predators' success in capturing their prey increases. Under these circumstances, artificially regulating a population's size may not be required. In addition, periodic physical events, such as droughts and floods, create disturbances that affect the survivorship of ungulates and birds through feedbacks among the three spheres—atmosphere, litho-hydrosphere, and biosphere.

In any case, slow and rapid changes of different spatial scales that initiate and maintain multiple conditions within an ecosystem become apparent only over several decades; hence there may be no a priori need to maintain one particular state. Beyond that, anthropogenic disturbances have direct (hunting) and indirect (transfer of disease to wildlife) effects on wildlife. Therefore, conservation must accommodate both infrequent and unpredictable events and long-term trends by planning for the timescale of those events—without aiming to maintain the status quo.[51]

At length, cumulative effects, gathering themselves below our level of conscious awareness, suddenly become visible. By then, it is too late to retract our decisions and actions even if the outcome they cause is decidedly negative with respect to our intentions. So it is that the cumulative effects of our activities multiply unnoticed until something in the environment shifts dramatically enough for us to see the effects through casual observation. That shift is defined by a threshold of tolerance in the ecosystem, beyond which the system as we knew it, suddenly, visibly, becomes something else.

In Africa, for example, the decline in the geographical distribution and numbers of elephants, as a result of expanding human activity, is recognized as one of the continent's serious conservation problems. Elephants and people coexist at various levels of human density, but when a threshold in human numbers and activity is reached, elephant populations disappear. This threshold is apparently related to a particular stage in the process of land being converted to agriculture, a situation in which farms become spatially dominant over the woodland that constitutes elephant habitat.[52]

However, if the African people had wanted to protect the elephants from extirpation in a particular area, repair of the elephant's habitat would need to have been initiated long before the threshold of vulnerability was reached. If, at this stage, habitat loss has seriously eroded the demographic potential of the species, halting the decline in population is limited more by demographic factors than by the amount of available habitat. Under this circumstance, it is not sufficient to conclude that mending the habitat will be sufficient to rescue a declining population.[53] Why? One reason has to do with our limited understanding of the effects of spatial scale on habitat quality.

Consider, for example, that although habitat edges are a ubiquitous feature of modern, fragmented landscapes (such as prairie remnants), researchers have an abiding tendency to focus their sampling designs on relatively small spatial scales. Their findings, therefore, can elucidate the influence of edge effects on animal communities over distances of only 66–820 feet, whereas edge effects can penetrate as far as six-tenths of a mile into a habitat patch. Such large-scale edge effects can lead to an 80 percent reduction in the population size of interior species, even in very large habitat fragments.

In fact, edge effects of this magnitude can drive the interior-dwelling species to local extinction, whether in a remnant of prairie or forest. With respect to forests, a global analysis of protected areas suggests that edge effects of six-tenths of a mile may compromise the ability of more than three-quarters of the world's forested reserves to maintain the uniqueness of their interior community structures.[54]

Returning to our example of the North American prairie, native grasslands fall into three categories: shortgrass, mixed grass, and tallgrass. They also vary in quality, as determined by the abundance of indigenous plants versus exotic species. With respect to the exotic species, two kinds of common organisms (invasive plants and invasive insects) illustrate the synergistic dynamics of prairie-grassland composition, structure, and function coupled with cumulative effects, lag periods, and thresholds.

Anyone who endeavors to repair a prairie remnant must determine whether invasive plants already exist in the area or have access to it. Either way, transportation to and from the area is of critical importance because roadsides are preferential migration corridors for exotic plants and can act as starting points for invasion into adjacent habitats. Because vehicles transport large amounts of seeds, it would be wise to carefully examine the existing roadsides for exotic plants and map their distribution; the rapid spread and interrupted patterns of a plant's distribution can indicate long-distance dispersal along roads. Dispersal of plants by vehicles can greatly accelerate invasion by exotic plants into a remnant, where they can induce rapid changes in the patterns of biodiversity by altering the species composition of the native plants and thus their remnant's overall physical structure and biophysical function.[55]

These changes occur within the realm of cumulative effects, lag periods, and thresholds, so by the time an invasive species is noticed, it may already be well established. In addition, many exotic plants have an advantage over indigenous prairie species because they do not require the obligate, symbiotic, mycorrhizal fungus in order to survive, whereas close to 100 percent of the native species in a prairie remnant do. In fact, the vast majority of indigenous plants in all terrestrial ecosystems require this mutualistic relationship, in which the host plant provides simple sugars from photosynthesis to the mycorrhizal fungus, which lacks chlorophyll and generally is not a competent saprophyte (a living plant that derives its nutrients from dead or decaying organic material). Fungal hyphae penetrate the tiny rootlets of the host plant to form a balanced, beneficial relationship with the roots. The fungus absorbs minerals, other metabolites, and water from the soil and translocates them into its host. Furthermore, nitrogen-fixing bacteria, which occur inside the mycorrhizae, use a fungal extract as food and in turn fix atmospheric nitrogen, which becomes available to both the fungus and its host plant in usable form.

In effect, mycorrhizal fungi serve as a highly effective extension of the host root system. Many of these fungi also produce growth regulators that induce production of new root tips and increase their useful life span. At the same time, a host plant prevents its mycorrhizal fungus from damaging its roots. Mycorrhizal colonization enhances resistance to attack by pathogens. Indeed, some mycorrhizal fungi also produce compounds that prevent pathogens from contacting the root system. Moreover, these fungi are dispersed throughout prairie and other ecosystems by such organisms as earthworms and small mammals, which eat the belowground fruiting-bodies and defecate the viable spore onto and within the soil as they move about.[56]

In addition to the likelihood of an exotic plant's being mycorrhizal-free, how closely related it is to the native species will largely determine not only how invasive it is but also how likely it is to undermine the repair of a relic piece of prairie. The more closely related an alien is to the indigenous plants, the less invasive it is likely to be, whereas the less related it is, the more invasive it will be. This relationship between the invader and the existing native community may explain why foreign species are not uniformly toxic in all novel habitats. Therefore, the degree of an invader's relatedness to the native biota may be a useful criterion for prioritizing whether to use it in a preventative mode or, if necessary, in an after-the-fact corrective one.[57]

A further advantage that some invasive exotic species have over native ones is their capability for independent seed production through self-fertilization or autonomous seed production. These plants usually have small, shallow flowers. Such plants are much more easily distributed than are species that are self-compatible, which means that the male gamete can fertilize the female gamete of the same plant, but the process requires an unrelated organism, such as a bee, to do the pollinating. Autonomous seed production increases an invasive plant's ability to extend its geographical range farther than can a plant that requires a bee or some other organism to accomplish pollination. Moreover, polycarpic plants, those that flower and set seeds many times before dying, have a vast competitive advantage over monocarpic species, those that flower and set seeds only once before dying.[58]

Understanding the behavior of invasive plants is critical. For example, in a study of fragments of tallgrass prairie, the ten most frequently occurring and abundant species of exotic plant were cool-season species, in contrast to those of the native-plant community, which was dominated by warm-season species. Timing is thus important for exotic species to succeed in the tallgrass prairie. Because it is biologically possible for an invasive species to out-compete a native one, it would be wise in any repair initiative to protect existing small, but relatively intact, fragments of tallgrass prairie as long-term refugia for indigenous species. Such refugia could be a source of genetically viable material for mending the prairie ecosystem should it be damaged by invasive exotics. In addition, an intact fragment could be the source of genetic variability to augment a prairie remnant in which ecological repair is ongoing.[59]

Some exotics, such as tall fescue, can introduce unwanted, virulent pathogens into a prairie remnant. Tall fescue, an indigenous grass of Eurasia and North Africa, is now widely spread in the United States. The challenge for prairie repair with this particular grass is its endophyte, which is a fungus that lives within the grass. This fungal endophyte is particularly troublesome because it causes fescue toxicosis—a fungal poison that can affect livestock.

Such hereditary fungal endophytes can increase the host plant's ability to compete, tolerate drought, and resist consumption by herbivores. Two endophyte-related mechanisms in tall fescue benefit the endophyte-infected plants growing in phosphorus-deficient soils. These mechanisms appear in the morphology of a grass's roots in the form of reduced diameters and longer root hairs. A chemical modification of the rhizosphere also results from the exudation of phenolic-like compounds. (The *rhizosphere* is the area surrounding the roots of plants wherein complex relations exist among the plant, the soil microorganisms, and the soil itself.) Although it's unclear whether these ecological benefits alter the dynamics of the endophyte-host relationship over time, the presence of herbivores—both mammals and insects—temporarily increases the frequency of the fungal infection in tall fescue.[60]

In one study, a normal, background level of mammalian grazing increased the endophyte frequency and thereby shifted the plant-species composition toward greater relative biomass of infected tall fescue. These results demonstrate that herbivores can drive plant-microbe dynamics and, in doing so, modify plant-community structure directly and indirectly. Part of the dynamics is low concentrations of copper in the tall fescue because of its fungal symbiont, which may then contribute to lower copper in livestock, which thus is a partial cause of fescue toxicity.[61]

In addition to tall fescue, cool-season grasses infected with endophytes have an extraordinary impact on the ecology of grasslands. As in tall fescue, these endophytes induce adaptive mechanisms (morphological, physiological, and biochemical) that help infected grasses avoid, tolerate, and recover from drought.[62]

Mineral nutrition (nitrogen, phosphorus, calcium), as mediated by the grass's endophyte, affects the production of ergot alkaloids. These alkaloids are produced by the ergot fungus, which looks like black smut on various grasses. The ergot alkaloids are at once potent neurotoxins and vasoconstrictors, which cause a condition called ergotism, or ergot poisoning, to which both livestock and people are susceptible.

The ergot fungus is common throughout North America, where it can infect such domestic crops as wheat and oats, a process I witnessed as a young man working on cattle ranches. I therefore include a brief description of ergot poisoning, or St. Anthony's fire, as the human type of the disease was known in medieval Europe. The outbreaks of "dancing mania" between the thirteenth and sixteenth centuries have sometimes been attributed to ergot poisoning, which was finally identified as the cause in the seventeenth century. It produces a common set of symptoms: gangrene with burning pain in the extremities or convulsions, hallucinations, severe psychosis, and death.

A ninth-century author described an outbreak of ergot poisoning: "A great plague of swollen blisters consumed the people by a loathsome rot so that their limbs were loosened and fell off before death." The cause of this great plague was the ingestion of toxic amounts of the alkaloids produced by the ergot fungus that infested rye, the growth of which was promoted by the cold, damp conditions common in France and Germany. Repeated epidemics occurred throughout the Middle Ages, when whole populations became infected from eating bread made from contaminated rye.[63]

As with invasive plants, it makes a difference where in the landscape one chooses a prairie remnant to repair. This is a critical consideration because many of the remaining patches of prairie are surrounded by intensive agriculture, with its insect pests. In southeast Minnesota, for instance, large numbers of corn-rootworm beetles invade prairie remnants from surrounding cornfields in late summer and attack the resident sunflowers.

The beetles feed extensively on sunflower pollen and occasionally on other flower parts, such as petals. Sunflowers located nearer cornfields sustain more floral damage than those farther from the cornfields. The beetles can also reduce the maturation of seeds, thereby interfering with the successful reproduction of sunflowers—and possibly other prairie composites that flower in late summer. The small size of most prairie remnants and the abundance of this flower-feeding beetle in landscapes dominated by corn agriculture may affect the sustainability of prairie-plant populations.[64]

A particularly dramatic example of the effect that an introduced insect can have on an ecosystem is the balsam woolly adelgid's role in the death of endemic, relict forests of Fraser fir trees at the southern limit of their distribution on the highest ridges of the southern Appalachian Mountains. Here, over a period of twenty-one years, the avian community populating the montane Fraser fir forests changed in response to the introduction of the balsam woolly adelgid insect.

A combined historical-geographic study was conducted at Mount Collins in the Great Smoky Mountains. Investigators looked at the distribution of birds over time within five southern Appalachian mountain ranges variably affected by the adelgid. Fraser fir was virtually eliminated on Mount Collins, and the canopy cover was reduced to half its previous level, as was the combined density for all breeding birds.

Birds that foraged in the canopy and midstory declined more significantly than those species that foraged on the tree's trunks and close to the ground. In addition, birds invading from open and disturbed forests diluted the boreal character of the original avifauna. In the other southern Appalachian mountain ranges, where habitat is not as extensive, the adelgid invasion resulted in even greater declines in the abundance of avian species, as well as having more pronounced effects on sensitive species. There also were more prominent invasions than in the past by species normally associated with forest succession.[65]

If we are to have a landscape with a desirable legacy to pass forward, we must protect existing biological, genetic, and functional diversity—including habitats and plant-community types—to foster the long-term ecological wholeness and biological richness of the patterns we create. Thus, to repair a prairie, or anything else for that matter, we have to understand the integration of it components.

Habitat Components and Animal Behavior

The arrangement of habitat components across a landscape is vastly different for animals in a prairie, forest (or anywhere else, for that matter) than for people in a city because we, the people, rearrange land on which we build our own shelters, to which we bring food and water from afar. Moreover, people can routinely store excess food for long periods in freezers, regardless of weather or climatic conditions. In contrast, nature creates habitable areas within a landscape (a snag of the right size, a fallen tree under which to dig, a cliff that ameliorates the ambient temperature, and so on), not the animals that need them. Animals must find their required ration of food and water on a daily basis, and many must locate existing or potential shelter as well. Another circumstance—the connectivity of habitat components in and across the landscape—is equal in importance.

Proximity to food, water, and shelter is critically important for the most sedentary and highly adapted species (such as salamanders). Proximity to food, water, and shelter is progressively less important for the most wide-ranging and adaptable species (such as coyotes, mountain lions, and elk), which can travel great distances in short periods of time. However, wide-ranging species that live in ever-more fragmented habitats require safe corridors for travel through hostile terrain from one habitat component to another. Let's take a quick look at a few of the possibilities.

WATER. A salamander or a frog cannot travel a mile to get a drink of water, but an elk, a bear, or a mountain lion can. By the same token, an eagle can fly a great distance to water, but most bats require drinking water within roughly a quarter mile of their daytime roosts or nursery colonies.

As an aside, an adaptation to the problem of getting water is that of the sandgrouse (a bird the size of a small chicken), which lives in the Egyptian and Nubian deserts. Sandgrouse nest well away from water, which minimizes the potential predation on their young, but their offspring require water nonetheless. Sandgrouse therefore fly great distances to water, soak their breast feathers in the precious liquid, and fly back to their nests to give their offspring a drink of the water stored in their feathers. But food is not such a problem because it is more widely distributed than water throughout the sandgrouse's desert habitat.

FOOD. Grazers, such as deer, require food and cover in close proximity because they are vulnerable to mountain lions and other predators while feeding. Thus, they venture from cover, eat rapidly, and return to cover, where, in relative safety, they chew their cud (regurgitated food), which corresponds to eating their food a second time. Deer also migrate from a summer range to a winter range and back again, meaning they need a good-quality corridor within which to travel.

Small forest animals, such as amphibians, shrews, and mice, use the open, downhill side of logs as a protective cover while navigating the surface of the forest floor in search of food—an idea emulated by the concrete tunnels open on the downhill side that protect traffic from snowdrifts and avalanches in the Austrian Alps and the higher elevations of Japan. Ants, however, create open highways in areas of dense, herbaceous vegetation, which forms the first stage of forest development; these highways are often crowded with hundreds of individuals going far afield from their colony in search of food. In addition, some rodents make aboveground runways from one fallen tree to another through the herbaceous ground cover, but they construct belowground burrows at other times in order to connect one place with another. Belowground burrow systems are also the preferred mode of travel for pocket gophers as they forage, although they construct aboveground burrow systems through the snow in winter.

Other animals, like river otters, seem to have a wanderlust and travel great distances to fulfill their requirements, which include a prodigious amount of playtime. The marten, a cousin of the otter, is also a traveler, but stays within forested areas, where its prey base is located; the marten is thus constrained in its movements by such habitat alterations as clear-cut logging because it will not readily venture into such open areas. Bobcats, which are stalkers, require sufficient cover in order to ambush their prey, but coyotes, which are chasers, require open areas in order to run their prey to ground.

SHELTER. Some small mammals have everything neatly packaged. Water voles live along streams in the higher elevations of western North America, where they use both the waterways and runways through herbaceous vegetation on the streams' banks to move about and obtain their vegetarian food. In addition, they burrow into the stream banks, where they build snug nests and have all their life's necessities in immediate proximity.

Birds and bats, however, spend time away from their living quarters (cavities in snags, nests in shrubs and trees, caves, and so on) while they forage and quench their thirst. Being aerially mobile, they are relatively free to move about and stitch their habitat requirements together with greater facility than some of their earthbound kin.

Still others, such as male mountain lions and elk, traverse great distances in search of food, water, and shelter. To them, "home is where their rump rests" on any given day.

SPACE. Not only is the arrangement of habitat components important to animals but so too is the habitat's extent. To illustrate, the water vole requires but a small area along a mountain stream in order to have a viable lifestyle. Compared with water voles, coyotes are exceedingly adaptable, relatively wide-ranging, independent animals that can seemingly survive anywhere, including in the suburbs of Los Angeles, California, where they help themselves to garden produce, scraps of human food, and food left out for pets. In addition, they are quite willing to eat neighborhood cats, chickens, ducks, and any other handy foods.

Roaming the country singly, in pairs, or as family groups, coyotes prey on a wide array of animals, beginning (as pups) with grasshoppers and graduating (as adults) to prey as large as adult mule deer and yearling domestic cattle or any other seasonal morsel they deem tasty. Coyotes are also adept at dining on fruits, a habit that has earned them the nickname melon wolves in some parts of their geographical distribution because they steal from farmers' fields.

As a generalist, the coyote can survive under a wide range of environmental conditions, from Texas to Alaska and from the Pacific Northwest to the Eastern Seaboard. Their arrival in Alaska and on the Eastern Seaboard relatively recently is due primarily to the clear-cutting of vast areas of dense forest. Our social activities have opened thousands of square miles for the coyote to inhabit because of its extraordinary adaptability. And because of its wide array of possible food items, the coyote can make use of a broad variety of energy sources.

Unlike coyotes, wolves are social animals that live in packs. Their group life places limits on their ability to use habitats. For example, a far greater number of coyotes than wolves can live in Yellowstone National Park because a pack of wolves acts as a single, large organism and therefore requires a much vaster area in which to hunt. A pair of coyotes can live on rabbits and fruits in season, but a pack of five to seven wolves, each of which is much larger than even a big coyote, acts collectively as a single, large animal that must continually secure prey the size of deer, elk, and North American moose in order to survive.

A pack of wolves has a much harder time staying fed than does a pair of coyotes. It takes far more time, energy, and trials for a pack of wolves to select, chase, and bring down large prey at any time of the year than it does for a coyote, which at certain times of the year can make do quite nicely on a diet of grasshoppers, meadow mice, and ripe berries.

As a specialist, the wolf is fitted to a narrow set of environmental circumstances and can survive only it if finds sufficient prey large enough to feed the pack as a whole. The wolf, therefore, has a limited range of prey items to which it is effectively adapted as sources of energy, and it can neither fit itself to a wide variety of conditions nor fit a wide variety of conditions to itself.

These requirements make the wolf, as a highly adapted social specialist, vulnerable to extinction because of humanity's continual encroachment on and fragmentation of its habitat, while the coyote, as a supremely adaptable, individualistic generalist, is likely to out-survive humanity itself. This outcome is particularly evident as the wolf's geographical range shrinks in the face of societal pressures on the landscape, while the coyote's geographical range increases in concert with those same pressures.

PRIVACY. With respect to privacy, the little water vole probably can most easily satisfy this requirement in that privacy is only a burrow away. But the medium-sized coyote is so adaptable and individualistic that it can find privacy almost anywhere. For a large pack of wolves in a progressively fragmented habitat, privacy is becoming increasingly rare, and this lack of privacy is part of the reason so much of their former geographical range is no longer suited for them to inhabit.[66] Beyond understanding the components of habitat, we need to know how they are configured into landscape patterns so we can emulate them in order to repair an ecosystem, such as a prairie.

Habitat Configuration, Size, and Quality

A little-considered prerequisite for determining the configuration, size, and quality of a habitat to be repaired is the body size of the largest species to occupy it. For example, assuming a prairie remnant can be maintained at various sizes without losing its ecological integrity, the question becomes which species can live there in viable numbers—pronghorn antelope, coyotes, rabbits, prairie chickens, or only gophers, meadow mice, and Henslow's sparrows.

Body size is perhaps the most important characteristic of an organism because it affects all physiological and ecological processes and is, therefore, a

fundamental influence on an organism's ability to survive and reproduce in different environments, which includes those modified by human activities. Some
species exhibit significantly different body sizes among macrohabitats; in other
words, individuals of the same species may have smaller bodies in fragmented
habitat than they do in contiguous habitats. The same may be true of individuals
living at different places along an elevational gradient.

Although the anthropogenic effects of altering landscapes may not be universal among species, they can be significant and rapid—developing in just a few
decades following habitat manipulation. Thus, habitat fragmentation may influence the biodiversity, species composition, and densities of local organisms, as
well as the body size of an indigenous species, which is one of its most fundamental and defining characteristics.[67]

The shape of a habitat patch also plays a strong role in determining the size of
the population of a given species it can accommodate in a fragmented landscape.
To illustrate, the rates of nest predation and brood parasitism of the grasshopper
sparrow, Henslow's sparrow, eastern meadowlark, and dickcissel were studied in
thirty-nine prairie fragments ranging in size from fifty-nine acres to roughly one
hundred thousand acres in five states in the mid-continental United States.

Throughout the region, the rates of nest predation were significantly influenced by the size of a habitat fragment. They were highest in prairie remnants of
less than 250 acres and lowest in patches greater than 2,475 acres. Rates of brood
parasitism by brown-headed cowbirds, however, were not consistently related to
fragment size, but were instead more strongly correlated with the regional abundance of cowbirds. Differences in the rates of nest predation between large remnants (54–68 percent of all nests lost to predators) and small patches (78–84
percent lost to predators) suggest that fragmentation of prairie habitats may be con-
tributing to regional declines of grassland birds. Such differences also point to the
advantage of repairing large areas of prairie to maintain viable avian populations.

Additional data on avian population density and nesting success collected in
thirteen prairie remnants of various sizes in southwestern Missouri revealed three
levels of sensitivity to an area. The most area-sensitive species, the greater prairie-
chicken, was absent from small fragments of prairie habitat. An intermediate
form of area sensitivity was apparent only in Henslow's sparrow, which had population densities scaled to the size of the prairie remnants.

At the third level, a species can be sensitive not only on a distributional scale
but also by having lower nesting success in small, rather than in large, habitat
patches of remaining prairie. The dickcissel, for example, was the only species in
southwestern Missouri that was area-sensitive on such a demographic level. These
data indicate that grassland-nesting species are subject both to the size of a habitat fragment, which determines area occupancy, and to population size, which
determines nesting success.

In addition to size, configuration of the overall area of the habitat core versus
the amount of edge effect is critical. The blockier the configuration is, the better

the biophysical integrity of the habitat's core. Conversely, the more convoluted, narrow, and linear the pattern, the greater the ratio of edge to interior, a configuration that magnifies the edge effect. One means of increasing the overall size of two adjacent patches of habitat is to establish a linking, biophysical corridor.

Although biological corridors may mitigate potential negative effects of inbreeding over long periods of time, so many variables are involved that corridors are not always the best method of conserving fragmented populations. Therefore, repairing a habitat by reconnecting disjunct populations within a habitat's core may be more effective than attempting to join isolated fragments, but reconnecting populations in the core may also require larger remnants to begin with than are realistically available. Not surprisingly, the foregoing habitat characteristics influence a habitat's quality, and small changes can have major effects on animal behavior and thus population dynamics.

For example, ecological studies indicate that the landscape context, such as the percentage of urbanization in the surrounding landscape, may significantly influence the abundance of plants and animals, as well as their distribution within remnant, indigenous habitats. With respect to small species, such as butterflies and bees, the type of grassland may have a significant effect on their richness and composition. In one study, tallgrass prairie supported much greater numbers of species than did shortgrass plots. Habitat quality also affected species richness and composition. As would be expected, areas of low quality generally supported fewer species than moderate- or high-quality sites. Although landscape context did not seem to have a clearly predictable impact on butterfly species richness or composition, it is clearly a major factor in accounting for the presence or absence of vertebrate species. These same kinds of wildlife-habitat relationships occur in other ecosystems, such as forests, tundra, and deserts, in every continent across the globe.[68]

With respect to the above-mentioned corridors, it makes no difference how large or small, adapted or adaptable a species is; all species require safe corridors in which to travel through hostile terrain. This necessity affects the requirements of a species with respect to how its habitat is configured, which, in large measure, determines its resultant quality.

Despite what skeptics say, observations of movements by naturally dispersing animals within and through fragmented landscapes can demonstrate the value of corridors more convincingly than can controlled experiments of animal movement. Such field observations relate directly to the species of animal and the reason it is moving (e.g., dispersing juveniles or seasonal migrations) and to the real landscapes where the animals must live. Moreover, evidence from well-designed studies indicates that corridors are valuable tools for the maintenance of biodiversity within fragmented landscapes. Therefore, whoever would destroy the last remnants of natural connectivity should be required to bear the burden of proof that destruction of the corridor in question will do no harm—now or in the future.[69]

The isolation of habitat patches is often cited as having a major impact on the dynamics of small populations occupying fragmented patches in a complex landscape.

To test this notion, field surveys of Bachman's sparrow were conducted where suitable habitat patches were not only in a linear configuration but also were isolated to varying degrees from the potential sources of dispersing birds. The results demonstrate that isolated patches of habitat in linear landscapes are less likely to be colonized than are more contiguous patches. Therefore, the configuration of habitat patches into a corridor can improve the ability of some species to find and settle in newly created habitats. Clearly, careful landscape design and planning can enhance habitat occupancy at a regional scale for species that do not disperse easily through landscapes.[70] Such planning is critical because landscape boundaries (edges of various kinds), although small in spatial extent, often have pronounced effects on the flow of ecological energy, especially when the area adjacent to a corridor is inhospitable to the species in question.

To visualize and appreciate the biophysical effects of a prairie habitat fragmented by barriers and edges of various sorts, one would have to be a gopher or a mouse and see it from their eye-view. Therefore, I employ here examples from a variety of ecosystems to illustrate the common effects of such broken-up habitats.

The trend, based on ecological literature, is to treat corridors and boundaries of various kinds as separate phenomena on the landscape. This conceptual approach misses a fundamental commonality, however, and that is their strong influence on the flow of energy through the directional control of organisms and processes, such as ecological feedback loops. Corridors and edges of various types exist at opposite ends of a permeability gradient and thus differ in their effects on the rates and direction that organisms, energy, and processes flow. The position of landscape structures along this gradient can determine its permeability, which depends on attributes of both the structure itself and the influence it has on the movement it allows.

By way of example, consider the influence of the international railroad line in Mongolia on the winter migration of Mongolian gazelles, which never cross the tracks, even though the best forage is on the opposite side. Although their movements mainly follow the railroad line in winter, the tracks form an effective barrier because they split the gazelles' habitat.

In this sense, the railroad is an impervious barrier to the gazelles' access to good-quality winter forage. What would happen to the railroad's permeability, and thus the flow of ecological energy, if a series of underpasses were constructed to allow the gazelles access to the side with the high-quality forage? What effect would the animals have on the plants and the plants on the animals in an exchange of energy—food for feces and urine? Could plant seeds be moved back forth under the railroad by wind and gazelles?

However, the permeability of a living structure, such as a riparian area, has different effects when it is part of a continuous forest as opposed to being a mere buffer zone between a stream and a recent clear-cut, which has a much more pronounced edge than a forest. In addition, a deciduous riparian area is one structural

habitat in summer when in full leaf but quite another in winter after leaf fall. And
the porosity of a corridor's edge between two habitats depends on the adaptability
of the species involved at any given season.[71]

If, for instance, a forested corridor were juxtaposed to a recent clear-cut covered in herbaceous vegetation, deer mice would make the corridor edges entirely
porous because they are highly exploratory, extremely adaptable, and thus equally
at home in either habitat. In contrast, the southern red-backed vole is a closed-
canopy, coniferous-forest specialist, and thus for it the edge is impervious.
Therefore corridors between intact forest habitats would maintain higher population connectivity than would landscapes without them.[72]

Another way to increase the permeability of the corridor-edge continuum is
to consider that the distance between or among patches of habitat may, in fact,
determine the relative effectiveness of corridors and other configurations. Habitat
"stepping stones," for example, could substantially reduce the isolation of remnants in fragmented landscapes for such species as butterflies and wetland birds.
When distances between patches are short compared with an animal's mobility
and the hostility of the terrain through which it must travel, a stepping-stone
approach may be the most effective way to promote dispersal. Alternatively, continuous corridors may have the highest value relative to other habitat configurations when longer distances separate patches of habitat in fragmented landscapes
and the species using the corridor is either a relative habitat specialist or a wide-
ranging species (such as the mountain lion) that prefers a more natural habitat.[73]
Yet another dimension to the porosity of habitat edges is the air.

The landscape at Los Tuxtlas, Mexico, was originally rainforest but is now
greatly fragmented and covered with pastures. To understand the ecological ramifications of the seeds that drop in feces from frugivorous bats under isolated trees
in the pastures, 652 bats of twenty species were captured, 83 percent of them fruit
bats. Of these, the most abundant species were the little yellow-shouldered bat,
Mexican fruit bat, Seba's short-tailed bat, and Toltec fruit-eating bat. The little
yellow-shouldered bat not only was the most abundant species but was also far
and away the most important dispenser of seeds in the pastures. Of the seeds,
89 percent were from zoochorous forest trees and shrubs, seven species of which
accounted for 79 percent of the total dispersed seeds: Mexican pepperleaf, semi-
epiphytic-strangler fig, trumpet tree, a free-standing fig tree, one tree with no
common name, and two shrubs with no common names.

As it turns out, both bats and birds are important to the onset of succession
in human-created pastures because they carry the seeds of pioneering and primary species (trees, shrubs, herbs, and epiphytes) and deposit them under isolated trees, thereby maintaining plant diversity while beginning to reconnect
forest fragments. Consequently, they contribute to the recovery of woody vegetation in disturbed areas within humid, tropical forests.[74] If, however, you, the
reader, either grew up in a city or now live in one, you may be familiar with habitat corridors in urban settings.

Birds in urban landscapes generally occupy parks, which are analogous to forest fragments, whereas tree-lined streets form linear corridors that connect the fragments within the urban matrix. To understand the species-habitat dynamics of an urban setting, a study conducted in Madrid, Spain, examined the effects of street location within the urbanscape, vegetative structure along the streets, and human disturbance (pedestrian and automotive) on bird species richness within the street corridors. In addition, the birds' temporal persistence, density of feeding and nesting guilds, and the probability of a street's being occupied by a single species were also accounted for.

The number of species increased from the least suitable habitats (streets without vegetation) to the most suitable habitats (urban parks), with tree-lined streets being an intermediate landscape element. Tree-lined streets that connected urban parks positively influenced the number of species within the streets' vegetation, as well as species persistence, population density, and the probability that the individual species would continue to occupy the streets. Human disturbance, however, exerted a negative influence on the same variables.

Wooded streets could potentially function as corridors that would allow certain species to fare well by supporting alternative habitats for feeding and nesting, particularly those birds that feed on the ground and nest in trees or tree cavities. Local improvements in quality and complexity of the vegetation associated with certain streets, as well as a reduction in the disturbance caused by people, could exert a positive influence on the regional connectivity of streets as a system of urban corridors for birds. Because of the differential use of corridors by species with various habitat requirements, streets as habitat corridors could be further improved by taking the requirements of different species into account.[75]

In fact, plants may be taking such requirements into account on their own. Although dispersal is a ubiquitous trait among living organisms, evolutionary theory postulates that the loss or death of propagules during dispersal episodes (the cost of dispersal) should select against it. As such, the cost of dispersal ought to be a strong selective force in fragmented habitats. To test this notion, patchy populations of the weed French hawksbeard were studied in small patches on sidewalks and around trees planted within the city of Montpellier, in southern France.

French hawksbeard is a Mediterranean composite in the aster family that spreads predominantly in cultural landscapes, including cities. This annual germinates with fall rains and persists during winter in rosette-leaf stage. Generalist insects, especially the domestic honeybee, pollinate the flower in town. Mating is largely by cross-fertilization with unrelated individuals and accounts for 80–90 percent of the pollination in the countryside and 60–90 percent in town.

French hawksbeard produces two types of seeds (i.e., it has *dimorphic seeds*): a small seed with a pappus and a large seed without it. (A *pappus* is a parachute-like structure attached to one end of the seed, which allows wind to disperse it.) The seed with a pappus favors dispersal over long distances, whereas the seed without a pappus falls to the ground near the parent plant.

Within the city, seeds with a dispersal parachute have a 55 percent greater chance of falling on concrete, which is unsuitable for germination, than do seeds without a parachute, which land at the base of the parent plant. The proportion of non-dispersing seeds is significantly higher in a city if the urban patches form a relatively fragmented environment, as opposed to contiguous populations in rural areas, which predominantly have dispersal parachutes. Because of the recent fragmentation of continuous habitat, this pattern of dispersal is consistent with rapid evolution, which occurs over five to twelve generations of selection as a result of the high cost of dispersal.[76] In this way, seeds are accumulating within urban habitat corridors that could increasingly serve such seed-eating birds as the European goldfinch.

Although corridors clearly work for animals, what about corridors for plants? To answer this question an investigation was launched into the role of corridors in seed dispersal of the rare, pond-dwelling, self-fertilizing buttercup in the Fontainebleau Forest of France. The connection of ponds through temporarily flooded natural corridors was found to facilitate the migration of seeds. As a result, a pond was more likely to be colonized by the rare buttercup when it was connected to other occupied ponds. Corridors are thus a critical element of a landscape's structure not only for animals but also for the persistence of certain species of plants living in fragmented habitats, where seed dispersal between and among habitat patches is essential.[77]

Mending the Prairie through Fire and Grazing

To return to the model of prairie repair that was discussed previously in this chapter, the reintroduction of fire and grazing by large ungulates, alone or in combination, has increasingly been recognized as central to the repair of North American mixed-grass and tallgrass prairies. Fire (a physical disturbance) and grazing (a biophysical disturbance) are major forces that shape the patterns of diversity among native and exotic species in many types of grassland; yet these disturbances have notoriously variable effects.

To examine these two types of disturbance in an ecosystem where exotic annuals dominate most grasslands and serpentine soil is the major refuge for native grassland species, studies were conducted in a mosaic of serpentine and non-serpentine soils in California. Both fire and grazing increased total species richness on both soils. However, on the one hand, fire enhanced the diversity of exotic species more on non-serpentine soils and increased native species richness disproportionately on serpentine soils. Grazing, on the other hand, increased native species richness on serpentine soils but not on non-serpentine soils. The reason cattle grazing increases the species richness of indigenous grasses, as shown by fertilization experiments, is that invasive grasses are limited by the amount of nitrogen in the soil.[78]

Several lines of evidence indicate that deposition of dry nitrogen from smog is responsible for the invasion of nonnative grasses. For example, the estimated rate at which nitrogen is deposited in grasslands south of San Jose is between 22 and 33 pounds of nitrogen/2.5 acres/year; whereas the grasslands on the San Francisco Peninsula experience nitrogen deposition of 9 to 13 pounds of nitrogen/2.5 acres/year. Moreover, cattle selectively graze on grasses over forbs, and their grazing leads to a net export of nitrogen—a nitrogen sink, if you will—as cattle are removed for slaughter. Although poorly managed cattle can significantly disrupt native ecosystems, in this case moderate, well-managed cattle grazing is essential for the maintenance of indigenous biodiversity in the face of invasive, exotic grasses and exogenous inputs of nitrogen from nearby urban areas.[79]

In the prairie remnants in the Loess Hills of Iowa, grazing by domestic livestock promoted the greatest overall species richness, whereas grazing and burning resulted in the lowest cover by woody plants. Burning, however, achieved the best overall increase in the cover and diversity of native species while simultaneously reducing exotic forbs and grasses, the latter being predominantly cool-season in habit, such as cheatgrass.[80]

In contrast, livestock grazing appears to be an exotic ecological force in grama grasslands of southeastern Arizona, one destructive to certain components of the native flora and fauna. The destructiveness of livestock grazing could result from the absence of extensive grazing by indigenous ungulates in the Southwest since the Pleistocene. Thus, the tolerance of particular grasslands to their use by domestic livestock may depend on their historic association with native grazing animals, as well as on the pervasiveness of certain exotics, like cheatgrass.[81]

Cheatgrass, an annual, was originally confined to roadways and abandoned farmlands but is now pervasive in the shrubsteppe ecosystem of the western United States, where it's expanding into shrub, ponderosa pine, and piñion-juniper ecosystems. The spread of cheatgrass causes fires to become more frequent and larger in scale, which has a dramatic effect on indigenous plants and animals. Therefore, although a mosaic of burning and grazing (alone and in combination) may provide the greatest landscape-level species richness, this strategy may also promote the persistence of exotic species, such as cheatgrass, in some areas because species-specific responses appear to be idiosyncratic.[82]

With respect to fire, entomologists have expressed concern that prescribed burning is incompatible with the conservation of insect diversity on small prairie sites. To evaluate this anxiety, a research project was conducted over seven seasons within small, isolated tallgrass-prairie remnants in northern Illinois, northwestern Indiana, and southeastern Wisconsin. The study focused on the responses of remnant-dependent and remnant-independent species to multiple fires. Sweep nets, light traps, sticky traps, and visual searches were used to gauge population responses and to track negatively affected populations to recovery.

Most species (93 percent) responded consistently to prescribed fires. Postfire responses, however, ranged from 26 percent positive to 40 percent negative for 151 species representing thirty-three families and seven orders. Three attributes—remnant-dependence, upland inhabitance, and nonvagility—were significant predictors of negative postfire response. (*Vagility* is the capacity or tendency of an organism or a species to move about or disperse in a given environment.) Among negatively affected populations, 68 percent recovered within a year, and all 163 populations tracked to recovery did so in two years or less. Therefore, a judicious use of rotational, cool-season burning can be a workable tool in helping to mend small, isolated prairie remnants.[83] In addition to fire, can grazing by large ungulates be used to repair a prairie remnant?

Many successful reintroductions of large mammalian herbivores have taken place throughout the world, but remarkably little attention has focused on how these actions affect native and exotic vegetation. One such herbivore is the tule elk, endemic to western California, which was on the brink of extinction in the mid-1800s but now has several stable populations.

The elk significantly altered the species composition of coastal grasslands, where the response of annual species (dominated heavily by exotic taxa) was dramatically different from that of perennial species. The herbivory increased the abundance of both indigenous and exotic annuals but either had no effect on or caused remarkable decreases in perennials. Moreover, the elk decreased the cover of native shrubs, a finding that indicates that herbivores can play an important role in maintaining open grasslands. In addition, elk dramatically reduced the abundance and biomass of velvet grass, which is a highly invasive exotic and thus a major problem in moderately moist perennial grasslands.[84]

Unlike the indigenous tule elk and bison, cattle are an exotic species that was introduced into North America more than two centuries ago. Whereas habitat and food items are partitioned among coexisting native herbivores, like the bison and pronghorn antelope, domestic cattle are much more like generalists in their use of habitat and their foraging. Moreover, livestock grazing is today perhaps the most ubiquitous use of land in the western United States.

As with grazing by native species, however, the biophysical results are highly variable. To understand some of the effects of livestock grazing, a study was carried out on the cover of plants, soil crusts, and plant-species richness at six sites with different potentially natural vegetation in the Chaco Culture National Historic Park in northern New Mexico. This park has a long history of human habitation and now is one of the largest grazing exclosures in the American West.

Species richness was higher under long-term protection than under current grazing at all sites. As with other studies, however, trends in the response of shrubs and grasses varied significantly among the sites, in that shrub cover increased with long-term protection at four upland sites, and grass cover increased with protection at four sites. The vegetative response to a discontinuation of grazing was determined partly by each site's ecological potential of both its edaphic and topographic characteristics.

Like any disturbance, the successful reintroduction of a large grazing mammal can be expected to have extremely complex effects on the plant community, thereby giving rise to both desirable and undesirable outcomes from the perspective of repairing a particular ecosystem. The nuances in vegetative response at Chaco, for instance, are thought to represent site-specific ecological variation and thus challenge the notions of a widespread invasion of shrubs, as is often inferred.[85] It is paramount, therefore, to pay careful attention to the existing heterogeneity of the ecological background in which one is proposing to work.

Nevertheless, the cover of perennial grass has declined in many arid types of grassland over the past two centuries, while shrub density has increased. These changes, which are characteristic of desertification, are thought to have occurred most often after prolonged periods of intense grazing by domestic livestock. At many such sites, however, the subsequent removal of livestock for up to twenty years did not increase the cover of grasses.

To understand the time required for grasses to recover in historically arid grasslands dominated by shrubs, vegetation was examined at two desertified sites that differed in the length of time they had been free of livestock. There was little noticeable difference between vegetation at the site from which livestock had been fenced out for twenty years and the shrub-dominated vegetation just outside the exclusion fence. Nevertheless, there was a significantly higher cover of perennial grasses in the area from which livestock had been removed thirty-nine years earlier, and all the increase had occurred within the last twenty years. It thus seems that perennial grasses in historic grassland ecosystems dominated by shrubs require a period of twenty or more years to recover from grazing once the domestic livestock have been eliminated.[86] The grasses' requirement of two decades or more for their recovery from livestock grazing is a beautiful illustration of the aforementioned dynamic shared by all ecosystems: cumulative effects, lag periods, and thresholds. It also demonstrates our limited powers of spontaneous observation, which can be thought of as the snapshot effect.

With respect to reintroducing large mammalian grazers into prairie remnants as part of the planned repair, Machteld Van Dierendonck and Michiel Wallis de Vries (both in the Department of Terrestrial Ecology and Nature Conservation, Agricultural University, Wageningen, The Netherlands) offer wise counsel based on experiences with the reintroduction of the Przewalski horse in the Mongolian steppe. For any planned reintroduction of any large mammalian grazer into an ecosystem being repaired, they recommend creating a framework for safeguarding the entire system with an integrated caretaking plan. Thus, each prairie site where such a reintroduction is anticipated requires a thorough assessment with respect to its suitability; the required information includes the area's current size, habitat types, land use, socioeconomics, relevant legislation, and potential problems. In addition, each prairie site must contain one or more facilities for acclimating genetically and physically healthy, socially adapted animals in biologically sound groups.

From the human viewpoint, an organizational structure that incorporates a vision statement, goals, and objectives should be established for each reintroduction site. In turn, these elements need to be developed into an effective caretaking plan, one that includes carefully monitoring of the population and its surrounding ecosystem, including research on and population control of the reintroduced ungulates to avoid unnecessary damage to the ecosystem. Finally, special attention needs to be given to local socioeconomic situations, community participation, and the training of a staff to manage the project.[87]

Returning to our example of the North American prairies, a graphic example in southeastern Oregon of the kind of problems Van Dierendonck and de Vries were talking about is illustrative of conflicting human desires that can erupt when communication and cooperation are lacking. A controversy arose between a rancher and a conservation organization while I was working in the rangelands of southeastern Oregon in the late 1970s. The subject of the debate was a population of rare plants that was discovered in an active grazing allotment. The conservationists wanted to enclose the plants within a fence to protect them from the cattle. The rancher, in turn, justified his grazing practices by arguing that the proposed "exclosure"—as he viewed the fence—was unnecessary because the plants and his cattle had coexisted for decades. The conservationists countered that the cattle not only are an introduced species but also are generalists in their eating habits and thus were a threat to the plants, whereas the indigenous grazers were more selective in their foraging behavior—like the bison of the plains.

Although the conservationists were correct about cattle being exotic organisms with catholic diets as compared to indigenous herbivores with selective diets, they did not mention their deep-seated dislike for cattle, which they regarded as unnatural intruders in the rangelands. Part of their dislike for the cattle was misplaced, however, because the cattle merely did whatever the rancher allowed them to do. Thus, as is often the case, the animals were blamed for the owner's lack of responsible behavior.

At length, I was drawn into the controversy. After examining the area, I proposed that a portion of the plant community be fenced to exclude the cattle, but not all of it because the rancher just might be correct in his assessment. The conservationists, however, did not even consider my proposal and fenced the whole area.

Although the situation seemed fine for a while, the cumulative effects of fencing out the livestock began taking an unseen toll. By the time the decline in the community's vigor became noticeable, the threshold of irreversibility had been crossed.

Had the conservationists been willing to test the rancher's notion, they would have been able to work with him to ensure the community's long-term survival. By unilaterally introducing the fence, however, they precipitated the demise of the very thing they wanted to save. The rare plants would probably still be flourishing if they had been given consideration commensurate with their special status.

Special Considerations

Three factors merit special attention when they are known to occur in an ecosystem slated for repair, namely microhabitats, mutualistic symbiotic relationships, and endemic species. Years ago, while working in the shrubsteppe ecosystem of southeastern Oregon, I became fascinated with the microhabitats created by large anthills and their thriving populations. The most striking aspect of this community, however, was the distinctive microclimate it created, which was especially noticeable in early spring.

As the snow melted, it did so on the south-facing slopes of the anthills, which left the north-facing slopes bedecked in snow. Once the snow was gone, the south-facing slopes warmed and were graced with early-spring vegetation, such as bur buttercups, whereas the north-facing slopes remained frosted or even frozen at times. As well, the ant's activities, from sunning themselves to foraging, occurred on the south-facing slopes and remained thus restricted until the sun's warmth more evenly affected their immediate environment.

A somewhat similar situation, particularly with respect to grasshoppers, occurs in the Natal Drakensberg Mountains of South Africa, where the hilltops act as thermal refugia from the cold-air drainage of winter. The increased insolation on the eastern and northern sides of the hilltops compared with the western and southern sides is particularly attractive to the grasshoppers. Crevices in the hill summits provide further micro-refugia in a matrix of thermally inhospitable land.[88]

Mutualistic symbiotic relationships need special attention because if one species is lost, so will the other be. For example, birds may hover over or perch on flowers to feed on their nectar, and their doing so cross-pollinates flowers of the same species on other plants when the birds visit them. However, the rat's tail plant, which is endemic to the South African cape, appears to be unique among bird-pollinated plants in that it has a sterile inflorescence axis whose sole function is to provide a perch for foraging birds. This structure enhances the plant's reproductive success by causing the malachite sunbird, its main pollinator, to adopt a position that is ideal for pollinating its unusual ground-level flowers.[89] This brings me to the importance of endemism per se.

Endemism is an important aspect of biodiversity that is often confined to small, isolated areas or to a severely limited number of species in an area or to both. The species involved are narrowly adapted and thus easily eliminated, and their elimination can have dramatic effects on the ecosystem of which they were a part, as illustrated by three disparate examples in New Zealand, the Balearic Islands of the western Mediterranean, and the central Pyrenees of France.

In New Zealand, two endemic mistletoes have declined considerably since 1840. Their decline is reputedly due to introduced herbivores but is coincident with a major decline in the densities of native birds. As it turns out, bellbirds and tuis are significant pollinators and seed dispersers not only of the endemic mistletoes but also of many other native plants. The continued existence of these mistletoes will require maintenance of native bird populations. A breakdown of such mutualistic relationships may have widespread consequences.[90]

The introduction of exotic species onto an island can have significant effects on the density of native populations and their distribution, as well as on their ecological and evolutionary feedback loops. Disruptions of this type can be dramatic, significantly reducing the reproductive success of native species and even causing their extinction, as happened on both Menorca and Mallorca islands in the Mediterranean as a result of the introduction of carnivorous mammals.

Prior to the release of carnivorous mammals, two endemic species—a perennial shrub and a frugivorous lizard—served each other's needs in a mutualistic symbiotic relationship. The Balearic Islands lizard is now extinct, and the shrub, the Balearic Islands daphne, is in danger of extinction on both Menorca and Mallorca islands. Fortunately, relict populations of both the lizard and the shrub, as well as their mutualism, still exist on a separate, isolated 158-acre islet, where the daphne is abundant.

The population of daphne with the greatest seedling recruitment is on the islet, where the lizards remain in abundance. In turn, the lizards appear to be the only dispersers of the shrub's seeds because they not only consume large amounts of the shrub's fruits without affecting either germination or seedling growth but also move the seed to sites suitable for the shrub's establishment. The disruption of such a specialized plant-vertebrate mutualism can set one or both partners on the road to extinction.[91]

In the central Pyrenees of France, the diversity of springtails, a tiny soil-dwelling insect, was studied at two sites, and a semi-natural beech forest was compared with a conifer plantation at each site. Although differences in the structure and composition of the springtail communities were observed in both the beech forest and the conifer plantation, these changes followed different patterns at the two sites. In both cases, however, the diversity of the springtail community was impoverished in the plantation, where endemic components of the community suffered a particularly severe loss in species richness and abundance. The non-endemic species, however, were less affected. Endemic biota represent the most valuable element in an ecosystem and thus are its most vulnerable component—one whose biological service to the system is seldom understood.[92]

Although not endemic by definition, rare species can also make significant contributions to the functioning of an ecosystem, but these contributions are often aggregated into data on common species and thus overlooked. In this case, prudence dictates that uncommon species be assumed to make positive contributions to the functioning of the ecosystem wherein they dwell.[93]

With the foregoing in mind, we would be wise to determine—as best we can—whether an ecosystem we are concerned about is functioning within healthy parameters. If, on the one hand, it is, then the question becomes how to sustain its processes and thus maintain its biophysical integrity. If, on the other hand, the ecosystem appears to be in decline, we need to figure out what action is necessary to mend its structure in order to revive and sustain its processes. In either instance, how do we know whether we are making changes that will lead to our desired outcomes?

Let's suppose that five prairie remnants are going to be repaired, each of which has a different soil type. It will be critical to the repair to understand how the perennial grasses behave during their recovery from grazing. The healing process is so slow and often subtle, however, that casual observation will not suffice. At this point monitoring becomes important because the process is specifically designed to observe and document subtle ecological changes.

Monitoring Your Efforts

Monitor has the same origin as *admonition*, which means a warning or caution, and is derived from the Latin *monitio*, a "reminder." With respect to repairing prairie remnants, monitoring means keeping watch over the process so that we can be warned if our behavior causes the ecosystem to deviate from the desired course. Thus, on the one hand, monitoring informs us of activities that may be too harsh and could offend the system, and, on the other, it helps us conserve the options embodied within the system for future generations and ourselves. Because monitoring, however helpful, is an illusive, abstract concept to many people, I use the personal pronoun *you* or *your* in various parts of the following discussion to add a measure of concreteness to the concept.

Good monitoring has seven steps: crafting a vision and goals; making a preliminary inventory; modeling your understanding; writing a caretaking plan with clearly defined objectives; observing implementation; verifying effectiveness; and validating the outcome(s).

Step one is crafting a carefully worded vision and attendant goals that state clearly and concisely what you want as a future condition. This necessary first step ensures that you know where you want to go, why you want to go there, and what you think the journey will be like. You use the vision and its goals to measure (monitor) all decisions, actions, and consequences so that you know whether in fact your journey is even possible as you imagined it and what the consequences of the journey might be. What, for example, do you want the prairie remnant to look like and function like when you have it repaired? What are your short-, medium, and long-term goals for the remnant?[94] Stating a vision and goals is paramount if the next steps in monitoring are to be successful.

Step two is taking a preliminary inventory—that is, carefully observing and understanding the initial situation, what is available here, now. Taking inventory requires asking three questions: What exists now in the prairie remnant of interest, before anything is purposefully altered? What condition is it in? What is the prognosis for its future repair? Even though preliminary monitoring may require asking multiple questions, the outcome is still a single realization.

Step three is modeling your understanding. This step involves configuring the current knowledge of the prairie remnant you want to repair into a conceptual model, an explicit map of your understanding. One or more computer models could augment such a map. Although it is critical in this exercise to assume at the outset that the map represents your best understanding of the remnant as a functional prairie ecosystem, it is equally critical to have the humility to assume that the map is flawed and that your understanding is incomplete. In this way, the viability of the vision and its goals, as well as the model itself, are continually tested and improved. As the model is improved, so is your knowledge of the prairie remnant with which you are working.

Step four requires that each prairie remnant have its own treatment plan in which all the particulars of its care, including objectives, are laid out. An *objective* is a specific statement of intended accomplishment. It is attainable, has a reference to time, is observable and measurable, and has an associated cost. The following are additional attributes of an objective: it starts with an action verb; it specifies a single outcome or result to be accomplished; it specifies a date for completion; it is framed in positive terms; it is as specific and quantitative as possible and thus lends itself to evaluation; it specifies only what, where, when, quantity, and duration and avoids mentioning the "why" and the "how"; and it is product oriented.

Once you have determined your objective(s), you not only will be able to but also must answer the following questions concisely: What do I want? Where do I want it? When do I want it? How much (or how many) do I want? For how long do I want it (or them)? If a component is missing, you may achieve your desire by default, but not by design.

As a simple illustration, imagine asking a product-oriented rancher what his main objective is for grazing his cattle on public lands. If he were completely honest, his answer would be: "I want all of the grass to fatten my cattle." If we break his statement into its components as questions, it looks like this: (1) What do you want? "The grass." (2) Where do you want it? "In my cattle." (3) When do you want it? "Now." (4) How much do you want? "All of it." (5) How long do you want it for? "As long as it lasts with continual grazing."

Only when you can answer all these questions concisely do you know where you want to go in repairing the relic prairie and the value of going there, and only then can you calculate the probability of arrival at your goal. Next you must determine the cost in both money and labor, make the commitment to bear it, and then commit yourself to keeping your commitment.

Step five is observing the implementation of a project on the ground by asking, Did we do what we said we were going to do? Although this type of monitoring is just documenting what was done to the prairie, it is critical. Without it, it may not be possible to figure out what went awry (if anything did or does), how or why it went askew, or how to remedy it. In addition, the next generation would have little or no idea of what you did to the parcel of land or why and thus no way to figure out what to do in order to remedy a problem (such as the appearance of an invasive plant) that has arisen since you either retired and moved away or died.

Step six is verifying the effectiveness of your actions on the ground; such verification assesses the implementation of your objectives, not your goals or vision. Whereas, on the one hand, your vision and its attendant goals describe the desired future condition of the prairie for which you are aiming, they are qualitative and are not designed to be quantified. Your objective, on the other hand, is quantitative and so is specifically designed to be quantifiable.

Monitoring to assess the effectiveness of an objective requires asking: Is the objective specific enough? Are the results clearly quantifiable and within specified scales of time? Systematically monitoring the effectiveness of your project with the aid of indicators provides information (feedback) that allows you to ascertain whether you are in fact headed toward the attainment of your vision, maintaining your current condition, or moving away from your vision.

A good indicator helps you to recognize potential problems in repairing a relic prairie and provides insight into possible solutions. What you choose to measure (say, the height of the various species of grasses), how you choose to measure it (with what instruments, in what season, and how often), and how you choose to interpret the grasses' recovery from livestock grazing would have a tremendous effect on how you construe the biological viability of the prairie in the long term, as well as on how you understand what happens over time.

Indicators close the circle of action by demanding that you come back to your beginning premise and ask (reflect on) whether, through your actions, you are better off now than when you started: if so, how; if not, why not; if not, can the situation be remedied; if so, how; if not, why not; and so on.

Here a caveat is necessary. Traditional one-dimensional indicators, which measure the apparent health of a single condition (sunflowers in one prairie remnant bordering a corn field), ignore the complex relationships among soil, water, and air quality, and the relationship of the remnant to its surrounding landscape. When each component is viewed as a separate issue and monitored in isolation, measurements tend to become skewed and lead to ineffective decisions with respect to grassland health. Therefore, if an accurate assessment of repair and sustainability is to be achieved, viable indicators need to be multi-dimensional and must measure the quality of relationships in the form of biophysical feedback loops among the components of the prairie being monitored.

Only with relevant indicators and a systemic way of tracking them is it possible to make a prognosis for the remnant's future based on your vision and goals (which state the desired condition) and on the collective objectives (which determine how the goals will be achieved and when). Likewise, relevant indicators and a systemic way of tracking them are a prerequisite to making the necessary target corrections because only now can you know what corrections to make.

Step seven is validating the outcome of on-the-ground-activities, which is considered by many to be research. This assessment involves testing the assumptions that went into the development of the objectives and the models they are based on.

Validating on-the-ground-actions may require asking such questions as Why didn't the results come out as expected? What does this mean with respect to our conceptual model of how the prairie works versus how it actually works? Will altering our approach make any difference in the outcome? If not, why not? If so, how and why? What target corrections do we need to make in order to bring our model in line with how the system actually works? Validation is necessary for determining the array of possible target corrections. In addition, monitoring for validation may have wide application for repairing other remnant grasslands.

Yet, even with the best of intentions, there is a weakness in monitoring that has be overcome if the results are to serve the purpose for which they were designed. This weakness is a lack of commitment to the long-term, costly feedback loop of information that must be relentlessly pursued if monitoring is to fulfill its role as an archive of purpose, action, and achievement. Whatever the stated purpose, the results of your local actions affect the whole world in some way.

How, you might wonder, is that possible. Well, place a pencil on a table and observe it. Now, pick up the pencil and put it down somewhere else. You have just changed the face of the whole Earth—never to be the same again because it's impossible to replace the pencil in the exact position that you had originally put it. You have just met your power to change the world, which you do every minute of every day. The question is, How do you change it—and why?

Ultimately, to make any ecosystem repair work over the long term, we must reevaluate our pet ideas, question the certainty of our knowledge, control the unchecked exploitation of the Earth (with its resultant pollution, especially that of the atmosphere), and control our human population. Unless we do, humanity will inevitably cause its own demise—despite the number of prairie acres that have been repaired.

For the sake of a healthy Earth, we need to understand—and accept with humility—that we are not in control of a single component of nature, much as we might wish it otherwise. Rather, we are in a mutual partnership, where the respect and care we give the Earth are reciprocated in kind.

7

Where Do We Go from Here?

The United Nations is by no means a perfect instrument, but it is a precious one. I urge you to seek agreement on ways [of] improving it, but above all of using it as its founders intended: to save succeeding generations from the scourge of war, to reaffirm faith in fundamental human rights, to re-establish the basic conditions for justice and the rule of law, and to promote social progress and better standards of life in larger freedom.—United Nations Secretary-General Kofi Annan, address to the United Nations General Assembly, New York, September 25, 2003

Although I could go on at length about what necessity dictates we adults of the world must do to fulfill our trusteeship of planet Earth as a biological living trust, I have, instead, selected two courses of action that we must simultaneously take if we are to bequeath a worthy legacy to those who are young and to those as yet unborn.

Beginning now, we must start (1) to critically examine our situation today and (2) to determine where society needs to be at the end of this century if people are to have any kind of dignified life with an overall sense of well-being. This scrutiny must be simultaneously focused on three areas: adaptations, behavioral changes, and imperative questions.

Adaptations

Adaptation is a process of deliberate change in response to social stress brought about by observable environmental conditions. The dominant adaptation to environmental change is primarily symptomatic, and thus actor-centered: individuals respond to specific circumstances in an effort to ameliorate perceived personal and social vulnerabilities. But individuals taking a more dynamic, systems-oriented approach honor the resilience of ecological systems by living within the long-term biophysical constraints of social-environmental sustainability.[1] If we adopt such an approach, we adapt immediately to current conditions, in part by repairing (as much as possible in the short term) the damage we have done to the ecosystems we rely on for a good quality of life so they can maintain their ability to serve our requirements. Such adaptation is a critical concept because we cannot move away from a negative idea (the fragmentation of habitat), which we continually try to do; we can only move toward a positive idea (the connectivity of habitat). We must, therefore, think about, plan for, and consciously move toward landscape connectivity if we are to counter the fragmentation of habitats.

Behavioral Changes

Adult members of every society must reevaluate their actions and make behavioral changes if humanity is to arrest the exponential acceleration of the destructive feedback loops we have so blindly set in motion (based largely on how we treat one another) while there is still time to repair the functional capacity of planet Earth as a global commons for the overall benefit of humankind.

Clearly, growth in the human population is confronting all of us with dilemmas of the most profound nature. We face such problems as the destruction of forests, the degradation of the land, rivalries over access to water, the depletion of an increasing number of natural resources, the inability to maintain levels of production, and outbreaks of disease. Burundi, in Africa, for instance, has a population of six million, one third of which is suffering from malaria. In Botswana and South Africa, a large percentage of the population has AIDS but no access to medicines because of the exceedingly high cost of pharmaceuticals.

In the United States alone, without these medicines and improved knowledge about maintaining good health, people in the past would have died when they were much younger, and the population would be around 140 million, rather than the 281 to 283 million it is today. When these improvements in health spread worldwide, the high rate of death in such places as Africa and the Indian subcontinent declined significantly, and the population boomed. Despite the extant problems of disease in many African nations, the population as a whole is expected to increase from eight hundred million to two billion by 2050.

If poor countries develop tastes for and can afford some of the industrialized nations' lifestyles, humanity is going to outstrip the available supply of goods and resources in a short time. In all likelihood, potable water will become the major source of competition and therefore the cause of wars. Food and the arable land on which to grow it shrinks each time one more person is born than dies, and wresting energy, such as wood from forests and fossil fuels from lands and seas, will greatly stress the ecosystems that produce it. With respect to forests, they are being summarily cut down for short-term monetary gains in countries that can least afford to lose them and the free ecological services they perform. As for fossil fuels, the question becomes one of how much pollution the environment can tolerate before it begins an irreversible downward spiral of degradation that humanity cannot easily survive.

Such an outcome is not necessary, of course, but to avoid it, men must raise the level of their consciousness with respect to the way they treat women; for example, they must give women the choice of how many children to have and when, and must, as well, value them for their abilities beyond satisfying males' sexual urges and bearing children—especially sons. Women must also be provided with safe access to legal abortions and adequate counseling about birth control, as well as not having to undergo the heinous practice of genital mutilation.

Moreover, men must see to it that gender equality is a given—and is taught—throughout the entire educational cycle, recognizing that although men and women differ biologically in some ways, they are equally human. Therefore, women deserve equality with men in all aspects of social life, including educational, social, political, and economic. Moreover, women must be given positions of political authority because they focus on relationships beyond violence and power-mongering and thus are desperately needed to balance male impatience and aggressive tendencies.

Many changes are required in order to restructure society as a whole so that women are equal partners in the human experiment of cohabitation. We the people, and men in particular, ignore the equality of women at our collective peril.

The future of humanity's place on the globe rests in our collective human willingness to make commitments and to be committed to keeping them, such as making the commitment to achieve a state of dynamic equilibrium with nature and a state of mind in which social-environmental sustainability is first and foremost in our social consciousness. Considering how we abuse one another and our environment, it seems patently obvious that we are a species at war against itself.

The primary reason humanity is at war against itself has to do with men because our attitude toward—and our apparent fear of—women is, perhaps, the world's major problem. To rectify this problem, we must, first, make women the authors of—not the subjects of—policies dealing with population by insisting they occupy at least half of all managerial and policy positions in the areas of family health and planning, human population dynamics and stabilization, and the environment. It is critical to the current population crisis that men ensure that women have the necessary authority to correct the inadequacies of their health, including reproductive health, and to address the inequitable distribution of food, water, and shelter. In addition, because men are responsible for genital mutilation, they must end it and stop the slave trade in girls and women as sex objects to satisfy their sexual urges. Men must also accede to demands for gender equality; they need to positively address the inequitable ownership of land, as well as to empower women in the ecological, economic, social, and political arenas.

Second, from the earliest grades in school, girls and boys must to be taught the importance of gender equality and their shared responsibility for limiting the size of the world's human population.

Third, worldwide, the attitudes of most men toward girls and women must change. If we men are at all serious about curbing the world's looming overpopulation, we, in both secular and religious life, will openly and honestly give women opportunities to be valued for things other than serving as slave labor, being sex objects, and bearing children. We will stop telling them what they can and cannot do with their bodies, which, after all, are entrusted to their consciences—not ours. We will allow women to choose when to have children and how many to have. And we will find the courage to accept and do our part in controlling the human population by having vasectomies. Until we are willing to do that, we have no right to speak!

All the above measures are necessary to lower birth rates and to bring the human population to a sustainable level within the global ecosystem's capacity to provide a dignified lifestyle for all people, as opposed to the mere existence so very many have now. It remains to be seen whether the world's men—especially those who control the world's organized religions with such dogmatic attitudes—are up to the task.[2] There is much riding on a positive outcome, and that outcome will, in turn, depend on the questions we ask.

Imperative Questions

It is becoming increasingly urgent that we recognize and articulate the questions we need to ask about our social treatment of the environment. When all is said and done, we will find that the consciousness with which we approach an issue resides in the questions we ask. To wit, the sustainability of human society lies in creating a sustainable environment, but politics often ignores long-term scientific inquiry and concentrates on immediate, competing social values. The more self-centered, product-oriented segment of humanity pays the most attention to these types of social values, despite the potentially detrimental effects on environmental sustainability, present and future. The question, therefore, is how to integrate science and sociology, intellect and intuition, spirituality and materialism in such a way that we can have a holistic perspective of both the world in which we live and the one we will bequeath to the children as their inherited circumstances.

The answer is that we must pay vastly more attention to the questions we ask. A good question, one that may be valid for a century or more, is a bridge of continuity among generations. We may develop a different answer every decade, but the answer does the only thing an answer can do: it brings a greater understanding of the question. An answer cannot exist without a question, and so the answer depends not on the information we derive from the illusion of having answered the question but rather on the question we asked in the first place.

In the final analysis, the questions we ask guide the evolution of humanity and society, and the questions we ask, not the answers we derive, determine the options we bequeath to the world's children. Answers are fleeting, here today and gone tomorrow, but questions may be valid for a century or more. Questions are flexible and open-ended, whereas answers are rigid, illusionary cul-de-sacs. The future, therefore, is a question to be defined—and ultimately determined—by questions.

Whether we as a species ultimately end up with an environment compatible to our existence or with one hostile to our existence depends on the level of consciousness we bring to the questions we ask. And before we can get fundamentally new answers, we must be willing to risk asking fundamentally new questions. This requirement presents an increasingly difficult task with respect to the sustainability of our environment because there is an ongoing and underlying disengagement from nature-based recreation by the American public, and because of this disengagement people are becoming increasingly separated from the biophysical systems that support them.[3]

Learning how to frame a good and effective question is paramount, both for the crafting of a collective vision for the future and for the process of monitoring the actions necessary to achieve the vision. A question is a powerful tool when used wisely because questions open the doors of possibility. For example, it was not possible to go to the moon until someone asked, Is it possible to go to the moon? At that moment, going to the moon became possible, albeit no one knew how. To be effective, each question needs to embody six characteristics: it must have a specific purpose, contain a single idea, be clear in meaning, stimulate thought, require a definite answer, and explicitly relate to previous information.

In a discussion about going to the moon, one might usefully ask, Do you know what the moon is? The specific purpose of this question is to find out whether the person knows what the moon is. Knowledge of the moon is the single idea contained in the question. The meaning of the question is clear: Do you, or do you not, know what the moon is? The question stimulates thought about what the moon is and may spark an idea of how one relates to it; if not, that can be addressed in a second question. The question, as asked, requires a definite answer, and the question relates to previous information.

Asking a question that focuses on right versus wrong is a hopeless exercise because it calls for human, moral judgment, and that kind of question is not a valid one to ask of either an ecosystem or science. A good question would be whether a proposed action is good or bad for taking care of a particular forest. To find out, you must inquire whether a good short-term economic decision is also a good long-term ecological decision and so a good long-term economic decision. Such questions are important because a good short-term economic decision can simultaneously be a bad long-term ecological decision and thus a bad long-term economic decision, one that generations of the future will have to pay for. The point is that one must ask before an answer can be forthcoming.

The old questions and the old answers have led us into the mess we are in today and are leading us toward the even greater mess we will be in tomorrow. We must therefore look long and hard at where we're headed with respect to the quality of the world we leave as a legacy. Only when we are willing to risk asking really new questions can we find really new answers.

With respect to new questions, those that will raise the level of social consciousness about the future of humanity, I have found an extremely insightful article—one that can go a long way in securing a dignified future for all generations. The paper is titled "The Identification of 100 Ecological Questions of High

Policy Relevance in the UK."[4] These questions are far more than merely ecological however. Indeed, they address the very heart and soul of social-environmental sustainability. What is more, they are eminently pertinent to all people everywhere—without exception—despite the geographical limitation of the title. Moreover, striving to answer them will raise the collective level of our consciousness above that responsible for the problems in the first place.

It is the heart-felt work of thirty-five authors who came together with a single idea—to leave this magnificent planet a little better than it now is, if I may speak for them. This set of questions is the best possible gift anyone could leave for our time and place in history. I commend them, and I thank them. As stated by author Alan Cohen, "True progress comes not through action, but through awakening." I would add the Vedic notion: "A man of God is one who is softer than the flowers, where kindness is concerned, and stronger than the thunder, where principles are at stake."

The questions we ask today about how to achieve social-environmental sustainability, the level of consciousness with which we derive the answers, and the courage (personal and political) to act in accord with our decisions will design the legacy we leave all generations. The time is now. It's all we have—or ever will.

So here is perhaps the most important question to ask now, right now: I'm just one person; what difference can I make? "If you think you are too small to make a difference," says author Anita Roddick, "you have never been in bed with a mosquito."

For example, you could ask: What can I do in my personal life to reduce pollutants in the air, which ultimately end up in the soil of my garden, where I grow vegetables and flowers? Can I curtail unnecessary driving by planning to do several errands in a given trip instead of one errand at a time? Can I walk or ride my bicycle at least part of the time? What can I do directly to protect the health of the soil in my garden? Can I become more consciously aware of the consequences of what I introduce into the soil or withhold from it?

With these questions in mind, close your eyes and imagine you have a garden over which the wind from the nearest industrial center is directly blowing. Imagine, too, a cloudburst, and watch the falling rain pick particles of pollution out of the air and deposit them in the soil of your garden. Now, follow the water, which has dissolved the pollution into itself, as it moves around and through the particles of soil.

Look, some of the water is evaporating back into the air, leaving its pollutants behind, where they concentrate in the remaining water. See, some of the dirty water is being taken up into the roots of the plants you will eat, which means that you will ultimately eat the pollutants. Now, watch as the rest of the polluted water passes through the soil, where it picks up other elements, which you introduced as fertilizer, and carries them downhill toward the nearest ditch.

Finally, arise and leave your garden. Go to the nearest ditch you know of or can find, and look at it closely. Breathe in consciously and smell the odors in the air and contemplate the pollution they may represent. Study closely the human garbage strewn along the ditch's bottom and sides. Follow the ditch for a ways and see if there is any place from which you would dare to drink the water and feel safe drinking it.

Remember, some of that water has come from someone's garden or lawn, be it a backyard plot of soil, such as your own, or a farmer's field—and someday, somehow, you will ingest it in one form or another. Can you think of a more sobering thought? Does this exercise give you a new perspective?

For most people, this type of visualization is simply an abstraction; so, let's make it concrete. Find a large piece of cardboard, about two feet by two feet. Get twelve red, twelve blue, and twelve clear pushpins. In addition, get two skeins of thin yarn, one red and one blue. Finally, get a sharp, pointed instrument of some kind to punch small holes with. Take the pins (thirty-six in all) and push them into one side of the cardboard in any configuration you want. Then punch a hole through the cardboard at the base of each pin. Next cut thirty-six twelve- to eighteen-inch pieces of yarn, twelve each of red and blue; tie a large-enough knot in one end of each to prevent it from fitting into a hole. Now, remove the red and blue pins and thread one each of the red and blue pieces of yarn through the holes by the appropriately colored pins. Gently pull all the yarn into the center of the underside of the cardboard and fasten the loose ends together with a small twist tie. What, you are probably wondering, is this all about?

The cardboard is the soil of your garden. The red pins and yarn represent toxic chemicals that you apply to your garden, such as herbicides and insecticides, which leach into the soil and, through it, into the nearest ditch—depicted by the bundled yarn contained within the twist tie. The blue pins and yarn represent petrochemical fertilizers, and the clear pins symbolize organic compost that is recycled in the soil.

The pieces of red and blue yarn represent the extent to which you are adding toxins to the nearest ditch, the stream into which it flows, and hence into the nearest river, which eventually finds its way into an estuary and the ocean. You have, in effect, added to the poisons that gather along the journey and concentrate in the ocean.

Understanding this problem, you decide to alter the way you tend your garden. Therefore, you replace the petrochemical fertilizers with compost. To demonstrate the effect of your shift in consciousness, replace all the blue pins and yarn with clear pins. Note that nothing is taking the place of the unused petro-chemicals in the ditch because there is nothing to leach into it; everything is recycled within the soil's infrastructure. Should you now decide to limit the toxic compounds as well, you could replace some of the red pins and yarn with clear pins. Once again, there is nothing leaching from the soil of your garden into the ditch and thus into the ocean.

If every gardener in the United States did what you have done, how much cleaner would the ditches, streams, rivers, estuaries, and oceans be? What if every gardener in North America did what you have done? What if all the farmers followed suit? How much cleaner and healthier would the oceans be—to everyone's benefit, regardless of which shore they lived on? Let's look at this phenomenon another way.

To understand the value and power of each gardener in the context of the collective thoughts and actions of all gardeners, pretend for a moment that we gardeners are snowflakes. We are part of the first snow of winter. Numbering in the millions, we fall one by one out of a quiet sky, touching our neighbors as we whirl and spin to Earth. As we fall, we magnify one another until we blot out the sky.

The pioneers, the first flakes to fall, land on warm soil and melt, disappearing without an apparent trace. But are they really lost? Have they really had no effect? Each flake that landed on the soil, only to melt and disappear, has given its coolness to the soil until, after enough flakes have landed and melted, the temperature of the soil drops noticeably.

Finally, because of the cumulative effect of all the flakes that have gone before, the soil has cooled enough for us—you and me—to survive as we land, and still the flakes fall one by one. It snows all night, and, by morning, a glittering, transformed world greets the rising sun. As far as the eye can see the world becomes winter white—one flake at a time—as we add our collective beauty to the wonder of the universe. We can do this; we can change, one person at a time. After all, how we individually think and behave is only a choice—our personal choice. And how we change will depend on the questions we ask, such as: If I change my behavior, can I help to cleanse the world's oceans to everyone's benefit—present and future? Remember, whether you are the first or the last person to change how they treat the soil of their garden, your effect on the oceans is the same—and your personal gift to the whole of the world is irreplaceable. Indeed, you have left this magnificent planet a little better for having been here. On behalf of all generations, I thank you.

APPENDIX:
COMMON AND SCIENTIFIC NAMES
OF PLANTS AND ANIMALS

Plants

Fungi

Asian chestnut-blight fungus	*Cryphonectria parasitica*
ergot fungus	*Claviceps purpurea*
tall fescue endophyte	*Neotyphodium coenophialum*

Lichens

witch's hair	*Alectoria sarmentosa*

Parasitic Plants

mistletoe, endemic to New Zealand	*Peraxilla colensoi*
mistletoe, endemic to New Zealand	*Peraxilla tetrapetala*

Algae

dinoflagellates	Dinoflagellata

Ferns

bracken fern	*Pteridium aquilinum*
fern	*Botrychium mormo*

Grasses

bluebunch wheatgrass	*Pseudoroegneria spicata*
corn (maize)	*Zea mays*
cheatgrass (a k a downy brome)	*Bromus tectorum*
crested wheatgrass	*Agropyron cristatum*
grama grasses	*Bouteloua* spp.
Idaho fescue	*Festuca idahoensis*
medusahead	*Taeniatherum caput-medusae*
prairie Junegrass	*Koeleria macrantha*
tall fescue	*Festuca arundinacea*

velvet grass   *Holcus lanatus*
wild oat   *Avena fatua*

Forbs

bur buttercup   *Ranunculus testiculatus*
camas   *Camassia* spp.
Canada thistle   *Cirsium arvense*
common teasel   *Dipsacus fullonum*
cotton   *Gossypium hirsutum*
edelweiss   *Leontopodium alpinum*
field bindweed (a k a morning glory)   *Convolvulus arvensis*
French hawksbeard   *Crepis sancta*
jasmines   *Jasminum* spp.
ox-eye daisy   *Leucanthemum vulgare*
Pyrenean yam   *Borderea chouardii*
Pyrenean yam   *Borderea pyrenaica*
rare buttercup   *Ranunculus nodiflorus*
rat's tail   *Babiana ringens*
sessile-flowered wake-robin   *Trillium sessile*
soybean   *Glycine max*
sunflowers   *Helianthus* spp.

Trees and Shrubs

Balearic Islands Daphne   *Daphne rodriguezii*
beeches   *Fagus* spp.
black cottonwood   *Populus trichocarpa*
black walnut   *Juglans nigra*
Canada yew   *Taxus canadensis*
coast redwood   *Sequoia sempervirens*
common snowberry   *Symphoricarpos albus*
cotoneasters   *Cotoneaster* spp.
Douglas-fir   *Pseudotsuga menziesii*
eastern hemlock   *Tsuga canadensis*
European beech   *Fagus sylvatica*
figs   *Ficus* spp.
Fraser fir   *Abies fraseri*
hawthorn   *Crataegus*
holly   *Ilex* spp.
junipers   *Juniperus* spp.
la palma real   *Attalea butyraceae*
Mahaleb cherry   *Prunus mahaleb*
Mexican pepperleaf   *Piper auritum*
Monterey, or radiata, pine   *Pinus radiata*

murumuru palm	*Astrocaryum murumuru*
narrowleaf cottonwood	*Populus angustifolia*
no common name	*Piper amalago*
no common name	*Piper yzabalanum*
no common name	*Solanum rudepanum*
northern red oak	*Quercus rubra*
Norway spruce	*Picea abies*
palma negra	*Astrocaryum standleyanum*
piñion pine	*Pinus edulis*
ponderosa pine	*Pinus ponderosa*
semi-epiphytic-strangler fig	*Ficus* spp.
trumpet tree	*Cecropia obtusifolia*
whistling-thorn acacia	*Acacia dreparalobium*
white cedar	*Thuja occidentalis*
white oak	*Quercus alba*
wild roses	*Rosa* spp.

Invertebrates

Bacteria

bacterium (food poisoning)	*Escherichia coli*
Bt	*Bacillus thuringiensis*
bubonic plague	*Yersinia pestis*

Trypanosomes

sleeping sickness	*Trypanosoma brucei*

Jellyfish

comb jellyfish	*Mnemiopsis leidyi*

Mollusks

Colorado River clam	*Mulinia coloradoensis*
common periwinkle	*Littorina littorea*
eastern gem clam	*Gemma gemma*
freshwater mussels	*Epioblasma* spp.
nutricola clam	*Nutricola confusa*
nutricola clam	*Nutricola tantilla*
pearl oyster	*Pinctada imbricata*
scallops	Pectinidae
turkey-wing mussel	*Arca zebra*
vesicomyid clams	*Calyptogena* spp. and *Vesicomya* spp.
zebra mussel	*Dreissena polymorpha*

Worms

earthworms	Annelida
European night crawler	*Lumbricus terrestris*
giant Palouse earthworm	*Driloleirus americanus*
zombie worm	*Osedax mucofloris*

Echinodermata

sea urchin	*Diadema setosum*

Crustaceans

Antarctic krill	*Euphausia superba*
blue crab	*Callinectes sapidus*
Caribbean spiny lobster	*Panulirus argus*
European green crab	*Carcinus maenas*
northern crayfish	*Orconectes virilis*
ostracodes	Ostracoda

Spiders

spiders	Arachnida

Insects

ants	Formicidae
Asian tiger mosquito	*Aedes albopictus*
balsam woolly adelgid	*Adelges piceae*
bark beetles	Coleoptera
black-headed ant	*Crematogaster nigriceps*
blueberry bee	*Osmia atriventris*
bumblebees	*Bombus* spp.
caddisflies	Trichoptera
carpenter ants	*Camponotus* spp.
click beetles	Elateridae
corn-rootworm beetle	*Diabrotica barberi*
cotton aphid	*Aphis gossypii*
crickets	Gryllidae
Douglas-fir beetle	*Dendroctonus pseudotsugae*
dragonflies	Odonata
European honeybee	*Apis melifera*
fig wasps	Agaonidae
fire ant	*Solenopsis invicta*
fleas	Siphonaptra
flea beetles	Chrysomelidae
flower flies (a k a hoverflies)	Syrphidae
giant waterbugs	Belostomatidae

golden buprestid	*Buprestis aurulenta*
grasshopper	*Melanoplus*
grasshoppers	Orthoptera
housefly	*Musca domestica*
Japanese beetle	*Popillia japonica*
lady beetle	*Propylaea japonica*
ladybird beetles	Coccinellidae
leaf beetles	Chrysomelidae
mayflies	Ephemeroptera
mimosa ant	*Crematogaster mimosae*
Pacific damp-wood termite	*Zootermopsis angusticollis*
Penzig's ant	*Tetraponera penzigi*
praying mantises	Mantodea
red-bellied checkered beetle	*Enoclerus sphegeus*
Saharan pebble beetles	Tenebrionidae
scale insects	Homoptera
Sjöstedt's ant	*Crematogaster sjostedti*
springtails	Collembola
termites	Isoptera
tiger beetles	Cicindelidae
tsetse flies	*Glossina* spp.
walking sticks	Phasmatodea
water boatman	Corixidae
water skippers	Gerridae
weevil	*Larinus planus*
whirligig beetles	Gyrinidae
yellow jacket	*Vespula pensylvanica*

Vertebrates

Fish

Banggai cardinalfish	*Pterapogon kauderni*
big-eye tuna	*Thunnus obesus*
coelacanth	*Latimeria chalumnae*
coelacanth	*Latimeria manadoensis*
flying fish	Exocoetidae
Pacific hagfish	*Eptatretus stoutii*
Pacific sleeper shark	*Somniosus pacificus*
yellowfin tuna	*Neothunnus macropterus*

Amphibians

cane toad	*Bufo marinus*
lungless salamanders	Plethodontidae

red-backed salamander   *Plethodon cinereus*

Reptiles

Balearic Islands lizard	*Podarcis lilfordi*
brown tree snake	*Boiga irregularis*
California glossy snake	*Arizona elegans*
Egyptian striped skink	*Mabuya vittata*
sea turtles	Chelonioidea
western skink (United States)	*Eumeces skiltonianus*

Birds

acorn woodpecker	*Melanerpes formicivorus*
Adélie penguin	*Pygoscelis adeliae*
American robin	*Turdus migratorius*
Bachman's sparrow	*Aimophila aestivalis*
bearded vulture	*Gypaetus barbatus*
bellbird	*Anthornis melanura*
Bewick's wren	*Thryomanes bewickii*
Brewer's sparrow	*Spizella breweri*
brown-headed cowbird	*Molothrus ater*
burrowing owl	*Athene cunicularia*
dickcissel	*Spiza americana*
eastern meadowlark	*Sturnella magna*
European goldfinch	*Carduelis carduelis*
fruit pigeons	Columbidae
golden eagle	*Aquila chrysaetos*
great green barbet	*Megalaima zeylanica*
greater prairie-chicken	*Tympanuchus cupido*
grasshopper sparrow	*Ammodramus savannaru*
Henslow's sparrow	*Ammodramus henslowii*
hornbills	Bucerotidae
hummingbird	Trochilidae
loggerhead shrike	*Lanius ludovicianus*
MacGillivary's warbler	*Oporornis tolmiei*
malachite sunbird	*Nectarinia famosa*
Nepalese sunbird	*Nectarinia nipalensis*
ovenbird	*Seiurus aurocapillus*
passenger pigeon	*Ectopistes migratorius*
prairie chicken	*Tympanuchus cupido*
rufous-sided towhee	*Pipilo erythrophthalmus*
sage sparrow	*Amphispiza belli*
sandgrouse	*Pterocles* spp.
scrub jay	*Aphelocoma californica*

sharp-tailed grouse	*Tympanuchus phasianellus*
song sparrow	*Melospiza melodia*
sooty tern	*Sterna fuscata*
tui	*Prosthemadera novaeseelandiae*
wedge-billed woodcreeper	*Glyphorynchus spirurus*
western meadowlark	*Sturnella neglecta*
western tanager	*Piranga ludoviciana*
white-crowned manakin	*Pipra pipra*
yellow warbler	*Dendroica petechia*
zebra finch	*Poephila guttata*

Mammals

African elephant	*Loxodonta africana*
American bison	*Bison bison*
Antarctic fur seal	*Arctocephalus gazella*
Asian elephant	*Elephas maximus*
Bailey's pocket mouse	*Chaetodipus baileyi*
baleen whales	Cetacea
banner-tailed kangaroo rat	*Dipodomys spectabilis*
big brown bat	*Eptesicus fuscus*
bobcat	*Felis rufus*
brown lemur	*Eulemur fulvus*
bush pig	*Potamochoerus larvatus*
California red-backed vole	*Clethrionomys californicus*
Cape buffalo	*Syncerus caffer*
collared peccary	*Pecari tajacu*
coyote	*Canis latrans*
deer mouse	*Peromyscus maniculatus*
desert pocket mouse	*Chaetodipus penicillatus*
domestic cattle	*Bos taurus*
duikers	*Cephalophus* spp.
Egyptian desert gerbil	*Gerbillus gerbillus*
erect man	*Homo erectus*
flying foxes	Chiroptera
forest buffalo	*Syncerus caffer nanus*
forest elephants	*Loxondonta africana cyclotis*
giraffe	*Giraffa camelopardalis*
gophers	*Thomomys* spp.
gray whale	*Eschrichtius robustus*
ground squirrels	*Spermophilus* spp.
hairy-winged bat	*Myotis volans*
Himalayan weasel	*Mustela sibirica*
human (modern)	*Homo sapiens*

jungle cat	*Felis chaus*
Key deer	*Odocoileus virginianus clavium*
little brown bat	*Myotis lucifigus*
little yellow-shouldered bat	*Sturnira lilium*
long-tailed weasel	*Mustela frenata*
lowland gorilla	*Gorilla gorilla gorilla*
mantled ground squirrel	*Spermophilus lateralis*
marshbuck (a k a sitatunga)	*Tragelaphus spekii*
marten	*Martes americana*
meadow mice	*Microtus* spp.
Merriam's kangaroo rat	*Dipodomys merriami*
Mexican fruit bat	*Artibeus jamaicensis*
Mongolian gazelle	*Procapra gutturosa*
montane vole	*Microtus montanus*
mule deer	*Odocoileus hemionus*
Nepalese golden jackal	*Canis aureus*
North American badger	*Taxidea taxus*
North American elk	*Cervus elaphus*
North American moose	*Alces alces*
North American opossum	*Didelphis viginiana*
Norway rat	*Rattus norvegicus*
Ord's kangaroo rat	*Dipodomys ordii*
Oregon creeping vole	*Microtus oregoni*
pangolins	*Manis* spp.
pocket gophers	Geomyidae
polar bear	*Ursus maritimus*
Polynesian rat (a k a Pacific rat)	*Rattus exulans*
pronghorn antelope	*Antilocapra americana*
Przewalski horse	*Equus ferus przewalskii*
puma (mountain lion)	*Felis concolor*
red river hog	*Potamochoerus porcus*
river otter	*Lutra canadensis*
Rocky Mountain elk	*Cervus elaphus nelsoni*
Santa Rosa beach mouse	*Peromyscus polionotus leucocephalus*
Seba's short-tailed bat	*Carollia perspicillata*
shrews	Soricidae
silver-haired bat	*Lasionycteris noctivagans*
southern red-backed vole	*Clethrionomys gapperi*
sperm whale	*Physeter macrocephalus*
Toltec fruit-eating bat	*Dermanura tolteca*
tule elk	*Cervus elaphus nannodes*
voles	*Microtus* spp.
water vole	*Microtus richardsoni*

western gray squirrel	*Sciurus griseus*
white-tailed deer	*Odocoileus virginianus*
wildebeests	*Connochaetes* spp.
wolf	*Canis lupus*
wood rats	*Neotoma* spp.

NOTES

PREFACE

1. Robert R. James, ed., *Winston S. Churchill: His Complete Speeches, 1897–1963* (New York: Chelsea House, 1974), 6:5592.
2. *The Teaching of Buddha* (Tokyo: Bukkyo Dendo Kyokai, 1966).
3. Quoted in Richard Alan Krieger, *Civilization's Quotations: Life's Ideal* (New York: Algora, 2002).

INTRODUCTION

1. David R. Brower, "Credo in the Form of a Poem," *Earth Island Journal*, March 22, 2001.
2. Line J. Gordon, Will Steffen, Bror F. Jönsson, and others, "Human Modification of Global Water Vapor Flows from the Land Surface," *Proceedings of the National Academy of Sciences* 102 (2005): 7612–7617.
3. The preceding two paragraphs are based on Richard A. Betts, Peter M. Cox, Susan E. Lee, and F. Ian Woodward, "Contrasting Physiological and Structural Vegetation Feedbacks in Climate Change Simulations," *Nature* 387 (1997): 796–799; Samuel Levis, Jonathan A. Foley, and David Pollard, "Large-Scale Vegetation Feedbacks on a Doubled CO_2 Climate," *Journal of Climate* 13 (2000): 1313–1325; N. Gedney, P. M. Cox, R. A. Betts, and others, "Detection of a Direct Carbon Dioxide Effect in Continental River Runoff Records," *Nature* 439 (2006): 835–838; and Shilong Piao, Pierre Friedlingstein, Philippe Ciais, and others, "Changes in Climate and Land Use Have a Larger Direct Impact Than Rising CO_2 on Global River Runoff Trends," *Proceedings of the National Academy of Sciences* 104 (2007): 15242–15247.
4. Quoted in Paul Hawken, "Natural Capitalism," *Mother Jones*, March/April 1997, 1.

CHAPTER 1   OF IGNORANCE AND KNOWLEDGE

1. Alan Oken, *Alan Oken's Complete Astrology* (New York: Bantam Books, 1980).
2. Carl T. Rowan, *Quote Cosmos*, http://www.quotecosmos.com/quotes/23323/view (accessed April 5, 2008).
3. Quoted in Gustave Mark Gilbert, *Nuremberg Diary* (New York: Da Capo Press, 1995), 278–279.
4. Satish Kumar, "Education for Sustainability," *Resurgence* 226 (September/October 2004): 3.
5. David W. Orr, "Walking North on a Southbound Train," *Conservation Biology* 17 (2003): 348–351.
6. Robert T. Lackey, "Axioms of Ecological Policy," *Fisheries* 31 (2006): 286–290.

7. "Floccinaucinihilipilification," *TheFreeDictionary by Farlex*, http://encyclopedia.thefreed-ictionary.com/floccinaucinihilipilification (accessed October 3, 2008).

8. *Online Oxford English Dictionary*, http://encyclopedia.thefreedictionary.com/Oxford+English+Dictionary (accessed April 5, 2008).

9. Tony Juniper, "Inspiring Change," *Resurgence* 227 (2004): 28–31.

10. Arnold Toynbee, *Civilization on Trial and the World and the West* (New York: Meridian Books, 1958).

11. Daniel I. Bolnick, Richard Svanbäck, Márcio S. Araújo, and Lennart Persson, "Comparative Support for the Niche Variation Hypothesis That More Generalized Populations Also Are More Heterogeneous," *Proceedings of the National Academy of Sciences* 104 (2007): 10075–10079.

12. Michael R. Burnett, Peter V. August, James H. Brown Jr., and Keith T. Killingbeck, "The Influence of Geomorphological Heterogeneity on Biodiversity I. A Patch-Scale Perspective," *Conservation Biology* 12 (1998): 363–370; William F. Nichols, Keith T. Killingbeck, and Peter V. August, "The Influence of Geomorphological Heterogeneity on Biodiversity II. A Landscape Perspective," *Conservation Biology* 12 (1998): 371–379; and Cheryl Palm, Pedro Sanchez, Sonya Ahamed, and Alex Awiti, "Soils: A Contemporary Perspective," *Annual Review of Environment and Resources* 32 (2007): 99–129.

13. From an advertisement in the *Arizona State University Magazine* (received in 2007) for David L. Pearson's 2001 book on tiger beetles published by Cornell University Press and written with Alfried P. Vogler.

14. The discussion of climate is based on John W. Williams, Stephen T. Jackson, and John E. Kutzbach, "Projected Distributions of Novel and Disappearing Climates by 2100 A.D.," *Proceedings of the National Academy of Sciences* 104 (2007): 5738–5742.

15. Wendell Berry, "The Road and the Wheel," *Earth Ethics* 1 (1990): 8–9.

16. The preceding story of the whistling-thorn acacia is based on Truman P. Young, Cynthia H. Stubblefield, and Lynne A. Isbell, "Ants on Swollen-Thorn Acacias: Species Coexistence in a Simple System," *Oecologia* 109 (1996): 98–107, and Todd Palmer, Maureen L. Stantan, Truman P. Young, and others, "Breakdown of an Ant-Plant Mutualism Follows the Loss of Large Herbivores from an African Savanna," *Science* 319 (2008): 192–195.

17. Jonathan Rowe, "The Hidden Commons," *Yes! A Journal of Positive Futures* (Summer 2001): 12–17.

18. Jianguo Liu and Eunice Yu, "Broken Homes Damage the Environment," *Proceedings of the National Academy of Sciences* 104 (2007): 20629–20634.

19. The discussion of roads and salt is based on R. B. Jackson, S. R. Carpenter, C. N. Dahm, and others, "Water in a Changing World," *Ecological Applications* 11 (2001): 1027–1045; Sujay S. Kaushal, Peter M. Groffman, Gene E. Likens, and others, "Increased Salinization of Fresh Water in the Northeastern United States," *Proceedings of the National Academy of Sciences* 102 (2005):13517–13520; and Robert B. Jackson and Esteban G. Jobbágy, "From Icy Roads to Salty Streams," *Proceedings of the National Academy of Sciences* 102 (2005): 14487–14488.

20. Roger M. Brown and David N. Laband, "Species Imperilment and Spatial Patterns of Development in the United States," *Conservation Biology* 20 (2006): 239–244.

21. The preceding two paragraphs are based on Eric A. Odell and Richard L. Knight, "Songbird and Medium-Sized Mammal Communities Associated with Exurban Development in Pitkin County, Colorado," *Conservation Biology* 15 (2001): 1143–1150; Buffy A. Lenth, Richard L. Knight, and Wendell C. Gilgert, "Conservation Value of Clustered Housing Developments," *Conservation Biology* 20 (2006): 1445–1456; and Vincent Devictor, Romain Julliard,

Denis Couvet, and others, "Functional Homogenization Effect of Urbanization on Bird Communities," *Conservation Biology* 21 (2007): 741–751.

22. K. Kumar, S. C. Gupta, S. K. Baidoo, and others, "Antibiotic Uptake by Plants from Soil Fertilized with Animal Manure," *Journal of Environmental Quality* 34 (2005): 2082–2085.

23. The foregoing discussion of the tradeoff of a commercial decision is based on Janet Raloff, "Lettuce Liability," *Science News* 172 (2007): 362–364, and Emma L. Sproston, M. Macrea, Iain D. Ogden, and others, "Slugs: Potential Novel Vectors of Escherichia coli O157," *Applied and Environmental Microbiology* 72 (2006): 144–149.

24. The foregoing discussion of kidney stones is based on Tom H. Brikowski, Yair Lotan, and Margaret S. Pearle, "Climate-Related Increase in the Prevalence of Urolithiasis in the United States," *Proceedings of the National Academy of Sciences* 105 (2008): 9841–9846; Margaret S. Pearle, Yair Lotan, and Tom H. Brikowski, "Predicted Climate-Related Increase in the Prevalence and Cost of Nephrolithiasis in the U.S.," *Journal of Urology* 179 (2008): 481–482; and Jeremy Manier, "Global Warming to Spark Rise in Kidney Stone Cases, Study Says Illinois Projected to See Increase by Year 2050," *Chicago Tribune*, July 14, 2008.

CHAPTER 2 OUR EVER-CHANGING LANDSCAPE PATTERNS

1. The preceding discussion of prehistoric England is based on I. G. Simmons, "The Earliest Cultural Landscapes of England," *Environmental Review* 12 (1988): 105–116; Philip Gibbard, "Palaeogeography: Europe Cut Adrift," *Nature* 448 (2007): 259–260; and Sanjeev Gupta, Jenny S. Collier, Andy Palmer-Felgate, and Graeme Potter, "Catastrophic Flooding Origin of Shelf Valley Systems in the English Channel," *Nature* 448 (2007): 342–345.

2. Brett Clark and Richard York, "Dialectical Materialism and Nature," *Organization & Environment* 18 (2005): 318–337.

3. Lester R. Brown, "A Copernican Shift," *Resurgence* 213 (2002): 14–15.

4. Andrew V. Suarez, Karin S. Pfennig, and Scott K. Robinson, "Nesting Success of a Disturbance-Dependent Songbird on Different Kinds of Edges," *Conservation Biology* 11 (1997): 928–935.

5. Denis A. Saunders, Richard J. Hobbs, and Chris R. Margules, "Biological Consequences of Ecosystem Fragmentation: A Review," *Conservation Biology* 5 (1991): 18–32.

6. Bruce C. Forbes, James J. Ebersole, and Beate Strandberg, "Anthropogenic Disturbance and Patch Dynamics in Circumpolar Arctic Ecosystems," *Conservation Biology* 15 (2001): 954–969.

7. John P. Smol, Alexander P. Wolfe, H. John B. Birks, and others, "Climate-Driven Regime Shifts in the Biological Communities of Arctic Lakes," *Proceedings of the National Academy of Sciences* 102 (2005): 4397–4402.

8. M. R. Besonen, W. Patridge, R. S. Bradley, and others, "A Record of Climate over the Last Millennium Based on Varved Lake Sediments from the Canadian High Arctic," *Holocene* 18 (2008): 169–180.

9. John P. Smol and Marianne S. V. Douglas, "Crossing the Final Ecological Threshold in High Arctic Ponds," *Proceedings of the National Academy of Sciences* 104 (2007): 12395–12397.

10. David M. Richardson, "Forestry Trees as Invasive Aliens," *Conservation Biology* 12 (1998): 18–26.

11. Kenneth V. Rosenberg, James D. Lowe, and André A. Dhondt, "Effects of Forest Fragmentation on Breeding Tanagers: A Continental Perspective," *Conservation Biology* 13 (1999): 568–583.

12. The discussion of habitat fragmentation and genetics is based on Ran Nathan and Gabriel G. Katul, "Foliage Shedding in Deciduous Forests Lifts Up Long-Distance Seed Dispersal by Wind," *Proceedings of the National Academy of Sciences* 102 (2005): 8251–8256; Cecile F. E. Bacles, Andrew J. Lowe, and Richard A. Ennos, "Effective Seed Dispersal across a Fragmented Landscape," *Science* 311 (2006): 628; and Alistair S. Jump and Josep Peñuelas, "Genetic Effects of Chronic Habitat Fragmentation in a Wind-Pollinated Tree," *Proceedings of the National Academy of Sciences* 103 (2006): 8096–8100.

13. Fred L. Bunnell, "Forest-Dwelling Vertebrate Faunas and Natural Fire Regimes in British Columbia: Patterns and Implications for Conservation," *Conservation Biology* 9 (1995): 636–644, and John M. Hagan, W. Matthew Vander Haegen, and Peter S. McKinley, "The Early Development of Forest Fragmentation Effects on Birds," *Conservation Biology* 10 (1996): 188–202.

14. Graciela Valladares, Adriana Salvo, and Luciano Cagnolo, "Habitat Fragmentation Effects on Trophic Processes of Insect-Plant Food Webs," *Conservation Biology* 20 (2006): 212–217.

15. The preceding discussion of roads is based on T. M. Spight, "The Water Economy of Salamanders: Evaporative Water Loss," *Physiological Zoology* 41 (1968): 195–203; J. R. Spotila, "Role of Temperature and Water in the Ecology of Lungless Salamanders," *Ecological Monographs* 42 (1972): 95–125; R. S. Vora, "Potential Soil Compaction Forty Years after Logging in Northeastern California," *Great Basin Naturalist* 48 (1988): 117–120; Rebecca A. Reed, Julia Johnson-Barnard, and William L. Baker, "Contribution of Roads to Forest Fragmentation in the Rocky Mountains," *Conservation Biology* 10 (1996): 1098–1106; T. Hels and E. Buchwald, "The Effect of Road Kills on Amphibian Populations," *Biological Conservation* 99 (2001): 331–340; David M. Marsh and Noelle G. Beckman, "Effects of Forest Roads on the Abundance and Activity of Terrestrial Salamanders," *Ecological Applications* 14 (2004): 1882–1891; M. J. Mazerolle, "Amphibian Road Mortality in Response to Nightly Variations in Traffic Intensity," *Herpetologica* 60 (2004): 45–53; David M. Marsh, Graham S. Milam, Nicholas P. Gorham, and Noelle G. Beckman, "Forest Roads as Partial Barriers to Terrestrial Salamander Movement," *Conservation Biology* 19 (2005): 2004–2008; and Raymond D. Semlitsch, Travis J. Ryan, Kevin Hamed, and others, "Salamander Abundance along Road Edges and within Abandoned Logging Roads in Appalachian Forests," *Conservation Biology* 21 (2007): 159–167.

16. Linda A. Dupuis, James N. M. Smith, and Fred L. Bunnell, "Relation of Terrestrial-Breeding Amphibian Abundance to Tree-Stand Age," *Conservation Biology* 9 (1995): 645–653.

17. The previous two paragraphs are based on Thomas W. Swetnam, "Forest Fire Primeval," *Natural Science* 3 (1988): 236–241; Wally W. Covington and M. M. Moore, *Changes in Forest Conditions and Multiresource Yields from Ponderosa Pine Forests since European Settlement* (unpublished report, submitted to J. Keane, Water Resources Operations, Salt River Project, Phoenix, AZ, 1991); and A. Joy Belsky and Dana M. Blumenthal, "Effects of Livestock Grazing on Stand Dynamics and Soils in Upland Forests of the Interior West," *Conservation Biology* 11 (1997): 315–327.

18. William S. Alverson, Donald M. Waller, and Stephen L. Solheim, "Forests Too Deer: Edge Effects in Northern Wisconsin," *Conservation Biology* 2 (1988): 348–358, and Sylvain Allombert, Steve Stockton, and Jean-Louis Martin, "A Natural Experiment on the Impact of Overabundant Deer on Forest Invertebrates," *Conservation Biology* 19 (2005): 1917–1929.

19. The discussion of witch's hair lichen is based on Per-Anders Esseen and Karl-Erik Renhorn, "Edge Effects on an Epiphytic Lichen in Fragmented Forests," *Conservation Biology* 12 (1998): 1307–1317.

20. Phillip G. Demaynadier and Malcolm L. Hunter, "Effects of Silvicultural Edges on the Distribution and Abundance of Amphibians in Maine," *Conservation Biology* 12 (1998): 340–352.

21. Stanley A. Temple and John R. Cary, "Modeling Dynamics of Habitat-Interior Bird Populations in Fragmented Landscapes," *Conservation Biology* 2 (1988): 340–347.

22. Katrin Böhning-Gaese, Mark L. Taper, and James H. Brown, "Are Declines in North American Insectivorous Songbirds Due to Causes on the Breeding Range?" *Conservation Biology* 7 (1993): 76–86; Amber J. Keyser, Geoffrey E. Hill, and Eric C. Soehren, "Effects of Forest Fragment Size, Nest Density, and Proximity to Edge on the Risk of Predation to Ground-Nesting Passerine Birds," *Conservation Biology* 12 (1998): 986–994; and Reiko Kurosawa and Robert A. Askins, "Effects of Habitat Fragmentation on Birds in Deciduous Forests in Japan," *Conservation Biology* 17 (2003): 695–707.

23. Chris Maser, *Mammals of the Pacific Northwest: From the Coast to the High Cascades* (Corvallis: Oregon State University Press, 1998), and William J. Zielinski and Steven T. Gellman, "Bat Use of Remnant Old-Growth Redwood Stands," *Conservation Biology* 13 (1999): 160–167.

24. Chris Maser, James M. Trappe, and Ronald A. Nussbaum, "Fungal–Small Mammal Interrelationships with Emphasis on Oregon Coniferous Forests," *Ecology* 59 (1978): 799–809; Douglas C. Ure and Chris Maser, "Mycophagy of Red-Backed Voles in Oregon and Washington," *Canadian Journal of Zoology* 60 (1982): 3307–3315; L. Scott Mills, "Edge Effects and Isolation: Red-Backed Voles on Forest Remnants," *Conservation Biology* 9 (1995): 395–403; and Maser, *Mammals of the Pacific Northwest*.

25. The previous two paragraphs are based on Timothy S. Brothers and Arthur Spingarn, "Forest Fragmentation and Alien Plant Invasion of Central Indiana Old-Growth Forests," *Conservation Biology* 6 (1992): 91–100.

26. The preceding three paragraphs are based on Whendee L. Silver, Sandra Brown, and Ariel E. Lugo, "Effects of Changes in Biodiversity on Ecosystem Function in Tropical Forests," *Conservation Biology* 10 (1996): 17–24; Robert John, James W. Dalling, Kyle E. Harms, and others, "Soil Nutrients Influence Spatial Distributions of Tropical Tree Species," *Proceedings of the National Academy of Sciences* 104 (2007): 864–869; and Benjamin Z. Houlton, Daniel M. Sigman, Edward A. G. Schuur, and Lars O. Hedin, "A Climate-Driven Switch in Plant Nitrogen Acquisition within Tropical Forest Communities," *Proceedings of the National Academy of Sciences* 104 (2007): 8902–8906.

27. Henning Steinfeld and Tom Wassenaar, "The Role of Livestock Production in Carbon and Nitrogen Cycles," *Annual Review of Environment and Resources* 32 (2007): 271–294.

28. The preceding two paragraphs are based on J. D. Majer and G. Beeston, "The Biodiversity Integrity Index: An Illustration Using Ants in Western Australia," *Conservation Biology* 10 (1996): 65–73, and Philip M. Fearnside, "Deforestation in Brazilian Amazonia: History, Rates, and Consequences," *Conservation Biology* 19 (2005): 680–688.

29. The previous two paragraphs are based on Ingolf Steffan-Dewenter, Michael Kessler, Jan Barkmann, and others, "Tradeoffs between Income, Biodiversity, and Ecosystem Functioning during Tropical Rainforest Conversion and Agroforestry Intensification," *Proceedings of the National Academy of Sciences* 104 (2007): 4973–4978.

30. William F. Laurance, Thomas E. Lovejoy, Heraldo L. Vasconcelos, and others, "Ecosystem Decay of Amazonian Forest Fragments: A 22-Year Investigation," *Conservation Biology* 16 (2002): 605–618.

31. The discussion of figs and fig wasps is based on Marie-Charlotte Anstett, Martine Hossaert-McKey, and Doyle McKey, "Modeling the Persistence of Small Populations of

Strongly Interdependent Species: Figs and Fig Wasps," *Conservation Biology* 11 (1997): 204–213; Drude Molbo, Carlos Machado, Jan Sevenster, and others, "Cryptic Species of Fig Pollinating Wasps: Implications for the Evolution of the Fig-Wasp Mutualism, Sex Allocation and Precision of Adaptation," *Proceedings of the National Academy of Sciences* 100 (2004): 5867–5872; and Ian Giddy, "Cloudbridge Nature Reserve," *Nature Notes*, no. 19 (2004), http://cloudbridge.org/fig-wasp.htm.

32. Marty S. Fujita and Merlin D. Tuttle, "Flying Foxes (Chiroptera: Pteropodidae): Threatened Animals of Key Ecological and Economic Importance," *Conservation Biology* 5 (1991): 455–463.

33. William F. Laurance, Henrique E. M. Nascimento, Susan G. Laurance, and others, "Rapid Decay of Tree-Community Composition in Amazonian Forest Fragments," *Proceedings of the National Academy of Sciences* 103 (2006): 19010–19014.

34. Andreas Hamann and Eberhard Curio, "Interactions among Frugivores and Fleshy Fruit Trees in a Philippine Submontane Rainforest," *Conservation Biology* 13 (1999): 766–773.

35. The previous two paragraphs are based on S. Joseph Wright, Horacio Zeballos, Iván Domínguez, and others, "Poachers Alter Mammal Abundance, Seed Dispersal, and Seed Predation in a Neotropical Forest," *Conservation Biology* 14 (2000): 227–239.

36. Jörg U. Ganzhorn, Joanna Fietz, Edmond Rakotovao, and others, "Lemurs and the Regeneration of Dry Deciduous Forest in Madagascar," *Conservation Biology* 13 (1999): 794–804.

37. The preceding two paragraphs are based on William D. Newmark, "Tropical Forest Fragmentation and the Local Extinction of Understory Birds in the Eastern Usambara Mountains, Tanzania," *Conservation Biology* 5 (1991): 67–78, and Jeffrey A. Stratford and Philip C. Stouffer, "Reduced Feather Growth Rates of Two Common Birds Inhabiting Central Amazonian Forest Fragments," *Conservation Biology* 15 (2001): 721–728.

38. The foregoing discussion of the Palouse prairie is based on W. Matthew Vander Haegen, Frederick C. Dobler, and D. John Pierce, "Shrubsteppe Bird Response to Habitat and Landscape Variables in Eastern Washington, U.S.A.," *Conservation Biology* 14 (2000): 1145–1160; Palouse Prairie Foundation, *The Palouse Prairie* (Moscow, ID: Palouse Prairie Foundation, 2002), www.palouseprairie.org; and R. Lee Lyman and Steve Wolverton, "The Late Prehistoric–Early Historic Game Sink in the Northwestern United States," *Conservation Biology* 16 (2002): 73–85.

39. Miguel A. Altieri, M. Kat Anderson, and Laura C. Merrick, "Peasant Agriculture and the Conservation of Crop and Wild Plant Resources," *Conservation Biology* 1 (1987): 49–58.

40. This paragraph and the previous two are based on Cathleen J. Wilson, Robin S. Reid, Nancy L. Stanton, and Brian D. Perry, "Effects of Land-Use and Tsetse Fly Control on Bird Species Richness in Southwestern Ethiopia," *Conservation Biology* 11 (1997): 435–447.

41. Gregory A. Jones, Kathryn E. Sieving, and Susan K. Jacobson, "Avian Diversity and Functional Insectivory on North-Central Florida Farmlands," *Conservation Biology* 19 (2005): 1234–1245.

42. The discussion of the conversion of forestland in Saskatchewan is based on Keith A. Hobson, Erin M. Bayne, and Steve L. Van Wilgenburg, "Large-Scale Conversion of Forest to Agriculture in the Boreal Plains of Saskatchewan," *Conservation Biology* 16 (2002): 1530–1541.

43. The discussion of agriculture and climate is based on John F. Morton, "The Impact of Climate Change on Smallholder and Subsistence Agriculture," *Proceedings of the National Academy of Sciences* 104 (2007): 19680–19685, and S. Mark Howden, Jean-François

Soussana, Francesco N. Tubiello, and others, "Adapting Agriculture to Climate Change," *Proceedings of the National Academy of Sciences* 104 (2007): 19691–19696.

44. Jeffrey C. Milder, James P. Lassoie, and Barbara L. Bedford, "Conserving Biodiversity and Ecosystem Function through Limited Development: An Empirical Evaluation," *Conservation Biology* 22 (2008): 70–79.

45. The discussion of sprawl is based on Volker C. Radeloff, Roger B. Hammer, and Susan I. Stewart, "Rural and Suburban Sprawl in the U.S. Midwest from 1940 to 2000 and Its Relation to Forest Fragmentation," *Conservation Biology* 19 (2005): 793–805.

46. Ken Thompson and Allan Jones, "Human Population Density and Prediction of Local Plant Extinction in Britain," *Conservation Biology* 13 (1999): 185–189, and Melinda G. Knutson, John R. Sauer, Douglas A. Olsen, and others, "Effects of Landscape Composition and Wetland Fragmentation on Frog and Toad Abundance and Species Richness in Iowa and Wisconsin, U.S.A.," *Conservation Biology* 13 (1999): 1437–1446.

47. Grzegorz Mikusiński and Per Angelstam, "Economic Geography, Forest Distribution, and Woodpecker Diversity in Central Europe," *Conservation Biology* 12 (1998): 200–208, and Z. Naveh, "Biodiversity to Ecodiversity: A Landscape Ecological Approach to Conservation and Restoration," *Restoration Ecology* 2 (1994): 180–189.

48. The preceding discussion of ice in the ocean is based on I. Marianne Lagerklint and James D. Wright, "Late Glacial Warming Prior to Heinrich Event 1; the Influence of Ice Rafting and Large Ice Sheets on the Timing of Initial Warming," *Geology* 27 (1999): 1099–1102; D. Genty, D. Blamart, B. Ghaleb, and others, "Timing and Dynamics of the Last Deglaciation from European and North African Δ^{13}c Stalagmite Profiles—Comparison with Chinese and South Hemisphere Stalagmites," *Quaternary Science Reviews* 25 (2006): 2118–2142; C. J. Mundy, D. G. Barber, and C. Michel, "Variability of Snow and Ice Thermal, Physical and Optical Properties Pertinent to Sea Ice Algae Biomass during Spring," *Journal of Marine Systems* 58 (2005): 107–120; and Volker Mohrholz, Jörg Dutz, and Gerd Kraus, "The Impact of Exceptionally Warm Summer Inflow Events on the Environmental Conditions in the Bornholm Basin," *Journal of Marine Systems* 60 (2006): 285–301.

49. The preceding discussion of the ocean is based on Martin V. Angel, "Biodiversity of the Pelagic Ocean," *Conservation Biology* 7 (1993): 760–772; I. Cacho, J. O. Grimalt, and M. Canals, "Response of the Western Mediterranean Sea to Rapid Climatic Variability during the Last 50,000 Years: A Molecular Biomarker Approach," *Journal of Marine Systems* 33–34 (2002): 253–272; Nathaniel B. Weston and Samantha B. Joye, "Temperature-Driven Decoupling of Key Phases of Organic Matter Degradation in Marine Sediments," *Proceedings of the National Academy of Sciences* 102 (2005): 17036–17040; C. Lin, X. Ning, J. Su, and others, "Environmental Changes and the Responses of the Ecosystems of the Yellow Sea during 1976–2000," *Journal of Marine Systems* 55 (2005): 223–234; M. A. Tobor-Kapon, J. Bloem, P.F.A.M. Romkens, and P. C. de Ruiter, "Functional Stability of Microbial Communities in Contaminated Soils near a Zinc Smelter (Budel, the Netherlands)," *Ecotoxicology* 15 (2006): 187–197; Georgi M. Daskalov, Alexander N. Grishin, Sergei Rodionov, and Vesselina Mihneva, "Trophic Cascades Triggered by Overfishing Reveal Possible Mechanisms of Ecosystem Regime Shifts," *Proceedings of the National Academy of Sciences* 104 (2007): 10518–10523; Z. V. Finkel, J. Sebbo, S. Feist-Burkhardt, and others, "A Universal Driver of Macroevolutionary Change in the Size of Marine Phytoplankton over the Cenozoic," *Proceedings of the National Academy of Sciences* 104 (2007): 20416–20420; H. Zhou, Z. N. Zhang, X. S. Liu, and others, "Changes in the Shelf Macrobenthic Community over Large Temporal and Spatial Scales in the Bohai Sea, China," *Journal of Marine Systems* 67(2007): 312–321; Moriaki Yasuhara, Thomas M. Cronin, Peter B. deMenocal, and others,

"Abrupt Climate Change and Collapse of Deep-Sea Ecosystems," *Proceedings of the National Academy of Sciences* 105 (2008): 1556–1560; James G. Mitchell, Hidekatsu Yamazaki, Laurent Seuront, and others, "Phytoplankton Patch Patterns: Seascape Anatomy in a Turbulent Ocean," *Journal of Marine Systems* 69 (2008): 247–253; Roman Stocker, Justin R. Seymour, Azadeh Samadani, and others, "Rapid Chemotactic Response Enables Marine Bacteria to Exploit Ephemeral Microscale Nutrient Patches," *Proceedings of the National Academy of Sciences* 105 (2008): 4209–4214; and Robert Ptacnik, Angelo G. Solimini, Tom Andersen, and others, "Diversity Predicts Stability and Resource Use Efficiency in Natural Phytoplankton Communities," *Proceedings of the National Academy of Sciences* 105 (2008): 5134–5138.

50. Charles A. Acosta, "Benthic Dispersal of Caribbean Spiny Lobsters among Insular Habitats: Implications for the Conservation of Exploited Marine Species," *Conservation Biology* 13 (1999): 603–612; "Seagrass," en.wikipedia.org/wiki/Seagrass (accessed May 5, 2008; updated October 27, 2008); and Department of Environmental Protection, Florida Marine Research Institute, "Seagrasses," www.dep.state.fl.us/coastal/habitats/seagrass/ (accessed May 5, 2008).

51. The previous two paragraphs are based on Fiorenza Micheli and Charles H. Peterson, "Estuarine Vegetated Habitats as Corridors for Predator Movements," *Conservation Biology* 13 (1999): 869–881.

52. Callum M. Roberts, "Rapid Build-Up of Fish Biomass in a Caribbean Marine Reserve," *Conservation Biology* 9 (1995): 815–826; Jane Lubcheco, Steven Gaines, Kirsten Grorud-Colvert, and others, "The Science of Marine Reserves," *Partnership for Interdisciplinary Studies of Coastal Oceans* (2007): 1–21; and Peter H. Taylor, "Coastal Connections," *Partnership for Interdisciplinary Studies of Coastal Oceans* 6 (2007): 1–17.

53. The preceding discussion of gray whales is based on S. Elizabeth Alter, Eric Rynes, and Stephen R. Palumbi, "DNA Evidence for Historic Population Size and Past Ecosystem Impacts of Gray Whales," *Proceedings of the National Academy of Sciences* 104 (2007): 15162–15167.

CHAPTER 3   HOW SPECIES ENRICH OUR LIVES AND THE WORLD

1. The preceding discussion of the parrot trade is based on Timothy F. Wright, Catherine A. Toft, Ernesto Enkerlin-Hoeflich, and others, "Nest Poaching in Neotropical Parrots," *Conservation Biology* 15 (2001): 710–720; Popko Wiersma, Agustí Muñoz-Garcia, Amy Walker, and Joseph B. Williams, "Tropical Birds Have a Slow Pace of Life," *Proceedings of the National Academy of Sciences* 104 (2007): 9340–9345; and W. Douglas Robinson, John D. Styrsky, Brian J. Payne, and others, "Why Are Incubation Periods Longer in the Tropics? A Common-Garden Experiment with House Wrens Reveals It Is All in the Egg," *American Naturalist* 171 (2008): 532–535.

2. Niclas Kolm and Anders Berglund, "Wild Populations of a Reef Fish Suffer from the 'Nondestructive' Aquarium Trade Fishery," *Conservation Biology* 17 (2003): 910–914.

3. The previous two paragraphs are based on William J. McShea and John H. Rappole, "Managing the Abundance and Diversity of Breeding Bird Populations through Manipulation of Deer Populations," *Conservation Biology* 14 (2000): 1161–1170.

4. Robert L. Beschta, "Cottonwoods, Elk, and Wolves in the Lamar Valley of Yellowstone National Park," *Ecological Applications* 13 (2003): 1295–1309; William J. Ripple and Robert L. Beschta, "Wolf Reintroduction, Predation Risk, and Cottonwood Recovery in Yellowstone National Park," *Forest Ecology and Management* 184 (2003): 299–313; William J. Ripple and Robert L. Beschta, "Wolves and the Ecology of Fear: Can Predation

Risk Structure Ecosystems?" *BioScience* 54 (2004): 755–766; William J. Ripple and Robert L. Beschta, "Wolves, Elk, Willows, and Trophic Cascades in the Upper Gallatin Range of Southwestern Montana, USA," *Forest Ecology and Management* 200 (2004): 161–181; and Robert L. Beschta, "Reduced Cottonwood Recruitment Following Extirpation of Wolves in Yellowstone's Northern Range," *Ecology* 86 (2005): 391–403.

5. Evan L. Preisser, Daniel I. Bolnick, and Michael F. Benard, "Scared to Death? The Effects of Intimidation and Consumption in Predator–Prey Interactions," *Ecology* 86 (2005): 501–509.

6. The foregoing discussion of pumas is based on Maurice G. Hornocker, "Winter Territoriality in Mountain Lions," *Journal of Wildlife Management* 33 (1969): 457–464, and Maurice G. Hornocker, "An Analysis of Mountain Lion Predation upon Mule Deer and Elk in the Idaho Primitive Area," *Wildlife Monograph* 21 (1970): 1–39.

7. The preceding story of the feral cats is from Chris Maser with Zane Maser, *The World Is in My Garden: A Journey of Consciousness* (Ashland, OR: White Cloud Press, 2005).

8. The previous two paragraphs are based on M. Nils Peterson, Roel R. Lopez, Edward J. Laurent, and others, "Wildlife Loss through Domestication: The Case of Endangered Key Deer," *Conservation Biology* 19 (2005): 939–944.

9. The foregoing discussion of domesticating animals is based on Chris Maser, *The Perpetual Consequences of Fear and Violence: Rethinking the Future* (Washington, DC: Maisonneuve Press, 2004).

10. Kate A. Brauman, Gretchen C. Daily, T. Ka'eo Duarte, and Harold A. Mooney, "The Nature and Value of Ecosystem Services: An Overview Highlighting Hydrologic Services," *Annual Review of Environment and Resources* 32 (2007): 67–98.

11. Unless otherwise noted, the following discussion of pollination is based on Janet N. Abramovitz, "Learning to Value Nature's Free Services," *Futurist* 31 (1997): 39–42; Stephen L. Buchmann and Gary Paul Nabhan, *The Forgotten Pollinators* (Washington, DC: Island Press, 1997); J. C. Biesmeijer, S.P.M. Roberts, M. Reemer, and others, "Parallel Declines in Pollinators and Insect-Pollinated Plants in Britain and the Netherlands," *Science* 313 (2006): 351–354; and Sandra Díaz, Sandra Lavorel, Francesco de Bello, and others, "Incorporating Plant Functional Diversity Effects in Ecosystem Service Assessments," *Proceedings of the National Academy of Sciences* 104 (2007): 20684–20689.

12. Quoted in Steve Newman, "Earthweek: A Diary of the Planet," Albany (OR) *Democrat-Herald*, Corvallis (OR) *Gazette-Times*, June 6, 1999.

13. Çaan H. Ekerciolu, Gretchen C. Daily, and Paul R. Ehrlich, "Ecosystem Consequences of Bird Declines," *Proceedings of the National Academy of Sciences* 101 (2004): 18042–18047.

14. The previous two paragraphs are based on Peter B. McIntyre, Laura E. Jones, Alexander S. Flecker, and Michael J. Vanni, "Fish Extinctions Alter Nutrient Recycling in Tropical Freshwaters," *Proceedings of the National Academy of Sciences* 104 (2007): 4461–4466.

15. Matthew E. S. Bracken, Sara E. Friberg, Cirse A. Gonzalez-Dorantes, and Susan L. Williams, "Functional Consequences of Realistic Biodiversity Changes in a Marine Ecosystem," *Proceedings of the National Academy of Sciences* 105 (2008): 924–928.

16. Louise H. Emmons, "Tropical Rain Forests: Why They Have So Many Species, and How We May Lose This Biodiversity without Cutting a Single Tree," *Orion* 8 (1989): 8–14.

17. The foregoing discussion of feedback loops in the African and South American rain forests is based on ibid.; William F. Laurance, Henrique E. M. Nascimento, Susan G. Laurance, and others, "Inferred Longevity of Amazonian Rainforest Trees Based on a Long-Term Demographic Study," *Forest Ecology and Management* 190 (2004): 131–143; P. Jordano, C. García, J. A. Godoy, and J. L. García-Castaño, "Differential Contribution of

Frugivores to Complex Seed Dispersal Patterns," *Proceedings of the National Academy of Sciences* 104 (2007): 3278–3282; and Elsa Youngsteadt, Satoshi Nojima, Christopher Häberlein, and others, "Seed Odor Mediates an Obligate Ant-Plant Mutualism in Amazonian Rainforests," *Proceedings of the National Academy of Sciences* 105 (2008): 4571–4575. Information about Al Gentry comes from Emmons, "Tropical Rain Forests."

18. Michael Soulé, "Allozyme Variation: Its Determinant in Space and Time," in *Molecular Evolution*, ed. F. J. Ayala, 60–77 (Sunderland, MA: Sinauer Associates, 1976); Richard Frankham, "Relationship of Genetic Variation to Population Size in Wildlife," *Conservation Biology* 10 (1996): 1500–1508; and M. A. Sanjayan, Kevin Crooks, Gerard Zegers, and David Foran, "Genetic Variation and the Immune Response in Natural Populations of Pocket Gophers," *Conservation Biology* 10 (1996): 1519–1527.

19. The preceding discussion of changes in the Amazon forest is based on Carol Savonen, "Ashes in the Amazon," *Journal of Forestry* 88 (1990): 20–25, and William F. Laurance and G. Bruce Williamson, "Positive Feedbacks among Forest Fragmentation, Drought, and Climate Change in the Amazon," *Conservation Biology* 15 (2001): 1529–1535; and Shilong Piao, Pierre Friedlingstein, Philippe Ciais, and others, "Changes in Climate and Land Use Have a Larger Direct Impact Than Rising CO_2 on Global River Runoff Trends," *Proceedings of the National Academy of Sciences* 104 (2007): 15242–15247.

20. James P. M. Syvitski, Charles J. Vörösmarty, Albert J. Kettner, and Pamela Green, "Impact of Humans on the Flux of Terrestrial Sediment to the Global Coastal Ocean," *Science* 308 (2005): 376–380; Christer Nilsson, Catherine A. Reidy, Mats Dynesius, and Carmen Revenga, "Fragmentation and Flow Regulation of the World's Large River Systems," *Science* 308 (2005): 405–408; and Bruce H. Wilkinson and Brandon J. McElroy, "The Impact of Humans on Continental Erosion and Sedimentation," *Geological Society of America Bulletin* 119 (2007): 140–156.

21. Chris Maser, "Abnormal Coloration in *Microtus montanus*," *Murrelet* 50 (1969): 39.

22. Donald Sparling, Gary M. Fellers, and Laura L. McConnell, "Pesticides and Amphibian Population Declines in California, USA," *Environmental Toxicology and Chemistry* 20 (2001): 1591–1595; Carlos Davidson and Roland A. Knapp, "Multiple Stressors and Amphibian Declines: Dual Impacts of Pesticides and Fish on Yellow-Legged Frogs," *Ecological Applications* 17 (2007): 587–597; Krista A. McCoy, Lauriel J. Bortnick, Chelsey M. Campbell, and others, "Agriculture Alters Gonadal Form and Function in the Toad *Bufo marinus*," *Environmental Health Perspectives* (2008): doi:10.1289/ehp.11536 (available at http://dx.doi.org/); and Tana V. McDaniel, Pamela A. Martin, John Struger, and others, "Potential Endocrine Disruption of Sexual Development in Free Ranging Male Northern Leopard Frogs (*Rana pipiens*) and Green Frogs (*Rana clamitans*) from Areas ofIntensive Row Crop Agriculture," *Aquatic Toxicology* 88 (2008): 230–242.

23. Shahid Naeem, "Species Redundancy and Ecosystem Reliability," *Conservation Biology* 12 (1998): 39–45.

24. The preceding discussion of plant-animal interactions is based on Alejandra I. Roldán and Javier A. Simonetti, "Plant-Mammal Interactions in Tropical Bolivian Forests with Different Hunting Pressures," *Conservation Biology* 15 (2001): 617–623.

25. The preceding discussion of bushmeat is based on David S. Wilkie, John G. Sidle, and Georges C. Boundzanga, "Mechanized Logging, Market Hunting, and a Bank Loan in Congo," *Conservation Biology* 6 (1992): 570–580; Emmanuel de Merode and Guy Cowlishaw, "Species Protection, the Changing Informal Economy, and the Politics of Access to the Bushmeat Trade in the Democratic Republic of Congo," *Conservation Biology* 20 (2006): 1262–1271; Guy Cowlishaw, Samantha Mendelson, and J. Marcus Rowcliffe,

"Structure and Operation of a Bushmeat Commodity Chain in Southwestern Ghana," *Conservation Biology* 19 (2005): 139–149; Richard E. Bodmer, John F. Eisenberg, and Kent H. Redford, "Hunting and the Likelihood of Extinction of Amazonian Mammals," *Conservation Biology* 11 (1997): 460–466; Carlos A. Peres, "Effects of Subsistence Hunting on Vertebrate Community Structure in Amazonian Forests," *Conservation Biology* 14 (2000): 240–253; Roldán and Simonetti, "Plant-Mammal Interactions in Tropical Bolivian Forests with Different Hunting Pressures"; Carlos A. Peres, "Synergistic Effects of Subsistence Hunting and Habitat Fragmentation on Amazonian Forest Vertebrates," *Conservation Biology* 15 (2001): 1490–1505; John E. Fa, Carlos A. Peres, and Jessica Meeuwig, "Bushmeat Exploitation in Tropical Forests: An Intercontinental Comparison," *Conservation Biology* 16 (2002): 232–237; Marc Thibault and Sonia Blaney, "The Oil Industry as an Underlying Factor in the Bushmeat Crisis in Central Africa," *Conservation Biology* 17 (2003): 1807–1813; David S. Wilkie, Malcolm Starkey, Kate Abernethy, and others, "Role of Prices and Wealth in Consumer Demand for Bushmeat in Gabon, Central Africa," *Conservation Biology* 19 (2005): 268–274; William F. Laurance, Barbara M. Croes, Landry Tchignoumba, and others, "Impacts of Roads and Hunting on Central African Rainforest Mammals," *Conservation Biology* 20 (2006): 1251–1261; A. Alonso, M. E. Lee, P. Campbell, and others, eds., "Gamba, Gabon: Biodiversity of an Equatorial African Rainforest," *Bulletin of the Biological Society of Washington*, no. 12 (2006): 1–448; Smithsonian Institution, *Gabon Biodiversity Program Briefing Paper 7* (2006): 1–32.

CHAPTER 4   THE NEVER-ENDING STORIES OF CAUSE, EFFECT, AND CHANGE

1. The foregoing discussion is based on Chris Maser and James M. Trappe, technical eds., *The Seen and Unseen World of the Fallen Tree*, USDA Forest Service General Technical Report PNW-164 (Portland, OR: Pacific Northwest Forest and Range Experiment Station, 1984); Chris Maser, *Forest Primeval: The Natural History of an Ancient Forest* (San Francisco: Sierra Club Books, 1989; rpt. Corvallis: Oregon State University Press, 2001); and Chris Maser, *Ecological Diversity in Sustainable Development: The Vital and Forgotten Dimension* (Boca Raton, FL: Lewis, 1999).

2. The foregoing discussion of whale carcasses is based on G. W. Rouse, S. K. Goffredi, and R. C. Vrijenhoek, "*Osedax*: Bone-Eating Marine Worms with Dwarf Males," *Science* 305 (2004): 668–671; T. G. Dahlgren, A. G. Glover, A. Baco, and C. R. Smith, "Fauna of Whale Falls: Systematics and Ecology of Anew Polychaete (*Annelida: Chrysopetalidae*) from the deep Pacific Ocean," *Deep Sea Research Part I: Oceanographic Research Papers* 51 (2004): 1873–1887; and Susan Milius, "Decades of Dinner: Underwater Community Begins with the Remains of a Whale," *Science News* 167 (2005):298–300.

3. Alice Friedemann, "Peak Soil: Why Cellulosic Ethanol, Biofuels Are Unsustainable and a Threat to America," *Cultural Change* (an online journal of the Sustainable Energy Institute, formerly Fossil Fuels Policy Action). http://culturechange.org/cms/index.php?option=com_content&task=view&id=107&Itemid=1 (accessed November 5, 2007).

4. A. Barrionuevo, "A Bet on Ethanol, with a Convert at the Helm," *New York Times*, October 8, 2006 (cited in Friedemann, "Peak Soil"); Ambuj D. Sagar and Sivan Kartha, "Bioenergy and Sustainable Development?" *Annual Review of Environment and Resources* 32 (2007): 131–167; Joel K. Bourne Jr., "Biofuels," *National Geographic Magazine*, August 2007, 41–58; and Seth Borenstein, "Food Scientists Say Stop Biofuels to Fight World Hunger," April 29, 2008, http://abcnews.go.com/Technology/Weather/ wireStory?id=4751187.

5. Wolfgang Haber, "Energy, Food, and Land—The Ecological Traps of Humankind," *Environmental Science and Pollution Research* 14 (2007): 359–365.

6. Natural Resources Conservation Service, United States Department of Agriculture, "Soil Quotations," http://soils.usda.gov/education/resources/k_12/quotes/ (accessed August 10, 2007).

7. J. R. McNeill and Verena Winiwarter, "Breaking the Sod: Humankind, History, and Soil," *Science* 304 (2004): 1627–1629.

8. Niles Eldredge, "Will Malthus Be Right?" *Time*, November 8, 1999, 102–103.

9. Jocelyn Kaiser, "Wounding Earth's Fragile Skin," *Science* 304 (2004): 1616–1618.

10. Eric Schmitt, "U.S. Population Has Biggest 10-Year Rise Ever," *New York Times*, April 3, 2001; Carol Savonen, "Population Growth: A Blessing or a Curse?" in *Looking for Oregon's Future* (Corvallis: Oregon State University Extension Service, 2001); and Jim Rydingsword, "Longevity, for Its Own Sake, Is Not Enough," Corvallis (OR) *Gazette-Times*, March 25, 2002.

11. Quoted in W. C. Lowdermilk, "Conquest of the Land through Seven Thousand Years," *Agricultural Information Bulletin* no. 99, Soil Conservation Service, United States Department of Agriculture (Washington, DC: United States Government Printing Office, 1975).

12. I. M. Young and J. W. Crawford, "Interactions and Self-Organization in the Soil-Microbe Complex," *Science* 304 (2004): 1637.

13. G. C. Daily, P. A. Matson, and P. M. Vitousek, "Ecosystem Services Supplied by Soil," in *Nature's Services: Societal Dependence on Natural Ecosystems*, ed. G. C. Daily, 113–132 (Washington, DC: Island Press, 1997); Gretchen C. Daily, Susan Alexander, Paul R. Ehrlich, and others, "Ecosystem Services: Benefits Supplied to Human Societies by Natural Ecosystems," *Issues in Ecology* 2 (1997): 1–16; Elizabeth Pennisi, "The Secret Life of Fungi," *Science* 304 (2004): 1620–1622; David A. Wardle, Richard D. Bardgett, John N. Klironomos, and others, "Ecological Linkages between Aboveground and Belowground Biota," *Science* 304 (2004): 1629–1633; and Young and Crawford, "Interactions and Self-Organization in the Soil-Microbe Complex."

14. Henry A. Wallace, Foreword to *Soils & Men: Yearbook of Agriculture*, ed. G. Hambridge (Washington, DC: United States Government Printing Office, 1938).

15. G. Hambridge, ed., *Soils & Men: Yearbook of Agriculture* (Washington, DC: United States Government Printing Office, 1938).

16. L. R. Oldeman, V. van Engelen, and J. Pulles, "The Extent of Human-Induced Soil Degradation," in *World Map of the Status of Human-Induced Soil Degradation: An Explanatory Note*, rev. 2d ed., Annex 5, ed. L. R. Oldeman, R.T.A. Hakkeling, and W. G. Sombroek (Wageningen, Netherlands: International Soil Reference and Information Centre, 1990).

17. Hans Jenny, *The Soil Resource, Origin and Behaviour* (New York: Springer-Verlag, 1980).

18. Elaine R. Ingham, "Organisms in the Soil: The Functions of Bacteria, Fungi, Protozoa, Nematodes, and Arthropods," *Natural Resource News* 5 (1995): 10–12, 16–17; and Andrew Sugden, Richard Stone, and Caroline Ash, "Ecology in the Underworld," *Science* 304 (2004): 1613–1620.

19. The discussion of soil microorganisms is based on Noah Fierer and Robert B. Jackson, "The Diversity and Biogeography of Soil Bacterial Communities," *Proceedings of the National Academy of Sciences* 103 (2006): 626–631.

20. Stephan Hättenschwiler and Patrick Gasser, "Soil Animals Alter Plant Litter Diversity Effects on Decomposition," *Proceedings of the National Academy of Sciences* 102 (2005): 1519–1524.

21. This paragraph is based on Albert Tietema, Claus Beier, Pieter H. B. de Visser, and others, "Nitrate Leaching in Coniferous Forest Ecosystems: The European Field-Scale

Manipulation Experiments NITREX (Nitrogen Saturation Experiments) and EXMAN (Experimental Manipulation of Forest Ecosystems)," *Global Biogeochemical Cycles* 11 (1997): 617–626, and Paul J. Squillace, Michael J. Morgan, Wayne W. Lapham, and others, "Volatile Organic Compounds in Untreated Ambient Groundwater of the United States, 1985–1995," *Environmental Science & Technology* 33 (1999): 4176–4187.

22. Paul D. Thacker, "Efforts to Stop Groundwater Pollution Disappoint," *Environmental Science & Technology* 40 (2006): 4815–4816.

23. Eugene P. Odum, *Ecology and Our Endangered Life Support Systems* (Stanford, CT: Sinauer Associates, 1989).

24. R. Lal, "Soil Carbon Sequestration Impacts on Global Climate Change and Food Security," *Science* 304 (2004): 1623–1627; Kees-Jan van Groenigen, Johan Six, Bruce A. Hungate, and others, "Element Interactions Limit Soil Carbon Storage," *Proceedings of the National Academy of Sciences* 103 (2006): 6571–6574; and Peter B. Reich, Sarah E. Hobbie, Tali Lee, and others, "Nitrogen Limitation Constrains Sustainability of Ecosystem Response to CO_2," *Nature* 440 (2006): 922–925.

25. Donald Worster, "Good Farming and the Public Good in Meeting the Expectations of the Land," in *Meeting the Expectations of the Land*, ed. Wes Jackson, Wendell Berry, and Bruce Colman (San Francisco: North Point Press, 1984), 39.

26. Lowdermilk, "Conquest of the Land Through 7,000 Years."

27. Friedemann, "Peak Soil."

28. Richard Manning, "The Oil We Eat," *Harper's*, February 2004, 37–45.

29. Evan Peacock, Wendell R. Haag, and Melvin L. Warren Jr., "Prehistoric Decline in Freshwater Mussels Coincident with the Advent of Maize Agriculture," *Conservation Biology* 19 (2005): 547–551.

30. Susan K. Jacobson, Kathryn E. Sieving, Gregory A. Jones, and Annamaria Van Doorn, "Assessment of Farmer Attitudes and Behavioral Intentions toward Bird Conservation on Organic and Conventional Florida Farms," *Conservation Biology* 17 (2003): 595–606; Liat P. Wickramasinghe, Stephen Harris, Gareth Jones, and Nancy Vaughan Jennings, "Abundance and Species Richness of Nocturnal Insects on Organic and Conventional Farms: Effects of Agricultural Intensification on Bat Foraging," *Conservation Biology* 18 (2004): 1283–1292; and Tatyana A. Rand and Svata M. Louda, "Spillover of Agriculturally Subsidized Predators as a Potential Threat to Native Insect Herbivores in Fragmented Landscapes," *Conservation Biology* 20 (2006): 1720–1729.

31. Maria D. Álvarez, "Illicit Crops and Bird Conservation Priorities in Colombia," *Conservation Biology* 16 (2002): 1086–1096.

32. Ibid.

33. Mary F. Willson, Joan L. Morrison, Kathryn E. Sieving, and others, "Patterns of Predation Risk and Survival of Bird Nests in a Chilean Agricultural Landscape," *Conservation Biology* 15 (2001): 447–456.

34. The preceding three paragraphs are based on Ricardo Steinbrecher, "What Is Wrong with Nature?" *Resurgence* 188 (1998): 16–19.

35. Antonio Regalado and Meeyoung Song, "Furor over Cross-Species Cloning," *Wall Street Journal*, March 19, 2002, and David Ehrenfeld, "Transgenics and Vertebrate Cloning as Tools for Species Conservation," *Conservation Biology* 20 (2006): 723–732.

36. The preceding four paragraphs are based on Lester R. Brown, "Can We Raise Grain Yields Fast Enough?" *World Watch* 10 (1997): 8–17.

37. The preceding paragraphs on nitrogen fixation are based on J. Allan Downie and Simon A. Walker, "Plant Responses to Nodulation Factors," *Current Opinion in Plant*

Biology 2 (1999): 483–489; Margaret E. McCully, "Roots in Soil: Unearthing the Complexities of Roots and Their Rhizospheres," *Annual Review of Plant Physiology and Plant Molecular Biology* 50 (1999): 695–718; Martin Parniske and J. Allan Downie, "Plant Biology: Locks, Keys and Symbioses," *Nature* 425 (2003): 569–570; Angie Lee and Ann M. Hirsch, "Signals and Responses: Choreographing the Complex Interaction between Legumes and Alpha- and Beta-Rhizobia," *Plant Signaling & Behavior* 1 (2006): 161–168; R. J. Yates, J. G. Howieson, W. G. Reeve, and others, "Host-Strain Mediated Selection for an Effective Nitrogen-Fixing Symbiosis between *Trifolium* spp. and *Rhizobium legumiosarum biovar trifolii*," *Soil Biology and Biochemistry* 40 (2008): 822–833; Susan Milius, "Out of Thin Air," *Science News* 173 (2008): 235–237; and Hassen Gherbi, Katharina Markmann, Sergio Svistoonoff, and others, "SymRK Defines a Common Genetic Basis for Plant Root Endosymbioses with Arbuscular Mycorrhiza Fungi, Rhizobia, and *Frankia* Bacteria," *Proceedings of the National Academy of Sciences* 105 (2008): 4928–4932.

38. Jennifer E. Fox, Jay Gulledge, Erika Engelhaupt, and others, "Pesticides Reduce Symbiotic Efficiency of Nitrogen-Fixing Rhizobia and Host Plants," *Proceedings of the National Academy of Sciences* 104 (2007): 10282–10287.

39. Katharine N. Suding, Scott L. Collins, Laura Gough, and others, "Functional- and Abundance-Based Mechanisms Explain Diversity Loss Due to N Fertilization," *Proceedings of the National Academy of Sciences* 102 (2005): 4387–4392, and Tim P. Barnett and David W. Pierce, "When Will Lake Mead Go Dry?" *Water Resources Research* 44 (2008), W03201, doi:10.1029/2007WR006704.

40. Tad W. Patzek, "Thermodynamics of the Corn-Ethanol Biofuel Cycle," *Critical Reviews in Plant Science* 23 (2004): 519–567; David Pimentel and Tad W. Patzek, "Ethanol Production Using Corn, Switchgrass, and Wood; Biodiesel Production Using Soybean and Sunflower," *Natural Resources Research* 14 (2005): 1, 65–76; Jason Hill, Erik Nelson, David Tilman, and others, "Environmental, Economic, and Energetic Costs and Benefits of Biodiesel and Ethanol Biofuels," *Proceedings of the National Academy of Sciences* 103 (2006): 11206–11210; Manfred Kroger, "Forum: Corn Is Food, Not Fuel," *Pittsburgh Post-Gazette*, April 8, 2007; Friedemann, "Peak Soil"; Lian Pin Koh, "Potential Habitat and Biodiversity Losses from Intensified Biodiesel Feedstock Production," *Conservation Biology* 21 (2007): 1373–1375; and Sid Perkins, "Groundwater Use Adds CO_2 to the Air," *Science News* 172 (2007): 301.

41. Discussion of the dead zone is based on Thomas O'Connor and David Whitall, "Linking Hypoxia to Shrimp Catch in the Northern Gulf of Mexico," *Marine Pollution Bulletin* 54 (2007): 460–463; Donald Scavia and Kristina A. Donnelly, "Reassessing Hypoxia Forecasts for the Gulf of Mexico," *Environmental Science & Technology* 41 (2007): 8111–8117; and Sarah C. Williams, "Dead Serious," *Science News* 172 (2007): 395–396.

42. The preceding two paragraphs are based on Hao Wei, Yunchang He, Qingji Li, and others, "Summer Hypoxia Adjacent to the Changjiang Estuary," *Journal of Marine Systems* 67 (2007): 292–303.

43. Reported in Perkins, "Groundwater Use Adds CO_2 to the Air,"

44. Ibid.

45. Suding, Collins, Gough, and others, "Functional- and Abundance-Based Mechanisms Explain Diversity Loss Due to N Fertilization," and Barnett and Pierce, "When Will Lake Mead Go Dry?"

46. Quoted in Alister Doyle, "'Water Wars' Loom? But None in Past 4,500 Years," Reuters, September 18, 2006, http://geo.oregonstate.edu/events/Press_2006/20060918_water-wars.pdf.

47. The preceding two paragraphs are based on Brown, "Can We Raise Grain Yields Fast Enough?"

48. The preceding discussion of ocean acidification is based on Helmuth Thomas and Venugopalan Ittekkot, "Determination of Anthropogenic CO_2 in the North Atlantic Ocean Using Water Mass Ages and CO_2 Equilibrium Chemistry," *Journal of Marine Systems* 27 (2001): 325–336; Jonathan Shaw, "The Great Global Experiment: As Climate Change Accelerates, How Will We Adapt to a Changed Earth?" *Harvard Magazine* 105 (2002): 34–43, 87–90; Kathy Tedesco, Richard A. Feely, Christopher L. Sabine, and Cathrine E. Cosca, "Impacts of Anthropogenic Co2 on Ocean Chemistry and Biology," *NOAA Archive of Spotlight Feature Articles*, 2005, http://www.oar.noaa.gov/spotlite/ spot_gcc.html; Lisa Stiffler, "Research in Pacific Shows Ocean Trouble," *Seattle Post Intelligencer*, March 31, 2006; Ruth Bibby, Polly Cleall-Harding, Simon Rundle, and others, "Ocean Acidification Disrupts Induced Defences in the Intertidal Gastropod *Littorina littorea*," *Biology Letters* 3 (2007): 699–701; Scott C. Doney, Natalie Mahowald, Ivan Lima, and others, "Impact of Anthropogenic Atmospheric Nitrogen and Sulfur Deposition on Ocean Acidification and the Inorganic Carbon System," *Proceedings of the National Academy of Sciences* 104 (2007): 14580–14585; Rosane Gonçalves Ito, Bernd Schneider, and Helmuth Thomas, "Distribution of Surface CO_2 and Air-Sea Fluxes in the Southwestern Subtropical Atlantic and Adjacent Continental Shelf," *Journal of Marine Systems* 56 (2005): 227–242; J. C. Blackford and F. J. Gilbert, "pH Variability and CO_2 Induced Acidification in the North Sea," *Journal of Marine Systems* 64 (2007): 229–241; Igor P. Semiletov, Irina I. Pipko, Irina Repina, and Natalia E. Shakhova, "Carbonate Chemistry Dynamics and Carbon Dioxide Fluxes across the Atmosphere-Ice-Water Interfaces in the Arctic Ocean: Pacific Sector of the Arctic," *Journal of Marine Systems* 66 (2007): 204–226; and O. Hoegh-Guldberg, P. J. Mumby, A. J. Hooten, and others, "Coral Reefs under Rapid Climate Change and Ocean Acidification," *Science* 318 (2007): 1737–1742.

49. Benjamin S. Halpern, Shaun Walbridge, Kimberly A. Selkoe, and others, "A Global Map of Human Impact on Marine Ecosystems," *Science* 319 (2008): 948–952.

50. The preceding discussion of Adélie penguins is based on Steven D. Emslie and William P. Patterson, "Abrupt Recent Shift in ^{13}C and ^{15}N Values in Adélie Penguin Eggshell in Antarctica," *Proceedings of the National Academy of Sciences* 104 (2007): 11666–11669.

51. Phillip S. Levin, Elizabeth E. Holmes, Kevin R. Piner, and Chris J. Harvey, "Shifts in a Pacific Ocean Fish Assemblage: The Potential Influence of Exploitation," *Conservation Biology* 20 (2006): 1181–1190.

52. Ransom A. Myers and Boris Worm, "Extinction, Survival or Recovery of Large Predatory Fishes," *Philosophical Transactions of the Royal Society of London: Biological Sciences* 360 (2005): 13–20; Peter Ward and Ransom A. Myers, "Shifts in Open-Ocean Fish Communities Coinciding with the Commencement of Commercial Fishing," *Ecology* 86 (2005): 835–847; and Kenneth T. Frank, Brian Petrie, Jae S. Choi, and William C. Leggett, "Trophic Cascades in a Formerly Cod-Dominated Ecosystem," *Science* 308 (2005): 1621–1623.

53. Benjamin S. Halpern, Karl Cottenie, and Bernardo R. Broitman, "Strong Top-Down Control in Southern California Kelp Forest Ecosystems," *Science* 312 (2006): 1230–1232; Chris L. J. Frid, S. Hansson, S. A. Rijnsdorp, and S. A. Steingrimsson, "Changing Levels of Predation on Benthos as a Result of Exploitation of Fish Populations," *Ambio* 28 (1999): 578–582; Chris L. J. Frid, Odette A. L. Paramor, and Catherine L. Scott, "Ecosystem-based Management of Fisheries: Is Science Limiting?" *Journal of Marine Science* 63 (2006): 1567–1572; Shelley C. Clarke, Jennifer E. Magnussen, Debra L. Abercrombie, and others, "Identification of Shark Species Composition and Proportion in the Hong Kong Shark Fin Market Based on Molecular Genetics and Trade Records," *Conservation Biology* 20 (2006): 201–211; and R. Q. Grafton, T. Kompas, and R. W. Hilborn, "Economics of Overexploitation Revisited," *Science* 318 (2007): 1601.

54. D. Pauly and V. Christensen, "Primary Production Required to Sustain Global Fisheries," *Nature* 374 (1995): 255–257; John A. Barth, Bruce A. Menge, Jane Lubchenco, and others, "Delayed Upwelling Alters Nearshore Coastal Ocean Ecosystems in the Northern California Current," *Proceedings of the National Academy of Sciences* 104 (2007): 3719–3724; F. Chan, J. A. Barth, J. Lubchenco, and others, "Emergence of Anoxia in the California Current Large Marine Ecosystem," *Science* 319 (2008): 920; and Ryan R. Rykaczewski and David M. Checkley Jr., "Influence of Ocean Winds on the Pelagic Ecosystem in Upwelling Regions," *Proceedings of the National Academy of Sciences* 105 (2008): 1965–1970.

55. Discussion of the Dust Bowl is based on Siegfried D. Schubert, Max J. Suarez, Philip J. Pegion, and others, "On the Cause of the 1930s Dust Bowl," *Science* 303 (2004): 1855–1859; Geoff Cunfer, "The Dust Bowl," EH.Net Encyclopedia, ed. Robert Whaples, August 19, 2004, http://eh.net/encyclopedia/article/Cunfer.DustBowl; and Wind Erosion Unit, United States Department of Agriculture at Kansas State University, "The Dust Bowl," www.usd.edu/anth/epa/dust.html (accessed February 3, 2008).

56. Sangwon Suh, "Are Services Better for Climate Change?" *Environmental Science & Technology* 40 (2006): 6555–6560, and Wouter Peters, Andrew R. Jacobson, Colm Sweeney, and others, "An Atmospheric Perspective on North American Carbon Dioxide Exchange: CarbonTracker," *Proceedings of the National Academy of Sciences* 104 (2007): 18925–18930.

57. The previous two paragraphs are based on Bridget F. O'Neill, Arthur R. Zangerl, Evan H. DeLucia, and Mar R. Berenbaum, "Longevity and Fecundity of Japanese Beetle (*Popillia japonica*) on Foliage Grown under Elevated Carbon Dioxide," *Environmental Entomology* 37 (2008): 601–607, and Jorge A. Zavala, Clare L. Casteel, Evan H. DeLucia, and May R. Berenbaum, "Anthropogenic Increase in Carbon Dioxide Compromises Plant Defense against Invasive Insects," *Proceedings of the National Academy of Sciences* 105 (2008): 5129–5133.

58. Feng Gao, San-Rong Zhu, Yu-Cheng Sun, and others, "Interactive Effects of Elevated CO_2 and Cotton Cultivar on Tri-Trophic Interaction of *Gossypium hirsutum*, *Aphis gossyppii*, and *Propylaea japonica*," *Environmental Entomology* 37 (2008): 29–37.

59. See note 56 in this chapter.

60. The discussion of the Bodélé Depression is based on I. Koren, Y. Kaufman, R. Washington, and others, "The Bodélé Depression: A Single Spot in the Sahara That Provides Most of the Mineral Dust to the Amazon Forest," *Environmental Research Letters* 1 (2006): 1–5.

61. Sid Perkins, "What Goes Up," *Science News* 172 (2007): 152–153, 156.

62. Kathleen C. Weathers, Mary L. Cadenasso, and Steward T. A. Pickett, "Forest Edges as Nutrient and Pollutant Concentrators: Potential Synergisms between Fragmentation, Forest Canopies, and the Atmosphere," *Conservation Biology* 15 (2001): 1506–1514.

63. Matthew Brown, "Federal Study: Pesticides Prevalent in National Wilderness," Corvallis (OR) *Gazette-Times*, February 27, 2008, and Roman Marks and Magdalena Bedowska, "Air-Sea Exchange of Mercury Vapor over the Gulf of Gdask and Southern Baltic Sea," *Journal of Marine Systems* (2001): 315–324.

64. Jacobson study described and Milford quote in A. Cunningham, "Not-So-Clear Alternative," *Science News* 171 (2007): 278.

65. R. L. France and N. C. Collins, "Extirpation of Crayfish in a Lake Affected by Long-Range Anthropogenic Acidification," *Conservation Biology* 7 (1993): 184–188.

66. W. James Gauderman, Edward Avol, Frank Gilliland, and others, "The Effect of Air Pollution on Lung Development from 10 to 18 Years of Age," *New England Journal of*

Medicine 351 (2004): 1057–1067, and Akira Toriba, Kazuichi Hayakawa, Christopher D. Simpson, and others, "Identification and Quantification of 1-Nitropyrene Metabolites in Human Urine as a Proposed Biomarker for Exposure to Diesel Exhaust," *Chemical Research in Toxicology* 20 (2007): 999–1007.

67. The discussion of brown clouds in this and the previous paragraph is based on Veerabhadran Ramanathan, Muvva V. Ramana, Gregory Roberts, and others, "Warming Trends in Asia Amplified by Brown Cloud Solar Absorption," *Nature* 448 (2007): 575–578.

68. Janet Raloff, "Elevated Pesticide Threatens Amphibians," *Science News* 168 (2005): 381.

69. The discussion of genetically engineered corn in this and the previous paragraph is based on David Quist and Ignacio H. Chapela, "Transgenic DNA Introgressed into Traditional Maize Landraces in Oaxaca, Mexico," *Nature* 414 (2001): 541–543; Anita Manning, "Gene-Altered DNA May Be 'Polluting' Corn," *USA Today*, November 29, 2001; and Mark Stevenson, "Mexicans Angered by Genetically Modified Corn," Corvallis (OR) *Gazette-Times*, December 30, 2001.

70. E. J. Rosi-Marshall, J. L. Tank, T. V. Royer and others, "Toxins in Transgenic Crop Byproducts May Affect Headwater Stream Ecosystems," *Proceedings of the National Academy of Sciences* 104 (2007): 16204–16208.

71. Luke Mehlo, Daphrose Gahakwa, Pham Trung Nghia, and others, "An Alternative Strategy for Sustainable Pest Resistance in Genetically Enhanced Crops," *Proceedings of the National Academy of Sciences* 102 (2005): 7812–7816, and Bruce E. Tabashnik, Aaron J. Gassmann, David W. Crowder, and Yves Carriére, "Insect Resistance to *Bt* Crops: Evidence versus Theory," *Nature Biotechnology* 26 (2008): 199–202.

72. Jennifer Clapp, "Unplanned Exposure to Genetically Modified Organisms," *Journal of Environment & Development* 15 (2006): 3–21.

73. Chris Maser, "The Humble Ditch," *Resurgence* 172 (1995): 38–40.

74. Janet Raloff, "Sharks, Dolphins Store Pollutants," *Science News* 170 (2006): 366, and Christian Sonne, Pall S. Leifsson, Rune Dietz, and others, "Xenoendocrine Pollutants May Reduce Size of Sexual Organs in East Greenland Polar Bears (*Ursus maritimus*)," *Environmental Science & Technology* 40 (2006): 5668–5674.

75. Quoted in Sid Perkins, "Back from the Dead?" *Science News* 172 (2007): 312, 314.

76. J.L.B. Smith, "A Living Coelacanthid Fish from South Africa," *Nature* (1939): 748–750.

77. Susan L. Jewett, "The Coelacanth: More Living Than Fossil," *Natural History Highlight of the Smithsonian National Museum of Natural History*, 2003, http://www.mnh.si.edu/highlight/coelacanth/.

78. Daniel Goodman, "How Do Any Species Persist? Lessons for Conservation Biology," *Conservation Biology* 1 (1987): 59–62.

79. Karen Hissmann, Hans Fricke, and Jürgen Schauer, "Population Monitoring of the Coelacanth (*Latimeria chalumnae*)," *Conservation Biology* 12 (1998): 759–765, and Janne S. Kotiaho, Veijo Kaitala, Atte Komonen, and Jussi Päivinen, "Predicting the Risk of Extinction from Shared Ecological Characteristics," *Proceedings of the National Academy of Sciences* 102 (2005): 1963–1967.

80. Peter Forey, "A Home from Home for Coelacanths," *Nature* 395 (1998): 319–320; Mark V. Erdmann, Roy L. Caldwell, and M. Kasim Moosa, "Indonesian 'King of the Sea' Discovered," *Nature* 395 (1998): 335; Jewett, "The Coelacanth"; and Perkins, "Back from the Dead?"

81. "Rare Fish Faces Extinction," Corvallis (OR) *Gazette-Times*, October 4, 1989.

82. Jose Gabriel Segarra-Moragues and Pilar Catalán, "Low Allozyme Variability in the Critically Endangered *Borderea chouardii* and in Its Congener *Borderea pyrenaica*

(Dioscoreaceae), Two Paleoendemic Relicts from the Central Pyrenees," *International Journal of Plant Science* 163 (2002): 159–166; María B. García, "Demographic Viability of a Relict Population of the Critically Endangered Plant *Borderea chouardii*," *Conservation Biology* 17 (2003): 1672–1680; Jose Gabriel Segarra-Moragues and Pilar Catalán, "Life History Variation between Species of the Relictual Genus *Borderea* (Dioscoreaceae): Phylogeography, Genetic Diversity, and Population Genetic Structure Assessed by RAPD Markers," *Biological Journal of the Linnean Society* 80 (2003): 483; and J. G. Segarra-Moragues, M. Palop-Esteban, F. Gonzalez-Candelas, and P. Catalán, "On the Verge of Extinction: Genetics of the Critically Endangered Iberian Plant Species, *Borderea chouardii* (Dioscoreaceae) and Implications for Conservation," *Molecular Ecology* 14 (2005): 969–982.

83. J. A. Thomas, M. G. Telfer, D. B. Roy, and others, "Comparative Losses of British Butterflies, Birds, and Plants and the Global Extinction Crisis," *Science* 303 (2004): 1879–1881.

84. F. Stuart Chapin III, Erika S. Zavaleta, Valerie T. Eviner, and others, "Consequences of Changing Biodiversity," *Nature* 405 (2000): 234–242; Gerardo Ceballos and Paul R. Ehrlich, "Mammal Population Losses and the Extinction Crisis," *Science* 296 (2002): 904–907; Lian Pin Koh, Robert R. Dunn, Navjot S. Sodhi, and others, "Species Coextinctions and the Biodiversity Crisis," *Science* 303 (2004): 1632–1634; and K. J. Gaston and R. A. Fuller, "Biodiversity and Extinction: Losing the Common and the Widespread," *Progress in Physical Geography* 31 (2007): 213–225.

85. The preceding four paragraphs are based on Chris D. Thomas, Alison Cameron, Rhys E. Green, and others, "Extinction Risk from Climate Change," *Nature* 427 (2004): 145–148; Janet Larsen, "The Sixth Great Extinction: A Status Report," March 2, 2004, Earth Policy Institute, http://www.earth-policy.org/Updates/Update35.htm (accessed January 10, 2008); and International Union for Conservation of Nature (IUCN), *Red List of Threatened Species*, www.redlist.org (accessed January 10, 2008).

CHAPTER 5   ACT LOCALLY AND AFFECT THE WHOLE WORLD

1. For a history of the Papal Bulls, see "The Inter Caetera, Papal Bull of May 4, 1493 by Alexander VI," http://www.kwabs.com/bull_of_1493.html (accessed November 18, 2008), and Manataka American Indian Council, "Revoking the Bull 'Inter Caetera' of 1493," http://bullsburning.itgo.com/ (accessed November 18, 2008).

2. The discussion of Easter Island is based on Michael Kiefer, "Fall of the Garden of Eden," *International Wildlife* (July-August 1989): 38–43; Terry L. Hunt and Carl P. Lipo, "Late Colonization of Easter Island," *Science* 311 (2006): 1603–1606; S. S. Barnes, E. Matisoo-Smith, and Terry L. Hunt, "Ancient DNA of the Pacific Rat (*Rattus exulans*) from Rapa Nui (Easter Island)," *Journal of Archaeological Science* 33 (2006): 1536–1540; and Terry L. Hunt, "Rethinking Easter Island's Ecological Catastrophe," *Journal of Archaeological Science* 34 (2007): 485–502.

3. The preceding discussion of the introduction of rats in the Aleutian Islands is based on Carolyn M. Kurle, Donald A. Croll, and Bernie R. Tershy, "Introduced Rats Indirectly Change Marine Rocky Intertidal Communities from Algae- to Invertebrate-Dominated," *Proceedings of the National Academy of Sciences* 105 (2008): 3800–3804.

4. Michael J. Gundale, "Influence of Exotic Earthworms on the Soil Organic Horizon and the Rare Fern *Botrychium mormo*," *Conservation Biology* 16 (2002): 1555–1561, and Rachel Ehrenberg, "Ecosystem Engineers," *Science News* 174 (2008): 13–14.

5. "Ah-Choo, Arizona No Longer Haven for Allergy Sufferers," Corvallis (OR) *Gazette-Times*, March 25, 1997.

6. Helmut Haberl, K. Heinz Erb, Fridolin Krausmann, and others, "Quantifying and Mapping the Human Appropriation of Net Primary Production in Earth's Terrestrial Ecosystems," *Proceedings of the National Academy of Sciences* 104 (2007): 12942–12947, and Jonathan A. Foley, Chad Monfreda, Navin Ramankutty, and David Zaks, "Our Share of the Planetary Pie," *Proceedings of the National Academy of Sciences* 104 (2007): 12585–12586.

7. Richard Gallagher and Betsy Carpenter, "Human-Dominated Ecosystems," *Science* 277 (1977): 485.

8. The discussion of the overuse of antibiotics, including the Lederberg quote, is based on "Overuse of Antibiotics Threatens World Health," Corvallis (OR) *Gazette-Times*, May 15, 1998.

9. Lyric Wallwork Winik, "Before the Next Epidemic Strikes," *Parade*, February 8, 1998, 6–9.

10. Peter M. Vitousek, Harold A. Mooney, Jane Lubchenco, and Jerry M. Melillo, "Human Domination of Earth's Ecosystems," *Science* 277 (1997): 494–499.

11. C. R. Allen, R. S. Lutz, and S. Demarais, "Red Imported Fire Ant Impacts on Northern Bobwhite Populations," *Ecological Applications* 5 (1995): 632–638; Anna Traveset and Nuria Riera, "Disruption of a Plant-Lizard Seed Dispersal System and Its Ecological Effects on a Threatened Endemic Plant in the Balearic Islands," *Conservation Biology* 19 (2005): 421–431; and Anthony Ricciardi, "Are Modern Biological Invasions an Unprecedented Form of Global Change?" *Conservation Biology* 21 (2007): 329–336.

12. Katharina Dehnen-Schmutz, Julia Touza, Charles Perrings, and Mark Williamson, "The Horticultural Trade and Ornamental Plant Invasions in Britain," *Conservation Biology* 21 (2007): 224–231.

13. Svata M. Louda and Charles W. O'Brien, "Unexpected Ecological Effects of Distributing the Exotic Weevil, *Larinus planus* (F.), for the Biological Control of Canada Thistle," *Conservation Biology* 16 (2002): 717–727; Gundale, "Influence of Exotic Earthworms"; Colin R. Townsend, "Individual, Population, Community, and Ecosystem Consequences of a Fish Invader in New Zealand Streams," *Conservation Biology* 17 (2003): 38–47; Gary J. Wiles, Jonathan Bart, Robert E. Beck Jr., and Celestino F. Aguon, "Impacts of the Brown Tree Snake: Patterns of Decline and Species Persistence in Guam's Avifauna," *Conservation Biology* 17 (2003): 1350–1360; John W. Chapman, Todd W. Miller, and Eugene V. Coan, "Live Seafood Species as Recipes for Invasion," *Conservation Biology* 17 (2003): 1386–1395; Mathieu Denoel, Georg Dzukic, and Milos L. Kalezic, "Effects of Widespread Fish Introductions on Paedomorphic Newts in Europe," *Conservation Biology* 19 (2005): 162–170; Michael J. Gundale, William M. Jolly, Thomas H. Deluca, "Susceptibility of a Northern Hardwood Forest to Exotic Earthworm Invasion," *Conservation Biology* 19 (2005): 1075–1083; Jaime Bosch, Pedro A. Rincón, Luz Boyero, and Iñigo Martínez-Solano, "Effects of Introduced Salmonids on a Montane Population of Iberian Frogs," *Conservation Biology* 20 (2006): 180–189; Jeffrey T. Foster and Scott K. Robinson, "Introduced Birds and the Fate of Hawaiian Rainforests," *Conservation Biology* 21 (2007): 1248–1257; Gregg Howald, C. Josh Donlan, Juan Pablo Galván, and others, "Invasive Rodent Eradication on Islands," *Conservation Biology* 21 (2007): 1258–1268; and Leandro A. Becker, Miguel A. Pascual, and Néstor G. Basso, "Colonization of the Southern Patagonia Ocean by Exotic Chinook Salmon," *Conservation Biology* 21 (2007): 1347–1352.

14. James T. Carlton, "Man's Role in Changing the Face of the Ocean: Biological Invasions and Implications for Conservation of Near-Shore Environments," *Conservation Biology* 3 (1989): 265–273.

15. F. Leprieur, O. Beauchard, S. Blanchet, and others, "Fish Invasions in the World's River Systems: When Natural Processes Are Blurred by Human Activities," *Public Library of*

Science, Biology 6 (2008): e28; R. Robinson, "Human Activity, Not Ecosystem Characters, Drives Potential Species Invasions," *Public Library of Science, Biology* 6 (2008): e39; Melih Ertan Çinar, Tuncer Katağan, Ferah Koçak, and others, "Faunal Assemblages of the Mussel *Mytilus galloprovincialis* in and around Alsancak Harbour (Izmir Bay, Eastern Mediterranean) with Special Emphasis on Alien Species," *Journal of Marine Systems* 71 (2008): 1–17; and Luigi Piazzi and David Balata, "The Spread of *Caulerpa racemosa* var. *cylindracea* in the Mediterranean Sea: An Example of How Biological Invasions Can Influence Beta Diversity," *Marine Environmental Research* 65 (2008): 50–61.

16. The previous discussion of the green crab and the gem clam is based on Edwin D. Grosholz, "Recent Biological Invasion May Hasten Invasional Meltdown by Accelerating Historical Introductions," *Proceedings of the National Academy of Sciences* 102 (2005): 1088–1091.

17. S. D. Porter and D. A. Savignano, "Invasion of Polygyne Fire Ants Decimates Native Ants and Disrupts Arthropod Community," *Ecology* 71(1990): 2095–2106, and Jonathan M. Bossenbroek, Ladd E. Johnson, Brett Peters, and David M. Lodge, "Forecasting the Expansion of Zebra Mussels in the United States," *Conservation Biology* 21 (2007): 800–810.

18. T. A. Zink, M. F. Allen, B. Heindl-Tenhunen, and E. B. Allen, "The Effect of a Disturbance Corridor on an Ecological Reserve," *Restoration Ecology* 3 (1995): 304–310, and J. H. Stiles and R. H. Jones, "Distribution of the Red Imported Fire Ant, *Solenopsis invicta*, in Road and Powerline Habitats," *Landscape Ecology* 335 (1998): 335–346.

19. Discussion of the detrimental effects of roads is based on C. Murcia, "Edge Effects in Fragmented Forests: Implications for Conservation," *Trends in Ecology & Evolution* 10 (1995): 58–62; W. Schmidt, "Plant Dispersal by Motor Cars," *Vegtatio* 80 (1989): 147–152; Richard T. T. Forman and L. Alexander, "Roads and Their Major Ecological Effects," *Annual Review of Ecology and Systematics* 29 (1998): 207–231; Richard T. T. Forman, "Estimate of the Area Affected Ecologically by the Road System in the United States," *Conservation Biology* 14 (2000): 31–35; Stephen C. Trombulak and Christopher A. Frissell, "Review of Ecological Effects of Roads on Terrestrial and Aquatic Communities," *Conservation Biology* 14 (2000): 18–30; Jonathan L. Gelbard and Jayne Belnap, "Roads as Conduits for Exotic Plant Invasions in a Semiarid Landscape," *Conservation Biology* 17 (2003): 420–432; Laurie A. Parendes and Julia A. Jones, "Role of Light Availability and Dispersal in Exotic Plant Invasion along Roads and Streams in the H. J. Andrews Experimental Forest, Oregon," *Conservation Biology* 14 (2000): 64–75; Radley Z. Watkins, Jiquan Chen, Jim Pickens, and Kimberley D. Brosofske, "Effects of Forest Roads on Understory Plants in a Managed Hardwood Landscape," *Conservation Biology* 17 (2003): 411–419; Susan G. Laurance, Philip C. Stouffer, and William F. Laurance, "Effects of Road Clearings on Movement Patterns of Understory Rainforest Birds in Central Amazonia," *Conservation Biology* 18 (2004): 1099–1109; Todd J. Hawbaker and Volker C. Radeloff, "Roads and Landscape Pattern in Northern Wisconsin Based on a Comparison of Four Road Data Sources," *Conservation Biology* 18 (2004): 1233–1244; David A. Steen and James P. Gibbs, "Effects of Roads on the Structure of Freshwater Turtle Populations," *Conservation Biology* 18 (2004): 1143–1148; Douglas Christen and Glenn Matlack, "The Role of Roadsides in Plant Invasions: A Demographic Approach," *Conservation Biology* 20 (2006): 385–391; and Mark C. Urban, "Road Facilitation of Trematode Infections in Snails of Northern Alaska," *Conservation Biology* 20 (2006): 1143–1149.

20. David S. Wilkie, Ellen Shaw, Fiona Rotberg, and others, "Roads, Development, and Conservation in the Congo Basin," *Conservation Biology* 14 (2002): 1614–1622.

21. Richard T. T. Forman and Robert D. Deblinger, "The Ecological Road-Effect Zone of a Massachusetts (U.S.A.) Suburban Highway," *Conservation Biology* 14 (2000): 36–46, and

David G. Haskell, "Effects of Forest Roads on Macroinvertebrate Soil Fauna of the Southern Appalachian Mountains," *Conservation Biology* 14 (2000): 57–63.

22. Daniel Simberloff, "New Tactics Could Halt Invasion of Harmful Species," Corvallis (OR) *Gazette-Times*, March 1, 1998.

23. Ibid.

24. Joseph P. Dudley, Joshua R. Ginsberg, Andrew J. Plumptre, and others, "Effects of War and Civil Strife on Wildlife and Wildlife Habitats," *Conservation Biology* 16 (2002): 319–329.

25. The preceding discussion of invasive plants and animals of based on Simberloff, "New Tactics Could Halt Invasion of Harmful Species."

26. The previous two paragraphs are based on Lisa Marinelli, "Hawaii Lawmakers Propose Plan to Protect Exotic Fish," Corvallis (OR) *Gazette-Times*, February 15, 1998.

27. Karen F. Schmidt, "'No-Take' Zones Spark Fisheries Debate," *Science* 277 (1997): 489–491, and N. Bax, J. T. Carlton, A. Mathews-Amos, and others, "The Control of Biological Invasions in the World's Oceans," *Conservation Biology* 15 (2001): 1234–1246.

28. J. E. Andrews, A. M. Greenaway, G. R. Bigg, and others, "Pollution History of a Tropical Estuary Revealed by Combined Hydrodynamic Modelling and Sediment Geochemistry," *Journal of Marine Systems* 18 (1999): 333–343; Mohamed A. Hamed and Ahmed M. Emara, "Marine Molluscs as Biomonitors for Heavy Metal Levels in the Gulf of Suez, Red Sea," *Journal of Marine Systems* 60 (2006): 220–234; and Benjamin S. Halpern, Kimberly A. Selkoe, Fiorenza Micheli, and Carrie V. Kappel, "Evaluating and Ranking the Vulnerability of Global Marine Ecosystems to Anthropogenic Threats," *Conservation Biology* 21 (2007): 1301–1315.

29. Jonathan Rowe, "The Hidden Commons," *Yes! A Journal of Positive Futures* (Summer 2001): 12–17.

30. Edmund Burke, quoted in David W. Orr, "Conservatives against Conservation," *Resurgence* 172 (1995): 15–17.

31. Karen A. Kidd, Paul J. Blanchfield, Kenneth H. Mills, and others, "Collapse of a Fish Population after Exposure to a Synthetic Estrogen," *Proceedings of the National Academy of Sciences* 104 (2007): 8897–8901.

32. David Korten, "What to Do When Corporations Rule the World," *Yes! A Journal of Positive Futures* (Summer 2001): 148–151.

33. Rowe, "The Hidden Commons."

34. D. G. Joakim Larsson, Cecilia de Pedro, and Nicklas Paxeus, "Effluent from Drug Manufactures Contains Extremely High Levels of Pharmaceuticals," *Journal of Hazardous Materials* 148 (2007): 751–755.

35. Ibid.

36. This paragraph is based on P. J. Bryant, C. M. Lafferty, and S. K. Lafferty, "Reoccupation of Laguna Guerrero Negro Baja California, Mexico, by Gray Whales," in *The Gray Whale Eschrictius robustus*, ed. M. L. Jones, S. L. Swartz, and S. Leatherwood, 375–386 (Orlando, FL: Academic Press, 1984); M. Andre, C. Kamminga, and D. Ketten, "Are Low-Frequency Sounds a Marine Hazard: A Case Study in the Canary Islands" (Underwater Bio-sonar and Bioacoustics Symposium, Loughborough University, UK, December 16–17, 1997); A. B. Morton and H. K. Symonds, "Displacement of *Orcinus orca* (L.) by high amplitude sound in British Columbia," *Journal of Marine Science* 59 (2002): 71–80; P.J.O. Miller, N. Biasson, A. Samuels, and P. L. Tyack, "Whale Songs Lengthen in Response to Sonar," *Nature* 405 (2000): 903; K. C. Balcomb and D. E. Claridge, "A Mass Stranding of Cetaceans Caused by Naval Sonar in the Bahamas," *Bahamas Journal of Science* 8 (2001): 1–12; R. D. McCauley, J. Fewtrell, and A. N. Popper, "High Intensity Anthropogenic Sound

Damages Fish Ears," *Journal of the Acoustical Society of America* 113 (2003): 638–642; Andrew J. Read, Phebe Drinker, and Simon Northridge, "Bycatch of Marine Mammals in U.S. and Global Fisheries," *Conservation Biology* 20 (2006): 163–169; and Lars Bejder, Amy Samuels, Hal Whitehead, and others, "Decline in Relative Abundance of Bottlenose Dolphins Exposed to Long-Term Disturbance," *Conservation Biology* 20 (2006): 1791–1798.

37. John P. Swaddle and Laura C. Page, "High Levels of Environmental Noise Erode Pair Preferences in Zebra Finches: Implications for Noise Pollution," *Animal Behaviour* 74 (2007): 363–368.

38. Brittany L. Bird, Lyn C. Branch, and Deborah L. Miller, "Effects of Coastal Lighting on Foraging Behavior of Beach Mice," *Conservation Biology* 18 (2004): 1435–1439; Ben Harder, "Light All Night," *Science News* 169 (2006): 170–172; and Robert F. Baldwin and Stephen C. Trombulak, "Losing the Dark: A Case for a National Policy on Land Conservation," *Conservation Biology*, 21 (2007): 1133–1134.

39. The previous two paragraphs are based on N. LeRoy Poff, Julian D. Olden, David M. Merritt, and David M. Pepin, "Homogenization of Regional River Dynamics by Dams and Global Biodiversity Implications," *Proceedings of the National Academy of Sciences* 104 (2007): 5732–5737.

40. My discussion of the Aswan High Dam is based largely on personal experience.

41. C. J. George, "The Role of the Aswan Dam in Changing Fisheries of the South-Western Mediterranean," in *The Careless Technology*, ed. M. T. Farvar and J. P. Milton, 179–188 (New York: Natural History Press, 1972).

42. R. E. Quelennec and C. B. Kruk, "Nile Suspended Load and Its Importance for the Nile Delta Morphology," in *Proceedings of the Seminar on Nile Delta Sedimentology*, ed. UNDP/UNESCO Project, Coastal Protection Studies, 130–144 (Alexandria, VA, 1976); C. Summerhayes and N. Marks, "Nile Delta Nature Evolution and Collapse of Continental Shelf Sediment System," in *Proceedings of the Seminar on Nile Delta Sedimentology*, ed. UNDP/UNESCO Project, Coastal Protection Studies, 162–190 (Alexandria, VA, 1976); and S. A. Toma and M. S. Salama, "Changes in Bottom Topography of the Western Shelf of the Nile Delta since 1922," *Marine Geology* 36 (1980): 325–339.

43. Young-Seuk Park, Jianbo Chang, Sovan Lek, and others, "Conservation Strategies for Endemic Fish Species Threatened by the Three Gorges Dam," *Conservation Biology* 17 (2003): 1748–1758, and G.-C. Gong, J. Chang, K.-P. Chiang, and others, "Reduction of Primary Production and Changing of Nutrient Ratio in the East China Sea: Effect of the Three Gorges Dam?" *Geophysical Research Letters* 33 (2006): L07610, doi:10.1029/2006GL025800.

44. Caryn C. Vaughn and Christopher M. Taylor, "Impoundments and the Decline of Freshwater Mussels: A Case Study of an Extinction Gradient," *Conservation Biology* 13 (1999): 912–920.

45. Carlie A. Rodriguez, Karl W. Flessa, and David L. Dettman, "Effects of Upstream Diversion of Colorado River Water on the Estuarine Bivalve Mollusc, *Mulinia coloradoensis*," *Conservation Biology* 15 (2001): 249–258.

46. The previous three paragraphs are based on R. G. Johnson, "Climate Control Requires a Dam at the Strait of Gibraltar," *Transactions* (American Geophysical Union) 78 (1997): 277–281, and N. Skliris, S. Sofianos, and A. Lascaratos, "Hydrological Changes in the Mediterranean Sea in Relation to Changes in the Freshwater Budget: A Numerical Modelling Study," *Journal of Marine Systems* 65 (2007): 400–416.

47. Johnson, "Climate Control Requires a Dam at the Strait of Gibraltar."

48. R. G. Johnson, "Ice Age Initiation by an Ocean-Atmospheric Circulation Change in the Labrador Sea," *Earth Planetary Science Letters* 148 (1997): 367.

49. Alan T. Hitch and Paul L. Leberg, "Breeding Distributions of North American Bird Species Moving North as a Result of Climate Change," *Conservation Biology* 21 (2007): 534–539.

50. The foregoing three paragraphs are based on Nicole Lemoine, Hans-Günther Bauer, Markus Peintinger, and Katrin Böhning-Gaese, "Effects of Climate and Land-Use Change on Species Abundance in a Central European Bird Community," *Conservation Biology* 21 (2007): 495–503.

CHAPTER 6   REPAIRING ECOSYSTEMS

1. Philip M. Fearnside, "Deforestation and International Economic Development Projects in Brazilian Amazonia," *Conservation Biology* 1 (1987): 214–221.

2. Francis E. Putz, Dennis P. Dykstra, and Rudolf Heinrich, "Why Poor Logging Practices Persist in the Tropics," *Conservation Biology* 14 (2000): 951–956.

3. Erwin H. Bulte, Richard Damania, and Ramón López, "On the Gains of Committing to Inefficiency: Corruption, Deforestation, and Low Land Productivity in Latin America," *Journal of Environmental Economics and Management* 54 (2007): 277–295.

4. Aldemaro Romero, "Death and Taxes: The Case of the Depletion of Pearl Oyster Beds in Sixteenth-Century Venezuela," *Conservation Biology* 17 (2003): 1013–1023.

5. The previous two paragraphs are based on G. J. Edgar and C. R. Samson, "Catastrophic Decline in Mollusc Diversity in Eastern Tasmania and Its Concurrence with Shellfish Fisheries," *Conservation Biology* 18 (2004): 1579–1588.

6. Shi & Provincial Forestry Bureau for Endangered Species, Import and Export Management Office of China, July 2007, unpublished data.

7. Norman Myers, "The Extinction Spasm Impending: Synergisms at Work," *Conservation Biology* 1 (1987): 14–21.

8. The preceding discussion of the trade in turtles and tortoises is based on M. D. Jenkins, *Tortoises and Freshwater Turtles: The Trade in Southeast Asia* (Cambridge, U.K.: TRAFFIC International, 1995); Le Dien Duc and S. Broad, *Investigations into Tortoise and Freshwater Turtle Trade in Vietnam* (Gland, Switzerland, and Cambridge, U.K.; Species Survival Commission, International Union for the Conservation of Nature, 1995); M. Lau, B. Chan, P. Crow, and G. Ades, "Trade and Conservation of Turtles and Tortoises in the Hong Kong Special Administrative Region, People's Republic of China," in *Asian Turtle Trade: Proceedings of a Workshop on Conservation and Trade of Freshwater Turtles and Tortoises in Asia*, ed. P. P. Van Dijk, B. L. Stuart, and A.G.J. Rhodin, Chelonian Research Monographs, no. 2, 39–44 (Lunenburg, MA: Chelonian Research Foundation, 2000); P. P. Van Dijk, B. L. Stuart, and A.G.J. Rhodin, eds., *Asian Turtle Trade: Proceedings of a Workshop on Conservation and Trade of Freshwater Turtles and Tortoises in Asia*, Chelonian Research Monographs, no. 2 (Lunenburg, MA: Chelonian Research Foundation, 2000); H. Shi and J. F. Parham, "Preliminary Observations of a Large Turtle Farm in Hainan Province, People's Republic of China," *Turtle and Tortoise Newsletter* 3 (2001): 2–4; R.H.P. Holloway, "Domestic Trade of Tortoises and Freshwater Turtles in Cambodia. Linnaeus Fund Research Report," *Chelonian Conservation and Biology* 4 (2003): 733–734; H. Shi, Z. Fan, F. Yin, and Z. Yuan, "New Data on the Trade and Captive Breeding of Turtles in Guangxi Province, South China," *Asiatic Herpetological Research* 10 (2004): 126–128; and Haitao Shi, James F. Parham, Michael Lau, and Chen Tien-His, "Farming Endangered Turtles to Extinction in China," *Conservation Biology* 21 (2007): 5–6.

9. The preceding two paragraphs are based on Donald Ludwig, Ray Hilborn, and Carl Walters, "Uncertainty, Resource Exploitation, and Conservation: Lesson from History," *Science* 260 (1993): 17, 36.

10. Jon Rosales, "Economic Growth and Biodiversity Loss in an Age of Tradable Permits," *Conservation Biology* 20 (2006): 1042–1050.

11. David Ehrenfeld, "The Environmental Limits to Globalization," *Conservation Biology* 19 (2005): 318–326.

12. The preceding three paragraphs are based on Jonna Engel and Rikk Kvitek, "Effects of Otter Trawling on a Benthic Community in Monterey Bay National Marine Sanctuary," *Conservation Biology* 12 (1998): 1204–1214; Peter J. Auster, "A Conceptual Model of the Impacts of Fishing Gear on the Integrity of Fish Habitats," *Conservation Biology* 12 (1998): 1198–1203; Cynthia H. Pilskaln, James H. Churchill, and Lawrence M. Mayer, "Resuspension of Sediment by Bottom Trawling in the Gulf of Maine and Potential Geochemical Consequences," *Conservation Biology* 12 (1998): 1223–1229; and Les Watling and Elliott A. Norse, "Disturbance of the Seabed by Mobile Fishing Gear: A Comparison to Forest Clearcutting," *Conservation Biology* 12 (1998): 1180–1197.

13. The foregoing discussion of whaling is based on Hal Whitehead, Jenny Christal, and Susan Dufault, "Past and Distant Whaling and the Rapid Decline of Sperm Whales off the Galápagos Islands," *Conservation Biology* 11 (1997): 1387–1396; Lyudmila Bogoslovskaya, *Inuit, Whaling, and Sustainability* (Lanham, MD: AltaMira Press, 1998); and Calestous Juma, *The Future of the International Whaling Commission: Strengthening Ocean Diplomacy*, report prepared for the International Whaling Commission (Cambridge, U.K.: International Whaling Commission, May 16, 2008).

14. Navjot S. Sodhi, Tien Ming Lee, Lian Pin Koh, and Dewi M. Prawiradilaga, "Long-Term Avifaunal Impoverishment in an Isolated Tropical Woodlot," *Conservation Biology* 20 (2006): 772–779.

15. Stuart W. Krasner, Howard S. Weinberg, Susan D. Richardson, and others, "Occurrence of a New Generation of Disinfection Byproducts," *Environmental Science & Technology* 40 (2006): 7175–7185; Susan Milius, "Birds Beware," *Science News* 170 (2006): 309–310.

16. The preceding three paragraphs are based on C. S. Holling and Gary K. Meffe, "Command and Control and the Pathology of Natural Resource Management," *Conservation Biology* 10 (1996): 328–337.

17. The preceding two paragraphs are based on David R. Foster and David A. Orwig, "Preemptive and Salvage Harvesting of New England Forests: When Doing Nothing Is a Viable Alternative," *Conservation Biology* 20 (2006): 959–970.

18. Robin Naidoo and Wiktor L. Adamowicz, "Effects of Economic Prosperity on Numbers of Threatened Species," *Conservation Biology* 15 (2001): 1021–1029, and Oliver R. W. Pergams, Brian Czech, J. Christopher Haney, and Dennis Nyberg, "Linkage of Conservation Activity to Trends in the U.S. Economy," *Conservation Biology* 18 (2004): 1617–1623.

19. U. Thara Srinivasan, Susan P. Carey, Eric Hallstein, and others, "The Debt of Nations and the Distribution of Ecological Impacts from Human Activities," *Proceedings of the National Academy of Sciences* 104 (2008): 1768–1773.

20. Ehrenfeld, "The Environmental Limits to Globalization."

21. Quoted in Paul Rauber, "A New Mobilization Is Just Beginning," *Sierra* (January/February 2004): 38–39.

22. George Monbio, "Why We Conform," *Resurgence* 221 (2003): 16–17.

23. Quoted in Susan Milius, "Wildfire, Walleyes, and Wine," *Science News* 171 (2007): 378–380.

24. Quoted in Sid Perkins, "Invasive, Indeed," *Science News* 172 (2007): 235–236.

25. The discussion of community assembly and incumbency is augmented by Nick B. Davies, "Territorial Defense in Speckled Wood Butterfly (*Pararge aegeria*): Resident Always Wins,"

Animal Behaviour 26 (1978): 138–147; Barry J. Fox, "Species Assembly and the Evolution of Community Structure," *Evolutionary Ecology* 1 (1987): 201–213; Camille Parmesan, Terry L. Root, and Michael R. Willig, "Impacts of Extreme Weather and Climate on Terrestrial Biota," *Bulletin of the American Meteorological Society* 81 (2000): 443–450; Peter Chesson, Renate L. E. Gebauer, Susan Schwinning, and others, "Resource Pulses, Species Interactions, and Diversity Maintenance in Arid and Semi-Arid Environments," *Oecologia* 141 (2004): 236–253; Tsuyoshi Takeuchi, "Matter of Size or Matter of Residency Experience? Territorial Contest in a Green Hairstreak, *Chrysozephyrus smaragdinus* (Lepidoptera: Lycaenidae)," *Ethology* 112 (2006): 293–299; and Katherine M. Thibault and James H. Brown, "Impact of an Extreme Climatic Event on Community Assembly," *Proceedings of the National Academy of Sciences* 105 (2008): 3410–3415.

26. David Stauth, "Streams May Depend on Violent Floods, Droughts," Corvallis (OR) *Gazette-Times*, January 24, 2003.

27. The preceding discussion of the indigenous population of the Americas and changes in the landscape is based on Martin A. Baumhoff and Robert F. Heizer, "Postglacial Climate and Archaeology in the Desert West," in *The Quaternary of the United States*, ed. J. E. Wright Jr. and D. G. Frey, 697–707 (Princeton, NJ: Princeton University Press, 1967); James B. Griffin, "Late Quaternary Prehistory in the Northeastern Woodlands," in *The Quaternary of the United States*, ed. J. E. Wright Jr. and D. G. Frey, 655–667 (Princeton, NJ: Princeton University Press, 1967); Clement W. Meighan, "Pacific Coast Archaeology," in *The Quaternary of the United States*, ed. J. E. Wright Jr. and D. G. Frey, 709–720 (Princeton, NJ: Princeton University Press, 1967); Robert L. Stephenson, "Quaternary Human Occupation of the Plains," in *The Quaternary of the United States*, ed. J. E. Wright Jr. and D. G. Frey, 685–696 (Princeton, NJ: Princeton University Press, 1967); Stephen Williams and James B. Stoltman, "An Outline of Southeastern United States Prehistory with Particular Emphasis on the Paleo-Indian Era," in *The Quaternary of the United States*, ed. J. E. Wright Jr. and D. G. Frey, 669–683 (Princeton, NJ: Princeton University Press, 1967); Martyn J. Bowden, "The Invention of American Tradition," *Journal of Historical Geography* (1992): 183–226; William M. Denevan, "The Pristine Myth: The Landscape of the Americas in 1492," *Annals of the Association of American Geographers* 82 (1992): 369–385; W. George Lovell, "Heavy Shadows and Black Night: Disease and Depopulation in Colonial Spanish America," *Annals of the Association of American Geographers* 82 (1992): 426–443; S. M. Wilson, "That Unmanned Wild Country: Native Americans Both Conserved and Transformed New World Environments," *Natural History* (May 1992): 16–17; Karl L. Butzer, "The Americas before and after 1492: An Introduction to Current Geographical Research," *Annals of the Association of American Geographers* 82 (1992): 345–368; Douglas MacCleery, "Understanding the Role the Human Dimension Has Played in Shaping America's Forest and Grassland Landscapes: Is There a Landscape Archaeologist in the House?" *Eco-Watch* 2 (1994): 1–12; and Hazel R. Delcourt and Paul A. Delcourt, "Pre-Columbian Native American Use of Fire on Southern Appalachian Landscapes," *Conservation Biology* 11 (1997): 1010–1014.

28. The foregoing discussion of dust in the alpine lakes is based on J. C. Neff, A. P. Ballantyne, G. L. Farmer, and others, "Increasing Eolian Dust Deposition in the Western United States Linked to Human Activity," *Nature Geoscience* 1 (2008): 189–195.

29. Stephen W. Barrett and Stephen F. Arno, "Indian Fires as an Ecological Influence in the Northern Rockies," *Journal of Forestry* 80 (1982): 647–651; James R. Habeck, "The Original Vegetation of the Mid-Willamette Valley, Oregon," *Northwest Science* 35 (1961): 65–77; Carl L. Johannessen, William A. Davenport, Artimus Millet, and Steven McWilliams, "The Vegetation of the Willamette Valley," *Annals of the Association of*

American Geographers 61 (1971): 286–302; John T. Curtis, *The Vegetation of Wisconsin* (Madison: University of Wisconsin Press, 1959); and Michael Williams, *Americans and Their Forests: A Historical Geography* (New York: Cambridge University Press, 1989).

30. Nigel C. A. Pitman, María Del Carmen Loyola Azáldegui, Karina Salas, and others, "Written Accounts of an Amazonian Landscape over the Last 450 Years," *Conservation Biology* 21 (2007): 253–262.

31. The preceding discussion of the passenger pigeon is based on A. W. Schorger, *The Passenger Pigeon* (Madison: University of Wisconsin Press, 1955); Errol Fuller, *Extinct Birds* (New York: Facts on File Publications, 1987); and Joshua W. Ellsworth and Brenda C. Mccomb, "Potential Effects of Passenger Pigeon Flocks on the Structure and Composition of Presettlement Forests of Eastern North America," *Conservation Biology* 17 (2003): 1548–1558.

32. Scott C. Doney and David S. Schimel, "Carbon and Climate System Coupling on Timescales from the Precambrian to the Anthropocene," *Annual Review of Environment and Resources* 32 (2007): 31–66.

33. Maximilian Auffhammer and Richard T. Carson, "Forecasting the Path of China's CO_2 Emissions Using Province Level Information," *Journal of Environmental Economics and Management* 55 (2008): (in press).

34. William H. Schlesinger, "Global Change Ecology," *Trends in Ecology & Evolution* 21 (2006): 348–351.

35. Ibid.

36. The discussion of disturbance regimes is based on Monica G. Turner, "Landscape Ecology: The Effect of Pattern on Process," *Annual Review of Ecological Systems* 20 (1989): 171–197.

37. Chris Maser and James R. Sedell, *From the Forest to the Sea: The Ecology of Wood in Streams, Rivers, Estuaries, and Oceans* (Delray Beach, FL: St. Lucie Press, 1994).

38. David J. Rapport, H. A. Regier, and T. C. Hutchinson, "Ecosystem Behavior under Stress," *American Naturalist* 125 (1985): 617–640, and David J. Rapport, "What Constitutes Ecosystem Health?" *Perspectives in Biology and Medicine* 33 (1989): 120–132.

39. Olivier Honnay and Hans Jacquemyn, "Susceptibility of Common and Rare Plant Species to the Genetic Consequences of Habitat Fragmentation," *Conservation Biology* 21 (2007): 823–831.

40. Robert M. Ewers and Raphael K. Didham, "The Effect of Fragment Shape and Species' Sensitivity to Habitat Edges on Animal Population Size," *Conservation Biology* 21 (2007): 926–936.

41. Leslie Allen, "Prairie Revival: Researchers Put Restoration to the Test," *Science News* 172 (2007): 376–377.

42. Judith X. Becerra, "The Impact of Herbivore-Plant Coevolution on Plant Community Structure," *Proceedings of the National Academy of Sciences* 104 (2007): 7483–7488.

43. Spencer T. Behmer and Anthony Joern, "Coexisting Generalist Herbivores Occupy Unique Nutritional Feeding Niches," *Proceedings of the National Academy of Sciences* 105 (2008): 1977–1982.

44. Quoted in Allen, "Prairie Revival."

45. Jasper van Ruijven and Frank Berendse, "Diversity-Productivity Relationships: Initial Effects, Long-Term Patterns, and Underlying Mechanisms," *Proceedings of the National Academy of Sciences* 102 (2005): 695–700.

46. Allen, "Prairie Revival."

47. Carly J. Stevens, Nancy B. Dise, J. Owen Mountford, and David J. Gowing, "Impact of Nitrogen Deposition on the Species Richness of Grasslands," *Science* 303 (2004): 1876–1879.

48. The preceding two paragraphs are based on J. Mathieu, J.-P. Rossi, P. Mora, and others, "Recovery of Soil Macrofauna Communities after Forest Clearance in Eastern Amazonia, Brazil," *Conservation Biology* 19 (2005): 1598–1605.

49. The preceding discussion of crested wheatgrass is based in part on Janice M. Christian and Scott D. Wilson, "Long-Term Ecosystem Impacts of an Introduced Grass in the Northern Great Plains," *Ecology* 80 (1999): 2397–2407, and C. M. D'Antonio and P. M. Vitousek, "Biological Invasions by Exotic Grasses, the Grass/Fire Cycle, and Global Change," *Annual Review of Ecology and Systematics* 23 (1992): 63–87.

50. The preceding three paragraphs are based on J. David Ligon and Peter B. Stacey, "Land Use, Lag Times and the Detection of Demographic Change: The Case of the Acorn Woodpecker," *Conservation Biology* 10 (1996): 840–846.

51. The previous two paragraphs are based on A.R.E. Sinclair, Simon A. R. Mduma, J. Grant, and others, "Long-Term Ecosystem Dynamics in the Serengeti: Lessons for Conservation," *Conservation Biology* 21 (2007): 580–590.

52. Richard E. Hoare and Johan T. Du Toit, "Coexistence between People and Elephants in African Savannas," *Conservation Biology* 13 (1999): 633–639.

53. Gregory R. Schrott, Kimberly A. With, and Anthony W. King, "Demographic Limitations of the Ability of Habitat Restoration to Rescue Declining Populations," *Conservation Biology* 19 (2005): 1181–1193.

54. The preceding two paragraphs are based on Robert M. Ewers and Raphael K. Didham, "Pervasive Impact of Large-Scale Edge Effects on a Beetle Community," *Proceedings of the National Academy of Sciences* 105 (2008): 5426–5429.

55. W. Schmidt, "Plant Dispersal by Motor Cars," *Vegtatio* 80 (1989): 147–152, and Moritz Von Der Lippe and Ingo Kowarik, "Long-Distance Dispersal of Plants by Vehicles as a Driver of Plant Invasions," *Conservation Biology* 21 (2007): 986–996.

56. The discussion of mycorrhizae is based on Zane Maser, Chris Maser, and Randy Molina, "Small-Mammal Mycophagy in Rangelands of Central and Southeastern Oregon," *Journal of Range Management* 41 (1988): 309–312; David Read, "The Ties That Bind," *Nature* 388 (1997): 517–518; David Read, "Plants on the Web," *Nature* 396 (1998): 22–23; Marcel G. A. van der Heijden, John N. Klironomos, Margot Ursic, and others, "Mycorrhizal Fungal Diversity Determines Plant Biodiversity, Ecosystem Variability and Productivity," *Nature* 396 (1998): 69–72; and M. C. Wicklow-Howard, "Mycorrhizal Ecology of Shrub-Steppe Habitat," in *Proceedings—Ecology and Management of Annual Rangelands*, ed. S. B. Monsen and S. G. Kitchen, 207–210 (Fort Collins, CO: Intermountain Research Station, USDA Forest Service, 1994).

57. Sharon Y. Strauss, Campbell O. Webb, and Nicolas Salamin, "Exotic Taxa Less Related to Native Species Are More Invasive," *Proceedings of the National Academy of Sciences* 103 (2006): 5841–5845.

58. William E. Kunin and Avi Shmida, "Plant Reproductive Traits as a Function of Local, Regional, and Global Abundance," *Conservation Biology* 11 (1997): 183–192, and Mark Van Kleunen and Steven D. Johnson, "Effects of Self-Compatibility on the Distribution Range of Invasive European Plants in North America," *Conservation Biology* 21 (2007): 1537–1544.

59. Anne C. Cully, Jack F. Cully Jr., and Ronald D. Hiebert, "Invasion of Exotic Plant Species in Tallgrass Prairie Fragments," *Conservation Biology* 17 (2003): 990–998.

60. Ibid.

61. S. B. Dennis, V. G. Allen, K. E. Saker, and others, "Influence of Neotyphodium Coenophialum on Copper Concentration in Tall Fescue," *Journal of Animal Science* 76 (1998): 2687–2693, and Keith Clay, Jenny Holah, and Jennifer A. Rudgers, "Herbivores Cause a Rapid Increase in Hereditary Symbiosis and Alter Plant Community Composition," *Proceedings of the National Academy of Sciences* 102 (2005): 12465–12470.

62. Dariusz P. Malinowski and David P. Belesky, "Adaptations of Endophyte-Infected Cool-Season Grasses to Environmental Stresses: Mechanisms of Drought and Mineral Stress Tolerance," *Crop Science* 40 (2000): 923–940.

63. Carol Hart, "Forged in St. Anthony's Fire," *Modern Drug Discovery* 2 (1999): 20–21, 23–24, 28, 31.

64. The preceding two paragraphs are based on Mark J. McKone, Kendra K. McLauchlan, Edward G. Lebrun, and Andrew C. McCall, "An Edge Effect Caused by Adult Corn-Rootworm Beetles on Sunflowers in Tallgrass Prairie Remnants," *Conservation Biology* 15 (2001): 1315–1324.

65. The preceding three paragraphs are based on Kerry N. Rabenold, Peter T. Fauth, Bradley W. Goodner, and others, "Response of Avian Communities to Disturbance by an Exotic Insect in Spruce-Fir Forests of the Southern Appalachians," *Conservation Biology* 12 (1998): 177–189.

66. The preceding discussion of habitat is based in part on Chris Maser, Andrew W. Claridge, and James M. Trappe, *Trees, Truffles, and Beasts: How Forests Function* (New Brunswick, NJ: Rutgers University Press, 2008).

67. Mark V. Lomolino and David R. Perault, "Body Size Variation of Mammals in a Fragmented, Temperate Rainforest," *Conservation Biology* 21 (2007): 1059–1069.

68. The preceding discussion of habitat configuration is based on James R. Herkert, Dan L. Reinking, David A. Wiedenfeld, and others, "Effects of Prairie Fragmentation on the Nest Success of Breeding Birds in the Mid-Continental United States," *Conservation Biology* 17 (2003): 587–594; Esa Huhta, Teija Aho, Ari Jäntti, and others, "Forest Fragmentation Increases Nest Predation in the Eurasian Treecreeper," *Conservation Biology* 18 (2004): 148–155; Sharon K. Collinge, Kathleen L. Prudic, and Jeffrey C. Oliver, "Effects of Local Habitat Characteristics and Landscape Context on Grassland Butterfly Diversity," *Conservation Biology* 17 (2003): 178–187; Ewers and Didham, "The Effect of Fragment Shape and Species' Sensitivity to Habitat Edges on Animal Population Size"; Matthew R. Falcy and Cristián F. Estades, "Effectiveness of Corridors Relative to Enlargement of Habitat Patches," *Conservation Biology* 21 (2007): 1341–1346; Tim Gerrodette and William G. Gilmartin, "Demographic Consequences of Changed Pupping and Hauling Sites of the Hawaiian Monk Seal," *Conservation Biology* 4 (1990): 423–430; and Maiken Winter and John Faaborg, "Patterns of Area Sensitivity in Grassland-Nesting Birds," *Conservation Biology* 13 (1999): 1424–1436.

69. Paul Beier and Reed F. Noss, "Do Habitat Corridors Provide Connectivity?" *Conservation Biology* 12 (1998): 1241–1252.

70. John B. Dunning Jr., Rene Borgella Jr., Krista Clements, and Gary K. Meffe, "Patch Isolation, Corridor Effects, and Colonization by a Resident Sparrow in a Managed Pine Woodland," *Conservation Biology* 9 (1995): 542–550.

71. The preceding discussion of corridors and boundaries (edges) is based on Craig S. Machtans, Marc-André Villard, and Susan J. Hannon, "Use of Riparian Buffer Strips as Movement Corridors by Forest Birds," *Conservation Biology* 10 (1996): 1366–1379; Linda M. Puth and Karen A. Wilson, "Boundaries and Corridors as a Continuum of Ecological Flow Control: Lessons from Rivers and Streams," *Conservation Biology* 15

(2001): 21–30; and Takehiko Y. Ito, Naoko Miura, Badamjav Lhagvasuren, and others, "Preliminary Evidence of a Barrier Effect of a Railroad on the Migration of Mongolian Gazelles," *Conservation Biology* 19 (2005): 945–948.

72. Stephen G. Mech and James G. Hallett, "Evaluating the Effectiveness of Corridors: A Genetic Approach," *Conservation Biology* 15 (2001): 467–474.

73. Paul Beier, "Determining Minimum Habitat Areas and Habitat Corridors for Cougars," *Conservation Biology* 7 (1993): 94–108; Susan M. Haig, David W. Mehlman, and Lewis W. Oring, "Avian Movements and Wetland Connectivity in Landscape Conservation," *Conservation Biology* 12 (1998): 749–758; and Nick Haddad, "Corridor Length and Patch Colonization by a Butterfly, *Junonia coenia*," *Conservation Biology* 14 (2000): 738–745.

74. The two previous paragraphs are based on Jorge Galindo-Gonzalez, Sergio Guevara, and Vinicio J. Sosa, "Bat- and Bird-Generated Seed Rains at Isolated Trees in Pastures in a Tropical Rainforest," *Conservation Biology* 14 (2000): 1693–1703.

75. The three previous paragraphs are based on Esteban Fernández-Juricic, "Avifaunal Use of Wooded Streets in an Urban Landscape," *Conservation Biology* 14 (2000): 513–521.

76. The preceding discussion of the French hawksbeard is based on P.-O. Cheptou, O. Carrue, S. Rouifed, and A. Cantarel, "Rapid Evolution of Seed Dispersal in an Urban Environment in the Weed *Crepis sancta*," *Proceedings of the National Academy of Sciences* 105 (2008): 3796–3799.

77. Florian Kirchner, Jean-Baptiste Ferdy, Christophe Andalo, and others, "Role of Corridors in Plant Dispersal: An Example with the Endangered *Ranunculus nodiflorus*," *Conservation Biology* 17 (2003): 401–410.

78. S. Harrison, B. D. Inouye, and H. D. Safford, "Ecological Heterogeneity in the Effects of Grazing and Fire on Grassland Diversity," *Conservation Biology* 17 (2003): 837–845.

79. Stuart B. Weiss, "Cars, Cows, and Checkerspot Butterflies: Nitrogen Deposition and Management of Nutrient-Poor Grasslands for a Threatened Species," *Conservation Biology* 13 (1999): 1476–1486.

80. Lars A. Brudvig, Catherine M. Mabry, James R. Miller, and Tracy A. Walker, "Evaluation of Central North American Prairie Management Based on Species Diversity, Life Form, and Individual Species Metrics," *Conservation Biology* 21 (2007): 864–874.

81. Carl E. Bock and Jane H. Bock, "Cover of Perennial Grasses in Southeastern Arizona in Relation to Livestock Grazing," *Conservation Biology* 7 (1993): 371–377.

82. W. D. Billings, "Ecological Impacts of Cheatgrass and Resultant Fire on Ecosystems in the Western Great Basin," in *Proceedings—Ecology and Management of Annual Rangelands*, ed. S. B. Monsen and S. G. Kitchen, 22–30 (Fort Collins, CO: Intermountain Research Station, USDA Forest Service, 1994), and S. B. Monsen, "The Competitive Influences of Cheatgrass (*Bromus tectorum*) on Site Restoration," in *Proceedings—Ecology and Management of Annual Rangelands*, ed. S. B. Monsen and S. G. Kitchen , 43–50 (Fort Collins, CO: Intermountain Research Station, USDA Forest Service, 1994).

83. The previous two paragraphs are based on Ron Panzer, "Compatibility of Prescribed Burning with the Conservation of Insects in Small, Isolated Prairie Reserves," *Conservation Biology* 16 (2002): 1296–1307.

84. Brent E. Johnson and J. Hall Cushman, "Influence of a Large Herbivore Reintroduction on Plant Invasions and Community Composition in a California Grassland," *Conservation Biology* 21 (2007): 515–526.

85. The previous three paragraphs are based on M. Lisa Floyd, Thomas L. Fleischner, David Hanna, and Paul Whitefield, "Effects of Historic Livestock Grazing on Vegetation at Chaco Culture National Historic Park, New Mexico," *Conservation Biology* 17 (2003): 1703–1711.

86. Thomas J. Valone, Marc Meyer, James H. Brown, and Robert M. Chew, "Timescale of Perennial Grass Recovery in Desertified Arid Grasslands Following Livestock Removal," *Conservation Biology* 16 (2002): 995–1002.

87. The two previous paragraphs are based on Machteld C. Van Dierendonck and Michiel F. Wallis de Vries, "Ungulate Reintroductions: Experiences with the Takhi or Przewalski Horse (*Equus ferus przewalskii*) in Mongolia," *Conservation Biology* 10 (1996): 728–740.

88. Michael J. Samways, "Land Forms and Winter Habitat Refugia in the Conservation of Montane Grasshoppers in Southern Africa," *Conservation Biology* 4 (1990): 375–382.

89. Bruce Andersom, William W. Cole, and Spencer C. H. Barrett, "A Plant Scores by Providing an Access Point for Visiting Sunbirds to Feed on Its Nectar," *Nature* 435 (2005): 41–42.

90. Alastair W. Robertson, Dave Kelly, Jenny J. Ladley, and Ashley D. Sparrow, "Effects of Pollinator Loss on Endemic New Zealand Mistletoes (Loranthaceae)," *Conservation Biology* 13 (1999): 499–508.

91. The two previous paragraphs are based on Anna Traveset and Nuria Riera, "Disruption of a Plant-Lizard Seed Dispersal System and Its Ecological Effects on a Threatened Endemic Plant in the Balearic Islands," *Conservation Biology* 19 (2005): 421–431.

92. Louis Deharveng, "Soil Collembola Diversity, Endemism, and Reforestation: A Case Study in the Pyrenees (France)," *Conservation Biology* 10 (1996): 74–84.

93. K. G. Lyons, C. A. Brigham, B. H. Traut, and M. W. Schwartz, "Rare Species and Ecosystem Functioning," *Conservation Biology* 19 (2005): 1019–1024.

94. To fully understand the visioning process, read Chris Maser, *Vision and Leadership in Sustainable Development* (Boca Raton, FL: Lewis, 1998).

CHAPTER 7   WHERE DO WE GO FROM HERE?

1. Donald R. Nelson, W. Neil Adger, and Katrina Brown, "Adaptation to Environmental Change: Contributions of a Resilience Framework," *Annual Review of Environment and Resources* 32 (2007): 395–419.

2. The discussion of population is based on Chris Maser, *The Perpetual Consequences of Fear and Violence: Rethinking the Future* (Washington, DC: Maisonneuve Press, 2004).

3. Oliver R. W. Pergams and Patricia A. Zaradic, "Evidence for a Fundamental and Pervasive Shift Away from Nature-Based Recreation," *Proceedings of the National Academy of Sciences* 105 (2008): 2295–2300.

4. William J. Sutherland, Susan Armstrong-Brown, Paul R. Armsworth, and others, "The Identification of 100 Ecological Questions of High Policy Relevance in the UK," *Journal of Applied Ecology* 43 (2006): 617–627.

GLOSSARY

abiotic—the nonliving chemical and physical components of the environment.

albedo effect—the electromagnetic radiation reflected back into space by the white surface of a growing ice sheet; *albedo* is Late Latin for whiteness, from the Latin *albus*, white.

autonomous, self-compatible plant—a plant that can fertilize itself without the services of an outside pollinator, such as a bee.

benthic—of or pertaining to a benthos, including the organisms that live there.

benthos—the biogeographic region at the lowest level of a body of water, such as an ocean or a lake, including the sediment surface and some subsurface layers.

biodiversity (see also "biological diversity")—the diversity of living organisms.

biological diversity—the condition of having a variety of biotic characteristics and traits (e.g., genes, species, and community types); life-history stages; structural forms (e.g., stratification, zonation, and the physical structures of plants); biotic patterns (e.g., reproductive, activity, food-web, social, and interactive); and functions (e.g., nutrient cycling, hydrological cycling, and provision of habitat).

biomass—the combined weight of all living organisms in a given area.

biophysical diversity—the diversity of living and nonliving components of an ecosystem.

biotic—composed of plants and animals.

boreal—of or pertaining to the northern part of the Northern Hemisphere.

calcareous—composed of, containing, or characteristic of calcium carbonate, calcium, or limestone; chalky.

calcium—a silvery, moderately hard, metallic element that constitutes approximately 3 percent of the earth's crust and is a basic component of bone, shell, and leaves.

calcium carbonate—a colorless- or white-crystalline compound that occurs naturally in chalk, limestone, marble, and other forms.

cambium—in woody vegetation, the layer of cells that lies between the secondary xylem and secondary phloem cell layers; through a process of cell division, the cambium produces the secondary xylem and the secondary phloem that are also known, respectively, as the wood and the innermost living bark.

canopy—the more or less continuous cover of branches and foliage collectively formed by the crowns of adjacent trees and other woody growth; layers of canopy may be called stories.

canopy closure—the progressive reduction of space between tree crowns as they spread laterally.

carbohydrate—any of a group of chemical compounds, including sugars, starches, and cellulose, containing carbon, hydrogen, and oxygen only.

carbon—a naturally abundant nonmetallic element that occurs in many inorganic and all organic compounds.

carbon flux—an abbreviated phrase referring to the net difference between the sequestration of carbon dioxide through photosynthesis and the respiration of carbon dioxide by such organisms as plants and microbes.

carnivore—a mammal that that preys or feeds on other animals, as opposed to eating plants.

carnivorous—flesh-eating.

carrying capacity—the maximum number of individuals of a species that an area can sustainably maintain without altering the integrity of the ecosystem that supports them.

chemoautotrophs—"chem" from the Greek *chemikos*, "of or pertaining to juices" + "auto" from the Greek *autos*, "self" + "troph" from the Greek *trophos*, "one who feeds"; hence, bacteria that derive energy from oxidizing inorganic compounds, such as hydrogen sulfide, ammonium, and ferrous iron.

chemotaxis (chemotactic, adj.)—a movement in which bodily cells, bacteria, and other single-cell organisms direct their movements according to certain chemicals in their environment, as when bacteria find food by swimming toward the highest concentration of food molecules or when they flee from poisons.

clear-cut—an area of land that was forested but now has all the trees cut down and removed.

clear-cutting—the act of cutting down all the trees in a forested area as an economic expedient in order to remove the timber to a mill.

climatic cycle—the cyclic changes in weather patterns in a geographical area over time.

climax—the culminating stage in plant succession for a given site where the vegetation is self-reproducing and thus has reached a stable condition through time.

colonization—the process or act of establishing a colony or colonies (also see "colony").

colony—a group of the same kind of plants or animals living together.

closed canopy—the condition that exists when the canopy created by trees or shrubs or both is dense enough to exclude most of the direct sunlight from the floor of the forest.

commensal—of or pertaining to the relationship between two different species from which one derives food and other benefits without negatively affecting the other, such as house sparrows benefiting from their association with humans in a town.

commons—something, such as an ocean or the air, that is everyone's birthright and so is "owned" in common by everyone and by no individual in particular.

community—a group of one or more populations of plants and/or animals using a common area; an ecological term used in a broad sense to include groups of plants and animals of various sizes and degrees of integration.

competition—a relationship among members of the same or different species such that use of a particular available resource by one party in the relationship reduces the amount of the resource that is available to the other party.

composition—the way in which something is made, especially the arrangement of its the parts.

compound—a substance composed of two or more chemical elements.

configuration—the shape or outline of a forest stand or plant community; the degree of irregularity in the edge between forest stands or communities, varying from simple to mosaic.

conifer (coniferous, adj.)—the most important order of the Gymnospermae, comprising a wide range of trees, mostly evergreens that bear cones and have needle-shaped or scale-like leaves; timber commercially identified as softwood.

coniferous forest—a forest dominated by cone-bearing trees (see also "conifer").

connectivity—the degree to which patches of habitat are connected by corridors that act as routes of migration for plants and animals to get from one viable patch to another.

continental shelf—a generally shallow, flat, submerged portion of a continent that extends to a point of steep descent to the floor of the ocean; generally the most productive part of the sea.

continuum—a continuous extent, succession, or whole, no part of which can be distinguished from neighboring parts except by arbitrary division.

copepod—any of numerous small marine and freshwater crustaceans of the order Copepoda.

cotyledon (or seed leaf)—a significant part of the embryo within the seed of a plant, which, upon germination, may become the embryonic, first leaves of the seedling.

crop—the vegetation growing on an area; in forestry it is thought of as the major woody growth forming the forest crop—any wood fiber that is harvestable.

crustacean—any of various predominantly aquatic arthropods of the class Crustacea, including lobsters, crabs, shrimps, and barnacles, characteristically having a segmented body, a chitinous exoskeleton, and paired, jointed limbs.

cryptic-species complex—a group of species that satisfies the biological definition of species—that is, they are reproductively isolated from each other but are distinguishable on a morphological basis; the only way to tell them apart is through such analysis as DNA sequencing.

cumulative effects—individual consequences of an action or repeated actions, which may or may not be observable, that reinforce one another as they occur over time until they cross a threshold and manifest as a stronger outcome than any of the individual consequences would be by themselves.

cycling—occurring in or passing through a cycle; moving in, or as if in, a circle.

dead zones—areas that are virtually depleted of dissolved oxygen, from which aquatic life either flees or suffocates.

decay—to decompose, to rot; in wood, the decomposition by fungi or other microorganisms resulting in softening, progressive loss of strength and weight, and changes in texture and color.

deciduous—falling off, shedding, or falling out at maturity, at certain seasons, or at certain ages; said of the leaves of certain trees, the antlers of deer, the first set of teeth (milk teeth) of most kinds of mammals.

decompose—to separate into component parts or elements; to break down; to decay or putrefy.

delta—the usually triangular deposit of alluvial material at the mouth of a river.

diatom—any of various minute one-celled or colonial algae of the class Bacillariophyceae, having siliceous cell walls consisting of two overlapping, symmetrical parts.

diversity—the relative degree of abundance of species of plants and animals, functions, communities, habitats, or habitat features per unit of area.

domestic mammal—a mammal that, through direct selection by humans, has certain inherent biological characteristics by which it differs from its wild ancestors.

dominant—species of plants or groups of plants that, by means of their numbers, coverage of an area, or size, influence or control the existence of associated species; also, individual animals that determine the behavior of one or more other animals in a way that establishes a social hierarchy.

duff—partially to fully decomposed organic material, generally located between litter and mineral soil.

dynamic—characterized by or tending to produce continuous change.

ecosystem—a community of all plants and animals and their physical environment, functioning together as an interdependent unit.

ecotone—the area influenced by the transition between plant communities or between successional stages or vegetative conditions within a plant community.

ectomycorrhiza—a mycorrhiza in which the fungus mantles the surface of the plant's feeder rootlet with fungal tissue, grows between the outer rootlet cells, and extends hyphal filaments into surrounding soil.

edaphic—related to or caused by particular soil conditions, including such factors as water content, acidity, aeration, and the availability of nutrients, especially as they affect living organisms.

eddy (s.), eddies (pl.)—a current, as of water or air, moving contrary to the direction of the main current, especially in a circular motion.

edge—the place where plant communities or where successional stages or vegetative conditions within a plant community come together.

endemic—native to or confined to a certain region; having a comparatively restricted distribution.

endophyte—a symbiont, often a bacterium or fungus, that lives within a plant for at least part of its life without causing apparent disease; endophytes are ubiquitous in all the species of plants, but the reciprocal relationships are not well understood.

endosymbiosis—a symbiotic relationship that takes place within an organism.

epifauna—animals that live on the surface of sediments or soils.

erosion—removal of soil or rocks from any place on the Earth's surface by weathering, dissolution, abrasion, wind, or water.

ethnobotany—*ethno*, "the study of people" + *botany*, "the study of plants"; thus, an inquiry into the complex relationship between plants and people in a given culture.

eutrophic—of a body of water, characterized by an increase of mineral and organic nutrients that reduce the dissolved oxygen, producing an environment that favors plant life over animal life.

eutrophication—the process whereby chemicals, typically compounds containing nitrogen or phosphorus, are introduced by humans into an aquatic system, where they act as excess nutrients that stimulate excessive growth of such plants as algae.

evapotranspiration—the combination of water lost to the atmosphere from the ground surface via evaporation from the capillary fringe of the groundwater table and the transpiration of groundwater by plants whose roots tap the capillary fringe of the groundwater table.

exotic plant species—plant species that meet one of the two following definitions: (1) they do not occur naturally in temperate or subtropical North America, or (2) they occur naturally in temperate or subtropical North America but are indigenous to a forest other than the one they are in.

fauna—animals collectively, especially the animals of a particular region or time.

fecal material—material discharged from the bowels; more generally, any discharge from the digestive tract of an organism.

feces—generally, any discharge from the digestive tract of an organism.

fixing nitrogen—taking gaseous, atmospheric nitrogen and altering it in such a way that it becomes available to and useable by the fungus and the host plant.

flora—plants collectively, especially the plants of a particular region or time.

forb—any herbaceous species of fleshy-leaved plants other than grasses, sedges, and rushes.

forest—generally, that portion of the ecosystem characterized by tree cover; more particularly, a plant community made up predominantly of trees and other woody vegetation that grows close together.

forest floor—the surface layer of a soil that supports forest vegetation.

fragmentation of habitat—the breaking up of contiguous habitat by intersecting it with roads, blocks of clear-cut forest, and so on.

fruiting body—the reproductive organ of a fungus.

function—the natural action of organisms and/or the nonliving components of a habitat.

fungal—caused by or associated with fungi (also see "fungi").

fungal hypha (s.), hyphae (pl.)—see "fungus" and "hyphae."

fungus (s.), fungi (pl.)—mushrooms, truffles, molds, yeasts, rusts, etc.; simply organized plants, unicellular or made of cellular filaments or strands called "hyphae," lacking chlorophyll, reproducing asexually and sexually through the formation of spores.

gene pool—narrowly, the genetic material of a localized interbreeding population; broadly, the genetic resources or materials of a species throughout its entire geographical distribution.

genetic diversity—the diversity of individuals of the same species, each with a different genetic makeup, as opposed to individuals that have been genetically selected to display a certain trait at the expense of others, such as a tree bred to grow fast but at the expense of being able to tolerate hot weather.

genus—the taxonomic group between family and species, containing one or more species that have certain characteristics in common; scientific names have two words, the first referring to genus and the second to species.

geomorphological—pertaining to geological structure: geomorphological features of the Black Hills, morphological features of granite, structural effects of folding and faulting of the earth's surface.

geomorphology—the geological study of the configuration and evolution of land forms.

germinate—to begin to grow, to sprout.

germplasm—a collection of genetic material for an organism. For nonarboreal plants, germplasm may be stored as a collection of seeds; for trees, the germplasm may be maintained by growing them in a nursery.

granivore—an herbivorous animal, which selectively eats the nutrient-rich seeds produced by plants.

granivorous—of or pertaining to a seed-eating animal (see "granivore").

grass—any species of plant in the family Poaceae, which is characterized by narrow leaves, hollow, jointed stems, and spikes or clusters of membranous flowers borne in smaller spikelets; such plants collectively.

habitat—the sum total of environmental conditions of a specific place occupied by a plant, an animal, or a population of such species.

heartwood—the inner layers of wood that, in a growing tree, have ceased to contain living cells and in which the reserve materials, such as starch, have been removed or converted into more durable substances.

herb (herbaceous, adj.)—a non-woody plant, as distinguished from a woody plant.

herbivorous—feeding chiefly on grasses or other plant material.

heterogeneity—the condition or state of being different in kind or nature.

Holocene epoch—the geological period that began approximately 11,550 years before the present and continued to about 300 years before the present, or to the 1700s.

home range—the area that an animal covers during its normal daily (twenty-four-hour) activities and that it does not defend against others of its own kind.

hydrological cycle—the way water falls as rain and/or snow, sinks into the soil and is either stored or flows below ground, runs over the surface of the soil in streams and rivers on their way to the sea, and evaporates into the atmosphere to be cycled again as rain and/or snow.

hypha (s.), hyphae (pl.)—filament or strand of a fungus thallus (non-reproductive, vegetative body) that is composed of one or more cylindrical cells; increases by growth at its tip; gives rise to new hyphae by lateral branching.

immigration—the act of moving into a new area.

impervious—incapable of being penetrated.

infaunal communities—communities composed of aquatic animals that live in the substrate of a body of water, especially in the soft bottom of an ocean.

inner bark—part of the living, growing tissue of a tree situated just under the non-living outer bark.

inorganic—not involving either living organisms or their products (see "organic").

inorganic materials—anything that does not involve organic products.

insectivore (insectivorous, adj.)—an animal or plant that eats insects and other invertebrate animals.

integration—coordination of parts into a functioning whole, as in a plant, animal, or human community.

integrator—an attribute that expresses the combined influences of a number of interacting variables.

integrity—the state of being unimpaired; soundness; completeness; unity.

landscape—a physiographic unit that is capable of sustaining several populations of a species.

larva (s.), larvae (pl.)—the newly hatched, earliest stage of any of various animals that undergo metamorphosis, such as insects, frogs, and salamanders, and differ markedly in form and appearance from the adult.

larval form—the form, or shape, of a larva.

legumes—members of the pea family in the broad sense, the Fabaceae.

lichen—a plant that is actually two plants in one; the outer plant is a fungus that houses an inner plant, an alga.

lignin—the chief non-carbohydrate constituent of wood, a polymer that functions as a natural binder and support for the cellulose fibers of woody plants.

litter (forest)—the uppermost layer of organic debris on the floor of a forest; essentially freshly fallen or slightly decomposed vegetable material, mainly foliate, or leaf, litter, but also twigs, wood, fragments of bark, flowers, and fruits.

litter fall—the fall of litter to the floor of the forest.

log—technically, a segment of tree stem that is cut to a predetermined length; generally, a tree stem that has fallen to the forest floor.

logging—the activity of cutting down trees and cutting their stems into predetermined lengths for sale to mills, where the stems are converted into lumber or other products.

long term—the timescale of the forest owner or manager as manifested by the objectives of the management plan, the rate of harvesting, and the commitment to maintain permanent forest cover; the length of time varies according to the context and ecological conditions and is a function of how long it takes a given ecosystem to recover its natural structure and composition following harvesting or disturbance or to produce mature or primary conditions.

macroevolution—any evolutionary change at or above the level of species.

macrofauna—invertebrates, such as beetles, ants, termites, spiders, earthworms, and mice.

macrohabitat—the larger habitat within which an organism dwells, such as a forest.

macroinvertebrates—aquatic invertebrates, including insects (e.g., larval mayflies, caddisflies, and dragonflies), crustaceans (such as crabs and lobsters), mollusks (such as snails and clams), and worms (such as flatworms), that inhabit a river channel, pond, lake, wetland, or ocean.

mammal—an animal that has hair on its body during some stage of its life and whose babies are initially nurtured by their mother's milk.

marine—of the ocean.

mature forest—a forest composed primarily of or dominated by mature trees in vigorous condition (also see "maturity").

mature tree—see "maturity."

maturity—in physiology, the stage at which a tree or other plant has attained full development and is in full production of seeds.

microbe—a microscopic organism.

microclimate—as I am using it here, the climate of an immediate area as determined by the topography and the vegetation of the area; the microclimate exerts a local influence over the prevailing, overall climate of the times, the macroclimate.

microhabitat—the small, specialized habitat in which an organism dwells, such as a log (microhabitat) within a forest (macrohabitat).

microorganism—a plant or animal of microscopic size, especially a bacterium or protozoan.

microscopic—too small to be seen by the unaided eye, large enough to be seen with the aid of a microscope; exceedingly small; minute.

microtopography—the features of a small place or region, such as one square foot, as opposed to the features of a large place or region, such as one square mile.

midden—a refuse pile of accumulated, uneaten scraps.

migration—a periodic movement away from or back to a given area.

mineral—any naturally occurring, homogeneous, inorganic substance that has a definite chemical composition and a characteristic crystalline structure, color, and hardness.

mineral soil—soil composed mainly of inorganic materials, with a relatively low amount of organic material.

mineralize—to convert to a mineral substance.

mollusk—any member of the phylum Mollusca, composed largely of marine invertebrates, including the edible shellfish, such as clams, and some one hundred thousand other species.

monocarpic plant—plants that flower and set seeds only once before dying.

montane—of the mountains.

mutualists—two species of organisms that derive a mutual benefit from living together, regardless of which species they are.

mycelium (s.), mycelia (pl.)—the vegetative part of a fungus, consisting of a mass of branching threadlike filaments or strands, called hyphae.

mycophagist—an animal that eats fungi.

mycophagous—fungi-eating.

mycorrhiza—the mutually beneficial symbiosis of specialized fungi with the feeder rootlets of plants; the fungus absorbs nutrients from the soil and shares them with the plant, while the plant produces sugars by photosynthesis and shares them with the fungus; the participating fungi and plants generally cannot survive without each other. (See also "ectomycorrhiza." and "vesicular-arbuscular mycorrhiza.")

mycorrhizal fungus—a fungus that forms mycorrhiza.

native species—a species that occurs naturally in a region; endemic to an area.

natural forest—forested areas in which many of the principal characteristics of the native ecosystems are present.

nitrogen (N_2)—a nonmetallic element constituting nearly four-fifths of the air by volume; a colorless, odorless, almost inert gas; it occurs in various minerals and in all proteins.

nitrogen fixation—the conversion of elemental nitrogen (N_2) from the atmosphere to organic combinations or to forms readily usable in biological processes.

nitrogen-fixing bacteria—bacteria that can take nitrogen gas out of the air and transform it into an organic compound that plants can use.

non-timber forest products—all forest products except timber; other materials obtained from trees, such as resins and leaves, as well as any other plant and animal products.

nutrient cycling—the circulation of elements, such as nitrogen and carbon, via specific pathways from the nonliving to the living portions of the environment and back again.

obligatory mutualists—two species of organisms in which for one partner to survive, both must survive, regardless of which species they are.

old forest—a forest that is past full maturity; the last stage in forest succession; a forest with two or more levels of canopy, heart rot, and other signs of physiological deterioration.

old-growth forest—a forest that is unroaded or lightly roaded, with no evidence of previous logging and of sufficient size and configuration to maintain ecological integrity, usually five hundred acres or larger.

old-growth stand—a stand of trees that is unroaded or lightly roaded, with no evidence of previous logging, usually ranging from fifteen to five hundred acres and of sufficient size and configuration to maintain specific ecological functions.

open canopy—a canopy condition that allows large amounts of direct sunlight to reach the ground.

organic—of, pertaining to, or derived from living organisms; of or designating a compound that contains the element carbon (see "inorganic").

organic matter in soil—materials derived from plants and animals, much of it in an advanced state of decay.

organism—any living individual of any species of plant or animal.

oxygen—a colorless, odorless, tasteless gaseous element constituting 21 percent (by volume) of the atmosphere; it combines with most elements, is essential for plant and animal respiration, and is required for nearly all combustion and combustive processes.

palearctic region—the biogeographic region of the Arctic, including the immediately adjacent, temperate regions of Europe, Asia, and Africa.

paleolimnology—the study of the archeological and geological properties of lakes; from the Greek *paleo*, "old" + *limne*, "lake" + *logos*, "to study."

pebble—a small stone eroded smooth.

pelagic—of, pertaining to, or living in open oceans or seas rather than in waters adjacent to land or inland waters.

peridotite—any of several coarse-grained, dark igneous rocks consisting mainly of olivine and other iron-magnesium minerals (see also "serpentine").

phenology—the study of plant and animal responses to seasonal changes in the timing of natural events; common examples include the date that migrating birds return, the dates various plants first flower, the date the first cabbage butterflies lay their eggs.

photosynthesis—the process by which chlorophyll-containing cells in green plants convert incident light to chemical energy and synthesize organic compounds from inorganic compounds, especially carbohydrates from carbon dioxide and water, with the simultaneous release of oxygen.

physiology (physiological, adj.)—the biological science of essential and characteristic life processes.

phytoplankton—plant organisms, generally microscopic, that float or drift in great numbers in fresh or salt water.

plankton—plant and animal organisms, generally microscopic, that float or drift in great numbers in fresh or salt water.

plant-community type—a vegetative complex with distinctive composition and boundaries that are recognizable in the field; the composition is a result of environmental influences on the site—such as the availability of seeds, soils, temperature, elevation, solar radiation, slope, aspect, and precipitation; denotes a general kind of climax vegetation, such as ponderosa pine or bunchgrass, from which several plant community types may be derived on the basis of characteristic lesser vegetation.

pleiotropy—from the Greek *pleio*, "many" + *trepein*, "influencing"; hence, the phenomenon of one gene producing many effects. (A common mistake is to use "pleiotrophic" instead of "pleiotropic.")

polycarpic plants—plants that flower and set seeds many times before dying.

polychaete—from the Greek *polys*, "many" + the New Latin *chaet*, "bristle"; hence, a many-bristled worm.

population—a group of individuals that are interfertile and that regularly contribute germ cells to the formation of fertile offspring.

predaceous—of or pertaining to an animal that feeds like a predator.

predator—any animal that kills and feeds on other animals.

primary productivity—the amount of green vegetation produced in a particular year.

process—a system of operations in the production of something; a series of actions, changes, or functions that brings about an end or result.

protein—any of a group of complex nitrogenous, organic compounds of high molecular weight that contain amino acids as their basic structural units and that occur in all living matter and are essential for the growth and repair of animal tissue.

recruitment—the juvenile individuals of a given species entering a population for the first time.

refugium (s.), refugia, (pl.)—a small island of habitat in which a species can survive and from which it can disperse when the surrounding habitat becomes suitable for it to live in.

rhizosphere—the area surrounding the roots of plants wherein complex relations exist among the plant, the soil microorganisms, and the soil itself.

riparian zone—an area identified by the presence of vegetation that requires free or unbound water or conditions more moist then normally found in the area.

rootlet—tiny roots that take up water and nutrients for a plant.

rootwad—the mass of roots, soil, and rocks that remains intact when a tree, shrub, or stump is up-rooted.

salinity—the saltiness of water.

salvage logging—the removal and sale of dead, dying, or deteriorating trees.

sand—loose, granular, gritty particles of worn or disintegrated rock, finer than gravel and coarser than dust.

sapling—a vigorously growing young tree more than a few feet high and an inch in diameter at breast height, with live bark and perhaps an occasional dead branch.

sapwood—the outer layer of wood in a growing tree that contains living cells and reserve materials, such as starch.

scrubland—an area of land that is uncultivated and covered with sparse, stunted vegetation.

secrete—to generate from bodily cells or fluids.

secretion—the process of secreting a substance, especially one that is not a product of bodily waste, from blood or cells.

sediment—material suspended in water; the deposition of such material onto a surface, such as a stream bottom, underlying the water.

seedling—a young tree grown from seed from the time of germination until it becomes a sapling; the division between seedling and sapling is indefinite and may be arbitrarily fixed.

self-compatible plant—a plant in which the male gamete can fertilize the female gamete of the same plant, but the process requires an unrelated organism, such as a bee, to do the pollinating.

serpentine—a type of peridotite, a rock found miles below the ocean floor on the bottom of the Pacific Plate; geologists call serpentine an "ultramafic" rock, where "mafic" refers to rocks that are relatively high in magnesium and iron but low in calcium, silica, and aluminum (see also "peridotite").

sessile—in zoology, permanently attached, unable to move, as an animal, such as an attached marine mussel, that is permanently attached to a substrate; in botany, attached by the base, as a flower or leaf, such as the sessile-flowered wake-robin, that lacks an intervening stalk between the flower and the trio of leaves; from the New Latin *sessilis*, "sitting."

shade-intolerant plant—a species of plant that does not germinate or grow well in shade.

shade-tolerant plant—a species of plant that both germinates and grows well in shade; when the shade is removed, however, the plant responds to the available light.

sheet flooding—flooding caused by comparatively shallow water flowing over a wide, relatively flat area that typically does not have the appearance of a well-defined watercourse; sheet flooding is especially dangerous because it is not often obvious, even when one is standing in such an area, that it could become inundated.

shrub—a plant with persistent woody stems and relatively low growth form; usually produces several basal shoots as opposed to a single stem; differs from a tree by its low stature and non-treelike form.

sibling species—species that are extremely similar in appearance but are nonetheless reproductively isolated from one another; often thought to be the result of fairly recent differentiation—in other words, they are relatively new products of the speciation process.

silt—a sedimentary material consisting of fine mineral particles intermediate in size between sand and clay.

siltload—silt that is carried in suspension, in contrast to soil, rocks, and other debris rolled along the bottom of a stream by the moving water.

silviculture—the art of producing and tending a forest by manipulating its establishment, composition, and growth to best fulfill the objectives of the owner, which may or may not include the production of timber.

snag—a standing dead tree from which the leaves and most of the branches have fallen; such a tree broken off but still more than twenty feet tall; also a large drifted tree that is stuck in the muddy bottom of a river or estuary and may or may not be visible above the surface.

soil—earth material so modified by physical, chemical, and biological agents that it will support rooted plants.

soil macrofauna—easily visible organisms that live in the soil or on its surface for at least some part of their life cycle, including invertebrates considered both as pests and/or as beneficial to the soil environment and plant production; ants, termites, earthworms, beetles, grubs, pill bugs, true bugs, cicadas, snails, millipedes, centipedes, crickets, wasps, pseudoscorpions, spiders, and especially moth and fly larvae all form part of the soil macrofauna community.

species—a unit of classification of plants and animals consisting of the largest and most inclusive array of sexually reproducing and cross-fertilizing individuals that share a common gene pool; the most inclusive Mendelian population; denotes the rank of a species taxon in the Linnaean hierarchy. In biological species their reproductive isolation from each other protects their genetic makeup. Degree of morphological difference is not an appropriate species definition. Unequal rates of evolution among different characteristics and a lack of information on the mating potential of isolated populations and the ecological barriers isolating them are the major difficulties in the demarcation of species. For an excellent treatise on the term *species*, see Ernst Mayr, "What Is a Species, and What Is Not?" *Philosophy of Science* 63 (1996): 262–277.

species composition—the species that occur on a site or in a successional or vegetative stage of a plant community.

species richness—the variety of species that inhabit a particular area.

structural diversity—the diversity in a plant community that results from the variety of physical forms of the plants within the community (such as the layering of vegetation into ground cover, shrub layer, as well as understory, midstory, and overstory trees).

subspecies—a subdivision (a division of lower rank) of a taxonomic species, usually based on characteristics that indicate variation as a result of geographical distribution.

succession—progressive changes in species composition and forest community structure caused by nonhuman processes over time.

successional stage—a stage or recognizable condition of a plant community and its attending animal community that occurs during its development from bare ground to climax.

summer range—a range, usually at higher elevation than a winter range, used by deer and elk during the summer; a summer range is usually much more extensive than a winter range.

suppression—the process of competition among trees in a young forest for soil nutrients, water, sunlight, and space in which to grow; more vigorous trees suppress the growth of weaker ones, often leading to their death.

suspended sediments—sediments that are suspended in and by the water in which they are carried and in which they may be kept from settling out by the motion or velocity of the water in which they are suspended.

symbiont—one of the organisms in a symbiotic relationship.

symbiosis—the living together of two or more dissimilar organisms in a close association; in a mutualistic symbiosis, all participants benefit, whereas in parasitic or disease symbiosis, one or some of the organisms benefit at the expense of the others.

symbiotic—of or pertaining to the close association of two or more different organisms that may be but are not necessarily of benefit to each other; sometimes obligatory for one or more of the organisms in the relationship.

terrestrial—inhabiting the land, rather than water, trees, or air.

thallophyte—a plant in any one of the phyla of fungi and algae.

thallus—the body of a plant that is not differentiated into leaves, stems, and roots; one- to many-celled, such as thallophytes; also, the non-reproductive part of a fungus.

threatened species—any species that is likely to become endangered within the foreseeable future throughout all or a significant portion of its range.

topography (typographic, adj.)—the features of a small place or region.

transpiration—the act of transpiring.

transpire—to release water from plant leaves.

trophic cascade—the effect that a change in the size of one population in a food web has on the populations below it.

trophic level—a stage in the food chain; it reflects the number of times energy has been transferred from one organism to another through feeding; for example, a grass is eaten by a mouse, which is then eaten by a weasel, which is then eaten by an owl.

truffle—in the broad sense, the fruiting body of fleshy, belowground fungi; in the strict sense, members of the genus Tuber, many of which are commercially harvested as food.

trunk—the stem of a tree.

tundra—a treeless area between the ice cap and the tree line of arctic regions, having permanently frozen subsoil and supporting low-growing vegetation, such as lichens, mosses, and stunted shrubs; also occurs above the tree line on high mountains, where it is known as "alpine tundra" as opposed to "arctic tundra."

upwelling—the phenomenon of cold water moving upward from the bottom and mixing with the warmer water at the surface.

vagility—the capacity or tendency of an organism or a species to move about or disperse in a given environment.

varve—couplets of coarse/fine layers of deposition frequently found in the sediments of glacial lakes, much like the annual growth rings in a tree.

vertebrate—an animal with a backbone.

vesicomyid clams—from the Latin *vesicoz*, "bladder" + the Greek *mydos*, "decay"; hence, clams that don't eat in the usual sense but rather get their nutrients from sulfide-metabolizing bacteria that live in their gills.

vesicular-arbuscular mycorrhiza—a mycorrhiza in which the fungus produces one or both of two structures within the cells of the plant rootlet: balloon-like cells (vesicles), which store energy in the form of lipids, and bush-like structures (arbuscules), in which nutrients are exchanged between fungus and root; the fungus extends hyphal filaments into surrounding soil but does not mantle the rootlet with fungal tissue. (See also "mycorrhiza" and "ectomycorrhiza.")

water catchment—a drainage basin that catches water—rain, snow, or both—and stores it in a slow-motion downward flow as it merges with ever-larger water catchments until it is finally accepted into the vessel of the sea.

water column—the vertical space of water that exists between the bottom and the surface of a body of water, such as a stream or lake.

water quality—the purity of water determined by a series of standard physiochemical parameters—turbidity, temperature, bacterial count, pH, and dissolved oxygen—or by biological parameters—community composition and functionality, as well as the incidence of disease.

water table—the surface in a permeable body of rock of a zone saturated with water.

winter range—a range, usually at lower elevation than a summer range, used by migratory deer and elk during the winter; usually better defined and smaller than a summer range.

woody—containing wood fibers.

woody debris—all woody material, from whatever source, that is dead and lying on the forest floor.

zoochore—a plant dispersed by animals.

zoochorous—of or pertaining to plants that are dependent on animals to disperse their seeds.

INDEX

Abraham, 93
acacias, whistling-thorn, 17
acidification, ecosystem responses to, 105–106
acorn woodpeckers, 172
adaptation: defined, 198; for social-environmental sustainability, 198–199
adelgid, balsam woolly, 178
Adélie penguins, 106
aesthetics, and biodiversity, 15
afforestation, 29
agribusiness, market-driven, 24
agricultural chemicals: in alpine ecosystem, 160; animal contact with, 77
agricultural lands: development of, 104; human alteration of, 46–49; loss of, 93
agricultural produce, increasing costs of, 78
agriculture: controlling foreign plants in, 128; conversion of forests to, 48; growing intensification of, 39, 99; intensified management in, 69; introduction of, 26–27; maize (corn), 98; origins of, 121. *See also* farms; irrigation
agro-ecosystems, protecting traditional, 46
agroforestry: cacao, 39; and introduction of alien species, 29
air currents, toxins carried by, 109. *See also* atmosphere
Aleutian Islands, 122
algae, 207: under deep snow, 51; in Palouse prairie ecosystem, 44
alien species: in Palouse prairie ecosystem, 44–45. *See also* exotic plants; invasive species
Allen, James, 8
Amazonia: bushmeat trade in, 83; deforestation in, 38, 39; economic projects in, 143; forest fragmentation in, 41; illicit crops in, 99
Amazonian forests: arboreal ants in, 72; hunting in, 82; life-sustaining dust for, 109; logging in, 169–170. *See also* tropical rain forests
amphibians, 179; ecological service of, 77; effect of microhabitat transitions on, 36; names for, 211–212; and pesticides, 77
analgesics, contamination with, 132
animal feed, rising prices for, 112–113
animals: advent of domesticating, 65–66; antibiotic overuse in, 124; as pets, 258; seeds dispersed by, 74–75; treatment of wild, 59

Annan, Kofi, 104, 198
Antarctic waters, overexploitation of, 106
ant diversity, and land use, 38–39
ant gardens, 72
anthills, microhabitats created by, 192
antibiotics: contamination with, 132; overuse of, 124; resistance to, 124
antidepressants, contamination with, 132
ants, 179; arboreal, 72; black-headed, 17; environmental services of, 72; fire, 126; mimosa, 17; Penzig's, 18; Sjöstedt's, 18; and whistling-thorn acacias, 18
Appalachian mountain ranges, adelgid invasion of, 178
aquarium trade, 58
aquatic ecosystems, Arctic, 29
Archer Daniels Midland, 91
Arctic, human alteration of, 28–29
Aristotle, 93
Asian tiger mosquito, 128
Aswan High Dam, 135: building of, 136; consequences of, 138; effect on Mediterranean of, 138; effect on soils of, 137; glaciation associated with, 139; Nubians displaced by, 138; removal of, 140
atmosphere, 1; degradation of, 142; increasing temperature of, 108; pollution of, 153
Australia, biodiversity in, 38–39
autogenic succession, 164–165
autonomous seed production, 176
Azáldegui, María, 161

Bachman's sparrow, 184
backup, use of term, 79
backup systems, maintenance of, 80–81
bacteria: names for, 209; nitrogen-fixing, 102; in Palouse prairie ecosystem, 44
Bailey's pocket mouse, 155
balance of nature, 154–156
balsam woolly adelgid, 178
Banggai cardinalfish, 58–59
banner-tailed kangaroo rat, 155
bark beetles, 87–88
bats: declining populations of, 99; in Gabon rain forest, 72; impact of habitat fragmentation on, 36; Mexican fruit, 185; and plant diversity, 185; plant reliance on, 41; Seba's short-tailed, 185; shelter for, 180; Toltec fruit-eating, 185; yellow-shouldered, 185

beauty in form, xii
bees: declining diversity of, 70; value of, 69
beetles: bark, 87–88; corn rootworm, 177–178; Japanese, 108; in tree decomposition, 87–88; wood-boring, 87
benthic communities: and biodiversity, 148; major disruptions in, 50
Berry, Wendell, 16, 98
biodiversity, 247; declines in, 71; effects of fire on, 187–188, 189; effects of grazing on, 187–188, 189; effects of mobile-fishing gear on, 148; growing loss of, 39; and interdependence of plants and animals, 74; lost, 80; maintaining, 14–15, 73; microbial, 53; protecting, 111, 178
biofuel, corn as, 98–105
biological corridors. See corridors
biological diversity, 80. See also biodiversity
biomass: defined, 247; standing, 81
biophysical principles, xv; cyclical nature of, 78; and science, 11; and social systems, 118; spirituality and, 4; study of, 7; and universal commonalities, 14
biosphere: defined, 1; degradation of, 142
biotic diversity, predictors of, 14. See also biodiversity
bird communities: impact of deer on, 60; as indicators of environmental impacts, 46
birds: on agricultural land, 47; declining species of, 115; in ecosystem repair, 164; effect of habitat fragmentation on, 36; environmental services of, 71; extinction-prone, 70–71; functionally extinct, 70–71; loss of large-bodied, 149; names for, 212–213; northward movement of, 140; and plant diversity, 185; population trends of, 141; seeds dispersed by, 73–75; shelter for, 180; and shrubsteppe communities, 45; tree seeds planted by, 30–31; tropical, 58; in urban settings, 186
birth rates, 201
bison, 167, 189
Black Blizzards, 107
black-headed ants, 17
blacklisting, of foreign species, 129
blight, 135
bobcats, 180
Bodega Bay area, 126
Bodélé Depression, 109
body size, importance of, 181–182
Bogor Botanical Gardens, 149
Bohai ecosystem, 53
Bohnsack, James, 130
bone-eating zombie worms, 90
Botswana, AIDS in, 199
bottom-fishing, in ocean, 148
Boulder County, Colorado, exurban lands in, 22
Brauman, Kate, 68
Britain: biodiversity in, 49; declining bat populations in, 99
British, in New World, 160
brood parasitism, 182
Brower, David, 2
brown clouds, 110
brown tree snake, 128–129

Bt corn, 112
BtRB corn, 112
bubonic plague, 138
Buddha, xiv
bunch-grasses, 43, 45
buprestid, golden, 87
Burke, Edmund, 130
burning, prescribed, and insect diversity, 188–189
burrow systems, 179
Burundi, malaria in, 199
bushmeat: effects of hunting for, 81–82; preferences for, 83–84; socioeconomic value of, 82; unsustainable hunting of, 83
bushmeat trade: in Congo, 84–85; in Ghana, 84
butterflies, declining species of, 115
"bycatch," 132

cacao agroforestry, 39
caddisfly, 156
calcium-secreting organisms, in world's oceans, 105
California glossy snake, 134
camas roots, 44
cane toads, 77
capital, natural, 151
capitalism, and pollution, 162
carbohydrates, found in trees, 87
carbon dioxide: atmospheric, 162; emissions, livestock's role in, 38; in groundwater, 104; oceanic effects of, 105–106
carbon flux, 52
cardinalfish, Banggai, 58–59
carrying capacity, 61, 248
Carson, Rachel, 89
Cassell, Gail, 124
catastrophic events, responses to, 156
cats: bobcats, 180; feral, 62; pumas, 60–61
cattle, as exotic species, 191
Caucasus, 26
cause and effect: irreversible, 115; law of, xiv
Chaco Culture National Historic Park, 189, 190
change, xii; as constant process, 120, 140, 158; fear of, 25; and role of backups, 79; uncertainty of, 149; as universal constant, 151. See also environmental change
Chang Jiang (Yangtze) River basin, 54, 103
Chapela, Ignacio, 111
cheatgrass, spread of, 188
chemoautotrophic state, of whale decomposition, 90
chemotaxis, 52
China: turtle farms in, 144; water diverted from farmlands to cities in, 144
Churchill, Winston, xiii, xiv
civilization, evolution of, xiv
clams: Colorado River, 137; eastern gem, 126; effect of dams on, 137
Clark, William, 160
clear-cutting: and coyote survival, 180; habitat fragmentation caused by, 33
climate change, 15; adapting agriculture to, 48; and acidification of oceans, 105; and CO_2 levels, 162; glaciation in, 139; Heimrich Event I, 50; irrefutable evidence

of, 140–141; and river flow, 3; and species extinction, 116; and tropical forests, 37–38; variability of, 15; Younger Dryas Event, 51
climates: disappearing, 15; effect on ecosystems of, 151; novel, 15
climax ecosystem, 157
cloning, cross-species, 100
coelacanths: discovery of, 113; endangered status of, 114; habitat of, 114
Cohen, Alan, 203
Colombia, illicit crops in, 99
Colorado River clam, 137
Columbus, Christopher, 159
Columbus, Ferdinand, 159
comb jellyfish, 53, 209
commonalities: climate change, 15; habitat fragmentation, 30; of life, 12–15
commons, the: abuse of, 130–131; global, 19; kinds of, 18–19; light pollution of, 134; noise pollution in, 133–134; silence in, 132–133; trespassing on, 131; visual pollution of, 135
communities, xii
competition, and fear, 117
compost, organic, 204
Confucius, 93
Congo, Republic of: bushmeat trade in, 82, 83; road density in, 127
connectivity: among forest fragments, 31; and landscape alteration, 28
consciousness: dynamic of, 5; raising, x, 199–200, 202–204
conservation, and economic growth, 151–152
copper, in tall fescue, 177
corn, 98; basic uses of, 113; as biofuel, 98–105; genetic engineering of, 100, 111–112; mass-scale ethanol production from, 91; monocultures, 103
corn production: fertilizer for, 101; pesticide runoff from, 103
corn rootworm beetles, 177–179
corporations, and the commons, 131
corridors, biological, 183; planning for, 184; for plants, 187; in urban settings, 185–186
cosmic time, 87
cotton crops, pesticides used on, 70
cottonwood trees: decline of, 156; and elk populations, 60
cotyledon, 30, 250
Courtney-Latimer, Marjorie, 113
coyotes: adaptability of, 180–181; open areas required by, 180
crab, European green, 126
Crawford, J. W., 94
crayfish, and acidification, 110
crested wheatgrass, 170–171
crop plants, specialization of, 69
crows, 66
crustaceans, names for, 210
cryptic-species complex, 40
Cubagua, Venezuela, pearl-oyster beds off coast of, 143
cultures: and biodiversity, 15; evolution of, 10–11; and protection of commons, 19
cycles: nature's ecological, 16; of universe, 15

dams: estimated average density of, 135; long-term ecological consequences of, 136
daphne, population of, 193
DDT: in coelacanth tissue, 115; in national parks, 109
dead zone: defined, 103; in oceans, 106
decision making: and exploiting natural resources, 147; leading toward sustainability, 154
decomposition: role of soil in, 95; of trees, 88–89
deep ecology, 119
deep-sea habitat, chemical pollution of, 114
deer, 179; impact on forests of, 35; Key, 64; overpopulation of, 59–60
Deere, John, 163
deforestation: in Amazonia, 38, 39, 169–170; changes in hydrological cycles caused by, 3; of Easter Island, 121–122; and illicit crops, 99; major cause of, 76; rate of, 47–48; rat-induced, 122; and river flow, 3
dengue fever, in U.S., 128
desertification, 190
desert pocket mouse, 155
de Vries, Michiel Wallis, 190, 191
dickcissel, 182
dieldrin, 109
diesel exhaust, 110
dignity, and biophysical principles, 4
dimorphic seeds, 186
dinoflagellates, and ocean temperatures, 51
diseases: dengue fever, 128; effect on North American indigenous population of, 158; evolution of chemical-resistant, 125; yellow fever, 128
disequilibrium, in nature, 155
disturbance: anthropogenic, 173; defined, 165; ecological, 165; human-introduced, 165
ditch, first, 120–121
diversity: insect, 188–189; kinds of, 80; microbial, 53; and productivity, 168. See also biodiversity
divorce, increasing rate of, 20
dolphins: effect of noise pollution on, 132; pollutants deposited in, 113
Doney, Scott, 162
Douglas firs, decomposition of, 87
dragonfly, 67
dredging, effects of, 148
driftwood, and biodiversity, 165
droughts: during Dust Bowl years, 107; ENSO, 76
dualism, in human psyche, 65
duff, 123, 251
dust, landscape altered by, 159–160
Dust Bowl years, 107

early warning system, for species invasion, 129
earthworms: ecosystem altered by, 123; giant Palouse, 44. See also worms
Easter Island, rat-induced deforestation of, 121–122
eastern gem clam, 126

Eban, Abba, 92
Echinodermata, 210
"ecological bypass," 163
ecological costs, 27
ecological cycles, 16
ecological restoration, 142
ecological services, impact of hunting on, 82
ecology: deep, 119; economics and, 27–28;
    shallow, 119
economics: and biodiversity, 15; and ecology,
    27–28; and genetically modified organisms,
    112; market, 131; traditional, linear, 96–97,
    101; of tropical ecosystems, 39; of village
    commons, 19
economic thinking, modern-day, 116
economic waste, 142
ecosystem dynamics: cause and effect in, 173;
    composition of, 1, 169; cumulative effects
    in, 172, 173, 175
ecosystem processes: fish in, 71; role of birds
    in, 71
ecosystem repair, 163; configuring landscape
    for, 181–187; microclimates in, 192;
    mutualistic symbiotic relationships in, 192;
    and social-environmental sustainability,
    198–199. See also prairie remnant repair
ecosystems, xii; arctic, 28; assessment of, 193;
    and biodiversity, 14–15; biologically
    sustainable, 164; climax, 157; deep sea, 51;
    endemic biota in, 193; external factors
    affecting, 151; fragmentation of, 28;
    function, 171, 172; healing, 168; historical
    manipulation of, 157–158; impact of
    invasive species on, 126–127; indicators of
    health of, 196; lag periods in, 175;
    midwestern, 167; model for repairing,
    168–169; pelagic, 52; pristine, 124;
    redundancy in, 78; self-regulation, 173; self-
    reinforcing feedback loops in, 71;
    shrubsteppe, 45, 188; structure of, 169;
    sustainability of, 153; thresholds in, 174, 175;
    tropical, 38; Yellow Sea, 54. See also prairie
    remnant repair; repairing ecosystems
edge effects: critical role of, 39; habitat core
    compared with, 182–183; and habitat
    fragmentation, 30, 32–33; measuring, 174
edges, habitat, 174; created by roads, 33; of
    forest, 34; overall complexity of, 33; sharp,
    35; between successional stages, 35
education: for adapting agriculture to climate
    change, 48; goals of, 7
E8₅ [ethanol/gasoline blend], 110
Ehrenfeld, David, 146, 152
elephants, decline of, 174
elk, 1801 overpopulation of, 59–60; tule, 189.
    See also ungulates
El Niño–Southern Oscillation (ENSO), 76
Emerson, Ralph Waldo, 68
Emmons, Louise, 71–72, 74
encephalitis, in U.S., 128
endemism, importance of, 192–193
endosulfan, 110
England, landscape of, 25. See also Britain
enrichment opportunist stage, of whale
    decomposition, 90

ENSO. See El Niño–Southern Oscillation
entail, defined, 130
environment: hidden functions of, 168;
    human relationships with, 68; imperative
    questions about, 201–205; social treatment
    of, 201; sustainable, 201
environmental change, adaptation to, 198
environmental services: pollination, 69;
    sustainability of, 69
epifauna, 148, 252
ergot poisoning, 177
erosion, in Arctic, 28–29. See also soil erosion
estrogens, synthetic, contamination with, 132
estuaries, hypoxic zones near, 103
estuarine habitats, 54
ethanol: as biofuel, 91; blend, 110
ethanol plantations, 103
ethics, and biodiversity, 15
ethnobotanical knowledge, 46
Europe: biodiversity in western, 49;
    grasslands of central, 169; population
    trends of birds in, 140–141
European beech, effect of habitat
    fragmentation on, 31–32
European green crab, 126
European night crawler, 123
eutrophication, 53, 252
evaporative transpiration, 3
evolution, of humanity, 201
evolutionary biology, 51
evolutionary process: human interference in,
    62; and rational impartiality, 63
exotic plants: advantages of, 175–176;
    invading old-growth forests, 37;
    relatedness to native biota of, 175
exotic species: dispersal of, 127; eradication
    of, 129; importation of, 129; introduced
    onto island, 192–193; of marine organisms,
    130; movement into new areas of, 128
experiences, life, 12
exploitation: and American myth, 160;
    controlling unchecked, 197. See also over-
    exploitation
extinction: sixth great, 116; susceptibility to, 42
extinction vulnerability: of birds, 70–71; of
    butterflies, 115; and increasing prosperity,
    152; plant-vertebrate mutualism in, 193; of
    wolves, 181
exurban development, and biodiversity, 49
eyesores, 135

families, separation of, 20, 23
farmland: loss of, 104; protection of, 45
farms: impact of global climate change on,
    48; Latin American subsidization of, 143;
    organic compared with inorganic, 99;
    small-scale heterogeneous compared with
    large-scale homogeneous, 46–47
fear, and competition, 117
Federal Bureau of Soils, 94
feedback loops, xi, xii, xiii; among plants and
    animals, 75; in deforested areas, 76; in
    development of soil, 96; and edge effects,
    33; maintenance of, 75 understanding of, 17
feedback loops, self-reinforcing, 18, 24;

during Dust Bowl years, 107; and genetic engineering, 112; obligatory mutualists in, 40–41; and results of increases in low-altitude ozone, 108; of root systems and soil erosion, 171
fencerows, elimination of, 23, 47
feral cats, 62
ferns, 207
fertilizers: increasing dependence on, 102; manufacturing, 104; petrochemical, 204; pollution created by, 78; reducing dependence on, 101
fescue: Idaho, 43; tall, 176–177; toxicosis, 176, 177
figs, 40
fig wasps, 40
finches, zebra, 132
fire ants, 126
fires: forest, 16; impact on prairie of, 167; variable effects of, 187
fire suppression, forest alterations caused by, 34
firs: Douglas, 87; Fraser, 178
fish: effects of dams on, 136; environmental services of, 71; invasive species among, 126; names for, 211; from national parks, 109
Fisher, Robert, 134
flash floods, caddisfly adaptation to, 156
floccinaucinihilipilification, 9–10
flooding, sheet, 155–156
flying foxes, plant reliance on, 41
Foley, Jonathan, 154
food: and animal behavior, 179–180; and habitat, 178–179; and population growth, 199; rising prices for, 112–113
food availability, in ecosystem dynamics, 173
food crops, for biofuel, 92
food poisoning, outbreak of, 23
food production, regional variability of, 3
food web, soil, 95
forbs, 208, 252
forest fires, and nature's cycles, 16
forest fragmentation, threat of, 41
forestry: agroforestry, 29, 39; commercial, 29; intensive, 30; invasive species problem in, 29–30; maintaining biodiversity, 32
forests: African, 85; Amazonian, 72, 82, 109, 169–170; ecosystems of, 87; edges of, 109; evolution of, 16; national, 128; natural restructuring of, 150; overhunting in, 81–82; passenger pigeons in, 161–162; and population growth, 199; temperate, 29–37, 95. See also tropical forests
fossil fuels, and population growth, 199
fossils, living, 115
foxhole, test of, 57
Fraser fir forest, adelgid invasion of, 178
freedom, xii; as inner state of consciousness, 3; limitation of, 1; relativity of, 1
French, in New World, 160
French hawksbead, 186
Friedemann, Alice, 91
functional diversity, 80
functions, beauty of, xii
fungi: dispersed throughout prairie, 175; mycorrhizal, 175; names for, 207; in

Palouse prairie ecosystem, 44; wood-rotting, 89
future, imperative questions about, 201–205

Gabon: Gamba Protected Areas Complex, 83; rain forest in, 72
gazelles, Mongolian, 184
gender equality, 200
gene flow, seed-mediated, 31, 75
generalists: coyotes as, 180; humans as, 124
generations, and sustainable environment, 201. See also sustainability
genetic diversity: defined, 80, 253; and feedback loops, 75. See also biodiversity
genetic engineering, 111–113; of corn, 100, 111–112; of crop plants, 69; effects of, 100
genetic erosion, 166
genetic variability, loss of, 125, 166
Gentry, Al, 72
germplasm, defined, 46, 253
Ghibe Valley of Ethiopia, 46–47
"ghost roads," 127–128
giant waterbug, 156
Gibraltar, proposed dam at, 139
glaciation, new, 139
global commons, planet Earth as, 199
global ecosystem, productive capacity of, xv. See also ecosystems
globalization, environmental impacts of, 147, 152
global warming, 19; effects on snow/sea-ice system, 51; jet stream affected by, 55; livestock's role in, 38; shifting winds associated with, 106–108; and tropical rain forests, 76; unanticipated result of, 24
Goering, Hermann, 7
golden buprestid, 87
gophers, pocket, 179
grain production, and ecological limiting factors, 101
grasses: bunch-grasses, 43, 45; names for, 207; perennial, 190; prairie Junegrass, 43; seagrasses, 54; wheatgrass, 43, 170–171. See also grasslands
grasshoppers: nutrient intake of, 167; variety of, 67
grasshopper sparrow, 182
grasslands: of Great Britain, 168–169; native compared with wheatgrass, 171; nitrogen deposited in, 188; role of herbivores in maintaining, 189; of southeastern Oregon, 170. See also grasses
gray whales, 55
grazers, 179. See also herbivores; ungulates
grazing: cattle, 187–188; livestock, 189; long-term, 34; and plant-microbe dynamics, 176–177; recovery of perennial grasses from, 190; selective, 167; variable effects of, 187
green crab, European, 126
greenhouse gases, 104; CO_2, 104, 105–106, 162; over Indian Ocean and Asia, 110
Greider, William, 153
Gross Domestic Product (GDP), calculation of, 70

Gross National Product (GNP), and threatened species, 152
groundwater: CO_2 in, 104; pollution of, 96. *See also* irrigation; water

Haber, Wolfgang, 91
habitat: and animal behavior, 178; configuration of landscape in, 181–187; connectivity of, 178–181, 199; and economic incentives, 149. *See also* edges, habitat
habitat fragmentation, 30, 199; and area occupancy, 182; and body size, 182; due to timber harvests, 35; edge effects of, 32–33; and forest birds, 36; forest-dependent understory species vulnerability to, 43; genetic changes associated with, 32; irregular patches in, 166, 167; in marine ecosystems, 54; and species extinction, 165–166; trophic processes affected by, 33
habitat patches. isolation of, 183–184. *See also* prairie remnants
hagfish, 89–90
harvests, and sustainability, 145
Hawaii, protection of tropical fish in, 129
hedgerows, 23
Heinrich, Hartman, 50
Heinrich Event I, 50
Henslow's sparrow, 182
herbicides, 204
herbivores: coexisting, 167; and plant-microbe dynamics, 176–177; reintroductions of, 189. *See also specific herbivores*
herders, nomadic, 65
history: repetition of, xiii; as social construction, 11
holistic approach, ix
Holocene epoch, 43–44
honeybees: losses of, 70; threats to, 69
horse, Przewalski, 190
housing: clustered, 22; environmental tradeoffs associated with, 49
hoverflies, decline in, 70
Hughes, James, 125
humanity, evolution of, 201
humans: clean air and survival of, 111; contamination with sewage of, 132; impact on landscapes of, 28, 49, 85; impact on litho-hydrosphere of, 110; as invasive species, 124; in nature, 157; prairies altered by, 43–45; as predators, 118
hunter-gatherer cultures: and biophysical principles, 4; Mesolithic, 26; other-centered cooperation in, 65; in Palouse prairie ecosystem, 44
hunting: and forest regeneration, 42; impact on mammal communities, 82, 83; and seed survival, 81
hydrological cycle, 3

ice age, 26
ice algae, and snow cover, 51
ice-rafted debris, 50
Idaho fescue, 43
ignorance, and knowledge, 5–6

impartiality, rational, 63
impermanence, Buddhist notion of, 1, 158
incumbency, advantage of, 156
Indian Ocean, contamination of, 132
indicators, of ecosystem's health, 196
indigenous populations: landscape alterations of, 160; land use of, 198–199; of North America, 158–159
individuality, expression of, 3
"informed denial," 20
injustice, environmental, 147
insect diversity, and prescribed burning, 188–189
insecticides, 204; banned, 109; genetically engineered, 112
insects: abilities of, 68; diversity of, 67–68; effect of increasing intensive agriculture on, 99; invasive species of, 108; names for, 210–211; in Palouse prairie ecosystem, 44
interactive systems, 151
interdependence, concept of, 2
Intergovernmental Panel on Climate Change, 154
"International Convention for the Regulation of [Commercial] Whaling," 148
International Development Association, 100
intuition, concept of, 12
invasive plants: advantages of, 175–176; on Palouse prairie, 45; successive, 37. *See also* exotic plants
invasive species: crested wheatgrass, 171; and economic development, 126; following catastrophic physical event, 155–156; humans as, 124; insects as, 178; modern vs. prehistoric, 125; movement into new areas of, 128; relatedness to native biota of, 175; and roads, 127; and special-interest groups, 129, 130
inventory, in remnant repair, 194
invertebrates: effects of early forest fragmentation on, 32; names for, 209–211; in Palouse prairie ecosystem, 44. *See also specific invertebrates*
"invisible present," 16, 133, 172, 173
irreversibility, threshold of, 191
irrigation: changes in hydrological cycles caused by, 3; origins of, 121; water diverted from, 104
islands, exotic species introduced onto, 192–193
isolation, and landscape alteration, 28

Jacobson, Mark Z., 110
James, William, xv
Japanese beetles, increasing populations of, 108
Jefferson, Thomas, 6
jellyfish, comb, 53, 209
Jenny, Hans, 95
jet stream, shift in, 55
Johnson, R. G., 138, 139
Junegrass, prairie, 43

kangaroo rat, 155
Key deer, 64

kidney-stone belt, 24
Killmann, Wulf, 138
Kleypas, Joanie, 105–106
Klostermaier, Klaus K., 11
Knapp, Allen, 167
knowledge: and ignorance, 5–6; language and, 8; questioning, 202; validity of, 6; as version of truth, 6–7
Korten, David, 131
krill, Antarctic, 106
Kumar, Satish, 7
Kyoto Protocol (1997), 146

Labrador-Sea-Water ventilation, 51
Lackey, Robert, 8
Lake Constance, changes in bird population of, 140–141
landscapes: early, 25–27; healing, 111 (see also prairie remnant repair); human alteration of, 28, 49, 85; urban, 186
land use: changes in, 3, 172; by early indigenous Americans, 98–99; in Latin America, 143; livestock's role in, 38; and runoff of precipitation, 77; in tropics, 38
language: and meaning, 8–11; power of, 10
Latin America, subsidies to farmers in, 143
laws, restricting entry, 129. See also trade
Leafy Green Marketing Agreement, 23
Lederberg, Joshua, 124
lemur, brown, 42
Leonardo da Vinci, 86
Lewis, Meriwether, 44, 160
lianas, in Gabon rain forest, 72
lichens: edge effects on, 35–36; names for, 207
life cycles, xii
life decisions, environmental impacts of, 24
lifestyles: mechanistic, 119; sustainability of, xiii
life support system, 157, 163
lights, glare from, 134
line-transect surveys, 82
lions, mountain, 180
litho-hydrosphere: degradation of, 142; human impact on, 110
Liu, Jianguo, 20
livelihood, 7
livestock: grazing, 189; pollution association with, 38
living standard, 7
Loango National Park, 83
lobsters, and seagrasses, 54
Loess Hills of Iowa, 188
logging: in Amazonian forests, 169–170; clear-cutting, 33, 180; exploitive, 22; forest alterations caused by, 34; lower-impact, 143; overhunting associated with, 82; in Republic of Congo, 127; salvage, 150
Los Tuxtlas, Mexico, 185
Lowdermilk, W. C., 97
lung cancer, 110

Macpherson, Gwen L., 104
macroevolution, 51
Madagascar, seed dispersal in, 42
Madrid, Spain, bird study in, 186

malachite sunbird, 192
Mallorca, 193
Malthus, Thomas, 93
mammals: declining species of, 115; hunting of, 82, 83; names for, 213–214; seeds dispersed by, 73–75
mammals, small: impact of habitat fragmentation on, 37; in Palouse prairie ecosystem, 44; rodents, 179
management, command-and-control, 150
Manning, Richard, 98
mantled ground squirrel, 31
marine sanctuaries, no-capture, 129–130
marine toads, 77
market, and the commons, 131
market strategies, for adapting agriculture to climate change, 48
marriage, and divorce, 20
material solutions, limitations of, 119
McKibben, Bill, 4
meaning: boundaries of, 8–9; and language, 8
mechanistic thinking, 119
media, and violence, 19
Mediterranean Sea, increased salinity of, 138–139
megafloods, and English landscape, 26
Menorca, 193
mercury, 109
Merriam's kangaroo rat, 155
Mexican fruit bat, 185
Mexico, and genetic engineering of corn, 111–112
Mexico, Gulf of, dead zone in, 103
Mexico City, 109
mice: and habitat fragmentation, 37; protective cover for, 179; Santa Rosa beach, 134; and sheet flooding, 155
microbial diversity, and ecosystem processes, 53. See also biodiversity
microclimate: creation of, 192; of tropical rain forests, 76
microhabitats, and ecosystem repair, 192
microorganisms: and fallen trees, 88–89; interactions with plants of, 102, 176–177; in root systems, 102; in soil, 95
Milford, Jana B., 110
military dictators, 7
Mill, John Stuart, 12
mimosa ants, 17
mineral nutrition, 177
Mississippi River, 103
mobile-scavenger state, of whale decomposition, 89
modeling, in remnant repair, 194–195
mollusks: names for, 209; in southeastern Tasmania, 144
Monbiot, George, 154
Mongolian gazelles, 184
monitoring: to assess effectiveness of objective, 196; of prairie remnant repair, 194
monkeys, in Gabon rain forest, 72
monocarpic plants, 176
monocultures: corn, 100, 103; move toward, 69
montane voles, 77
Monterey pine, 30

mortality, decrease in rate of, 199
mosquito, Asian tiger, 128
Moukalaba-Doudou National Park, 83
mountain lions, 180
mouse, deer, 37. *See also* mice
Muehlen, Werner, 69
Musick, John, 115
mussels: effect of dams on, 137; turkey-wing, 143; zebra, 126
mutualistic symbiotic relationships: on Balearic Islands, 193; importance of, 192
mutualists, obligatory, 40
myth, grand American, 160

Naeem, Shahid, 78–79
Naess, Arne, 119
Nantahala National Forest, 34
Natal Drakensberg Mountains, 192
national forests, roads in, 128
nationalization, and American myth, 160
national parks, and airborne contaminants, 109
nations, and ignorance, 6–7
natural resources, overexploitation of, 147. *See also* resources
nature: balance of, 154–156; similarities in, 13
Nelson, Gaylord, 1
Neolithic period, 26–27
neotropical forests, overhunting in, 81–82
neotropical parrots, nest poaching of, 58
nephrolithiasis, 24
nesting habitats, impact of agriculture on, 100
nest predation: and habitat fragments, 182; rate of, 28
Nevada, water shortage in, 104
New Stone Age, 26–27
New World: European alteration of, 158; European invasion of, 118
New Zealand, mistletoes of, 192
night crawler, European, 123
nitrogen: atmospheric, 102; deposition from smog of, 187–188; and plant diversity, 168–169
nitrogen cycle, and climate change, 37–38
"nod factors," 102
noise, pollution of, 132
nomadic herders, 65
North America, landscape altered in, 162–163
Nubian culture: and Aswan High Dam, 138; extinction of, 140
nutrients: recycling of, 71; species-specific niches for, 167

oak, white, 161
observation, limited powers of, 190
oceans: bottom-fishing in, 148; CO_2 in, 105–106; direct human influence on, 106; greatest threats to, 131; human alteration of, 50–55; invasive species in, 125–126; noise pollution in, 132; warming of, 107
Odum, Eugene, 96
oil companies, ecological impact of, 83
Oken, Alan, 6
Ord's kangaroo rat, 155

Oregon creeping vole, impact of habitat fragmentation on, 37
Oregon Trail, 160
organic compost, 204
Orr, David W., 8, 10, 20
ostracode fauna, 51
otters, river, 179
outcomes, validation of, 196–197
ovenbird, 32
overcapitalization, 150
overexploitation, 142, 144; of Antarctic waters, 106; of natural resources, 147; and political power, 143
overfishing, 106; consequences of, 71; regime shifts triggered by, 53; species vulnerable to, 55
overhunting, results of, 81–82
overpopulation, role of women in, 200–201. *See also* population growth
ownership: concept of, xii; of water, 120
ozone, increases in low-altitude, 108

Pacific sleeper sharks, 90
pairing success, in habitat fragmentation, 32
Palearctic region, 136
paleolimnological records, 29
Palouse prairie ecosystem, 43
pappus, 186
parasitic plants, 207
parrots: neotropical, 58; trade in, 58
passenger pigeons: destruction of, 161–162; population estimates for, 161
Patancheru, India, pharmaceutical manufacturers in, 131–132
patterns, xi
PCBs: in coelacanth tissue, 115; in national parks, 109
Pearl River basin, 103
Pearson, David, 14
pelagic species, 52
penguins, Adèlie, 106
Penzig's ants, 18
perception, and language, 8–9
periwinkle, 105
Perry, David, 97
persistence, nature's criteria for, 115
personal growth, xiii
pesticides: amphibian populations exposed to, 77; and pollinators, 70; pollution created by, 78
pests, perception of, 69
petrochemical fertilizers, 204
pharmaceuticals, contamination with, 132
philosophy, and care for soil, 93
Phoenix, Arizona, pollen contamination of, 123
phytoplankton: changes in assemblages of, 50; in global-carbon cycle, 53; and productivity of fisheries, 52
pine trees: as invasive species, 30; Monterey, 30; ponderosa, 30, 31, 34
Pitman, Nigel, 161
plant invasions, pathway for, 125. *See also* invasive species
plant-microbe dynamics, 176–177
plant-microbe symbiosis, 102

plants: Arctic migration of, 28–29; bird-pollinated, 191; declining species of vascular, 115; in ecosystem repair, 164; genetically engineered, 112; names for, 207; parasitic, 207. *See also* vegetation
Plastic Lake, Ontario, 110
pleiotropic effects, detection of, 100
Pleistocene epoch, 26
poaching, effect on seed dispersal of, 42
pocket gophers, 179
pocket mice, 155
polar bears, 113
politicians, and "informed denial," 20
politics: and American myth, 160; and genetically modified organisms, 112; role of women in, 200
pollen grains, in Phoenix, Arizona, 123
pollinators, wild, 70
pollution: and air quality, 10; of atmosphere, 153; boundaries for, 131; corn production associated with, 113; fighting against, 203–205; genetic, 112; light, 134; noise, 132–133; overfishing and, 53; and population growth, 199; thinking about, 120; visual, 135; of world's oceans, 53
polycarpic plants, 176
polychaetes, 90
polychlorinated biphenyls (PCBs), 109, 115
ponderosa pine, 31, 34
population density, and invasive species, 126
population growth, xiv, 92, 124; birth rates in, 201; increase in, 199; and reproductive rights, 200; and resource management, 149–150; in U.S., 93
population size: and genetic variation, 75; and nesting success, 182
Portal, Arizona, sheet flooding at, 155–156
prairie: disappearance of, 163; grasses, 163; human alteration of, 43–45; tallgrass vs. shortgrass, 183
prairie, North American: biodiversity of, 167; crested wheatgrass on, 170–171; native grasslands of, 174
prairie Junegrass, 43
prairie remnant repair, 168, 177; modeling of, 194; role of monitoring in, 194–197; soil type in, 193–194; taking inventory for, 194; through fire and grazing, 187–191; treatment plan for, 195; verification of effectiveness, 195–196; vision for, 194
prairie remnants: effect of barriers and edges on, 174, 184; genetic variability found in, 176; size of, 182
predator-prey interactions, 173; forced redistribution resulting from, 61; intimidation in, 60
predators: depletion of marine, 53; humans as, 118
Prigogine, Ilya, 140
privacy, and animal behavior, 181
private property, and intact ecosystems, 162
"privatizing," 131
protein, trees as source of, 87
Przewalski horse, 190

pumas, hunting patterns of, 60–61
Pyrenees of France, springtails in, 193

quality of life: determination of, 164; and light pollution, 134; and noise pollution, 133
questions, imperative, 201–205
Quist, David, 111

radiata pine, 30
railroads, effect on migration routes of, 184
rain forests, 72; destruction of, 39; seed dispersal for, 73–74. *See also* tropical rain forests
ranching, landscape altered by, 159–160
Randall, Jack, 130
rare species, contributions of, 193
ratchet effect, 145–146
rats, 155
rat's tail plant, 192
Read, Herbert, 117
reality: construction of, 6; and intuition, 12; and language, 10
reclamation, of national rivers, 135
recreation, nature-based, 202
Red List survey, 116
redundancy, use of term, 79
reef fish: and aquarium trade, 58; marine reserves for, 55
reef stage, of whale decomposition, 90
refrigerators, energy used by, 21
regime shifts: defined, 50; and overfishing, 53
reintroduction, planned, 190–191
relationships: changing, 120; commensal, 22; continuum of, 87; practice of, 2
reliability engineering, 78–79
religion, and care for soil, 93
religious fanatics, 7
repairing ecosystems, 163; biophysical dynamics of, 169; model for, 168–169; and nature's dynamics, 164–168; purpose of, 164. *See also* prairie remnant repair
reproductive rate, in habitat fragmentation, 32. *See also* population growth
reptiles, 212
resources: changes in allocation of, 48; management of, 149; overexploitation of, 142–146, 147
restoration, 2; concept of, 152; as oxymoron, 160–161; process of, 152–153; rethinking concept of, 153–157
reversibility, potential for, 149
rhizosphere, 176, 259
rhythms, xi, 15
rice, 98
riparian ecosystems, global biodiversity in, 135
river flows, homogenization of, 135
river otters, 179
rivers: and changing land use, 3; reclamation of national, 135
roads: ecological effects of, 33, 127–128; habitat fragmentation caused by, 33; and hunting, 82; negative effects of, 127
roadsides, as migration corridors for exotic plants, 174

rock salt, deicing with, 21–22
Roddick, Anita, 203
rodents, 179. *See also specific rodents*
Roggeveen, Jacob, 122
Roosevelt, Franklin D., 92
root systems: dynamics of, 175; functions of,
    101–102
Ropke, Wilhelm, 131
Rowan, Carl, 7
Rowe, Jonathan, 19, 131
runoff of precipitation, 77
rural development, and biodiversity, 49.
    See also farmlands
rural sprawl, environmental effects of, 49

safety, and light pollution, 134
Sahara Desert, 109
salamanders: effect of forest roads on, 33–34;
    effect of microhabitat transitions on, 36;
    and habitat, 178–179
Salas, Karina, 161
salt, deicing with, 21–22
sanctuaries, no-capture marine, 129–130
San Juan Mountains, domestic livestock
    introduced into, 159
Santa Rosa beach mice, 134
saprophyte, defined, 175
sardines, of Nile Delta, 136
Saskatchewan, Canada, deforestation in, 47–48
Schimel, David, 162
schistosomiasis, dam-induced spread of, 136,
    137
Schlesinger, William H., 162, 164
schools, goals of, 7
science: as discipline of disproof, 18; and
    economic thinking, 116; and feedback
    loops, 18; overreliance on, 118; politicizing,
    147; social construction of, 12; as social
    contract, 11; subjectivity of, 49
scientific research, 12
scientific thinking, reductionist, 79
seabed, ecological restoration of, 147
seagrasses, importance of, 54
sea nesting turtles, 134
Seba's short-tailed bat, 185
sediments, marine, 52
seed dispersal: in city, 186–187; effect of
    poaching on, 42; long-distance, 31; by
    mammals, 73–75; for tropical rain forests,
    73–74
self-conquest, xiv
self-fertilization, 176
serpentine soil, 187, 260
sewage, human, contamination with, 132
sharks, 113; overexploitation of, 106; Pacific
    sleeper, 90
sheet flooding, 155–156
shellfish, anthropogenic impact on, 144
shelter, and animal behavior, 180
Shenstone, William, 9
shrews, 179
shrubs: increasing density of, 190; names for,
    208
shrubsteppe ecosystem: cheatgrass in, 188;
    fragmentation of, 45

sibling species, 40
Simberloff, Daniel, 128, 129
similarities: and differences, 13; identifying,
    13–14
Sjöstedt's ants, 18
slave trade, 200
sleeping sickness, tsetse fly transmitted, 46
smallholder farmers, impacts of global
    climate change on, 48
Smith, J.L.B, 113
snag, 88, 261
snake: brown tree, 128–129; California glossy,
    134
snowberry, 43
snowboarding worm, 90
snowflake, metaphor of, 205
snow/sea-ice system, 51
social insanity, xv
social justice, xiii
society: evolution of, 201; and ignorance, 6–7
soil: care of, 92–98; and cyclical nature of
    forest, 16; degradation of, 94; dynamics of,
    95; fertility of, 98; health of, 96, 97; impact
    of roads on, 127; importance of, 94;
    serpentine, 187, 260; topsoil, 38
soil erosion: consequences of, 93; of Easter
    Island, 122; effect of new restrictions on,
    23; impact of grazing on, 35
soil-water regime, alteration of, 22
songbirds, migratory, decline in, 36
South Africa, AIDS in, 199
southern red-back vole, 185
space, and animal behavior, 180
Spanish, in New World, 118, 160
sparrow: Bachman's, 184; grasshopper, 182;
    Henslow's, 182
special-interest groups, and invasive species,
    129, 130
specialist: defined, 124; wolf as, 181
species: apparent disappearance of, 114;
    backup replacement of, 79; estimating
    historic size of, 55; genetic erosion of, 166;
    and genetic variability, 125; loss of, 55;
    nonnative, 126. *See also* invasive species
species dispersal, and landscape patterns, 166
species diversity: and fertilizer use, 102; in
    tropical rain forests, 72
species extraction, sociopolitical factors
    in, 85
species redundancy, in ecosystem
    processes, 78
sperm whales, 148
spiders, 210
spinach, contamination of, 23
springtails, 193
squirrels: in Gabon rain forest, 72; mantled
    ground, 31
Steinbrecher, Ricardo, 100
streets, as biological corridors, 186. *See also*
    roads
subsistence farmers, impacts of global
    climate change on, 48
suburban development, and biodiversity, 49
suburban sprawl, environmental effects of, 49
succession, autogenic, 164–165

superorder, loss of, 115
surface runoff, 3
survival, human, and clean air, 111
sustainability, x; achieving, 154; and backup
    systems, 81; of ecosystems, 153; of
    ecosystem services, 69; of fisheries, 53; of
    food supply, 104; in generational time
    scales, 28; and harvest rate, 145; of human
    society, 201; and human survival, 111;
    imperative questions about, 201–205; loss
    of, 148; social-environmental, xiv, 164, 198,
    200, 203; of water catchments, 22
symbiosis, of obligatory mutualists, 40
symptomatic thinking, 119
systems thinking, ix

Takoradi, Ghana, 84
tall fescue, 176–177
Taylor Grazing Act (1934), 160
technological specialization, 93–94
technology: and economic thinking, 116; and
    feedback loops, 18; impact of, 66;
    limitations of, 119; overreliance on, 118
telegraph, ecological impact of, 162
temperate forests: human alteration of,
    29–37; rate of decomposition in, 95
temperature, in plant-microbe dynamics, 177
thermohaline circulation, 50
Three Gorges Dam, 136
tile drains, in corn production, 103
time: generational scales, 28; perception of,
    15; in plant-microbes dynamics, 176
toads, 77. See also amphibians
Toltec fruit-eating bat, 185
topsoil, ecological integrity of, 38
Toronto, rock salt contamination of water
    in, 22
tortoises, exploitation of, 144. See also turtles
tourism industry, and protection of tropical
    fish, 129
Toynbee, Arnold, 12
trade: aquarium, 58; bushmeat, 84–85; and
    invasive species, 128
tradeoffs: of commercial decision, 23–24; of
    personal decisions, 20–23; of social
    decisions, 24
trade winds, cycle of, 107
transgenic technology, 111
travel, and invasive species, 128
trawling, effects of, 148
treatment plan, for remnant repair, 195.
    See also prairie remnant repair
tree communities, impact of changes in, 41
tree-farm management, 22
trees: cottonwood, 156; death of, 87;
    decomposition of, 87, 91; drifting, 165;
    fallen, 88; in Gabon rain forest, 72; names
    for, 208; pioneering species of, 73 planted
    by birds, 30–31; shade-tolerant, 88;
    zoochorous, 42. See also specific trees
trophic cascade, 122, 263
trophic levels, 33, 263
tropical birds, 58. See also specific birds
tropical fish, protection of, 129
tropical forests: dry compared with rain, 73;

human alteration of, 37–43; neotropical,
    81–82
tropical rain forests: deforestation of, 76;
    destruction of, 39; seed dispersal for,
    73–74; species in, 76
trypanosomes, 209
tule elk, 189
tuna, overexploitation of, 106
turkey-wing mussel, 143
turtles: exploitation of, 144; farm-raised,
    144–145; road mortality of, 127; sea nesting,
    134; wild-caught, 145

uncertainty: of continual change, 149; of
    ecological cycles, 147
ungulates: effect of road construction on, 82;
    impact of railroad line on, 184; overpopu-
    lation of, 59–60; in Palouse prairie
    ecosystem, 44; reintroduced, 190–191.
    See also specific ungulates
United Nations, 198
universality, expression of, 3
universities, goals of, 7
unsustainable practices, of turtle farming,
    144–145
urbanization: effect on wildlife of, 49–50; and
    food supplies, 92; and habitat fragments,
    181; and invasive species, 126
urban landscapes, biological corridors in, 186
urban sky glow, 134
urban sprawl, 20: impact on wild animals of,
    64; noise pollution associated with, 133
Usambara Mountains, 43

vagility, 189, 263
validation, of outcomes, 196–197
Van Dierendonck, Machteld, 190
variability: loss of genetic, 125, 166; systems
    governed by, 1
varve thickness, in Arctic, 29
vasectomies, 201
vegetation: CO_2 absorbed by, 107–108; and
    fallen trees, 88; in forest, 87; species
    composition of, 171; in state of dynamic
    imbalance, 151. See also plants
vehicles, dispersal of plants by, 175
verification, of effectiveness of prairie repair,
    195–196
vertebrates: effects of early forest
    fragmentation on, 32; names for, 211–215;
    and plant-vertebrate mutualism, 193
    (see also humans; mammals)
village commons, 18–19. See also
    commons, the
violence: and media, 19; and politics, 200
vision: and imperative questions, 202; in
    remnant repair, 194
visual pollution, 135
voles: impact of habitat fragmentation on, 37;
    montane, 77; Oregon creeping, 37;
    southern red-back, 185; water, 180

Wafer, Lionel, 159
war, leadership for, 7
wasps, fig, 40

water: and animal behavior, 179; diverted from farmlands to cities, 179; "ownership" of, 120; and population growth, 199; shortages of, 104; wasted, 21. *See also* irrigation

waterbug, giant, 156

water catchments, sustainability of, 22

water infiltration, impact of grazing on, 35

water skippers, 67

water voles, 180

wealthy nations, impact on poor nations of, 152

wedge-billed woodcreeper, 43

well-being, and biophysical principles, 4

Western Airborne Containments Assessment Project, 109

Western Europe, biodiversity in, 49

west Nile fever, 126

whales: decomposition of, 89–91; effect of noise pollution on, 132; gray, 55; sperm, 148

whaling industry, 148

wheat, 98

wheatgrass; bluebunch, 43; crested, 170–171

wheel, introduction of, 66

whistling-thorn acacias, feedback loops of, 17

white-crowned manakin, 43

whitelisting, of foreign species, 129

white oak, 161

white-tailed deer, impact on forests of, 35

wildlife, treatment of, 59

wild rose, 43

wind, and global warming, 106–108

witch's hair, edge effect on, 35–36

wolves: melon, 180; reintroduced to Yellowstone National Park, 60; space requirements of, 180–181

women: political leadership of, 200; reproductive rights of, 200

wood-boring beetles, 87

woodpeckers: acorn, 172; and urbanization, 50

words: perceived meaning of, 9; power of, 10

World Bank, 100, 104

Worldwatch Institute, 101

worms: bone-eating zombie, 90; earthworms, 44, 123; names for, 210; snowboarding, 90

Worster, Donald, 97

Yangtze River, effect of Three Gorges Dam on, 136

yellow fever, in U.S., 128

Yellow Sea, China's, 54

yellow-shouldered bat, 185

Yellowstone National Park, 181

Young, I. M., 94

Younger Dryas Event, 51

Yu, Eunice, 20

zebra finches, 132

zebra mussels, 126

zoochorous trees, 73

ABOUT THE AUTHOR

Chris Maser is the senior author of *Trees, Truffles, and Beasts: How Forests Function* (Rutgers University Press, 2008). He has more than forty years of experience in ecological research, including broad international experience; he has spent many years developing new ways of understanding how forests are structured and function, as well as new ways of understanding social-environmental sustainability. He has written over 250 papers (mostly in scientific journals) and some twenty-five books on these topics. His books, as well as some of his papers, are in academic and public libraries in every state in the United States and all but one province in Canada, as well as in seventy other countries. He has lived, worked, and/or lectured in Austria, Canada, Chile, Egypt, France, Germany, Japan, Malaysia, Nepal, Slovakia, Switzerland, and much of the United States. (See his website if you want more information: www.chrismaser.com.)

Printed in the United States
148570LV00003B/2/P

9 780813 545592